ALL ABOUT GORILLAS

Other books by DAVID P. WILLOUGHBY

The Super-Athletes
The Empire of Equus
Growth and Nutrition in the Horse

**SOUTH BRUNSWICK AND NEW YORK: A. S. BARNES AND COMPANY
LONDON: THOMAS YOSELOFF, LTD.**

ALL ABOUT GORILLAS

David P. Willoughby

© 1978 by David P. Willoughby

A. S. Barnes and Co., Inc.
Cranbury, New Jersey 08512

Thomas Yoseloff Ltd
Magdalen House
136–148 Tooley Street
London SE1 2TT, England

Library of Congress Cataloging in Publication Data

Willoughby, David P
  All about gorillas.

  Bibliography: p.
  Includes index.
  1. Gorillas. I. Title.
QL737.P96W47   599'.884   76-50215
ISBN 0-498-01845-8

PRINTED IN THE UNITED STATES OF AMERICA

To the memory of my parents,
David and **Mary Willoughby,**
who introduced me to the
wonder-world of Natural
History

Contents

	Acknowledgments	9
	Introduction	11
1	Evolution and Classification of the Primates	15
2	Is "Bigfoot" a Descendant of GIGANTOPITHECUS?	25
3	Early Accounts of the Gorilla	35
4	Races, Natural Habitats, and Geographical Distribution of the Gorilla	46
5	Size, Physique and Growth, and Development of the Gorilla	57
6	Two Other Manlike Apes—The Chimpanzee and the Orangutan	107
7	An Anthropometric Comparison of the Three Higher Apes and Man	128
8	Intelligence, Brain Weight, Physical Powers, Longevity	142
9	Habits and Family Life of the Gorilla	159
10	Some Distinguished Captive Gorillas	172
11	Care and Feeding of Gorillas in Captivity	211
12	Gorillas in Fiction and Fantasy	223
13	A Census of Zoo Gorillas	232
14	The Gorilla's Chances for Survival	250
	Select Bibliography	256
	Index	258

Acknowledgments

In my endeavor to assemble data on the measurements of young and growing zoo gorillas, I have been assisted greatly by the cooperation of numerous zoo directors and primate keepers, each of whom recorded for me the ages and weights (and in some cases girths and other measurements) of the young gorillas currently in their custody. Too, I have received information on the measurements of a number of wild-shot adult gorillas from hunters or custodians of such animals. Especially helpful in these respects have been the following persons: Alfreda L. Acosta, Los Angeles Zoo; Edward H. Bean, Chicago Zoological Park; the late Mrs. Belle J. Benchley, San Diego Zoological Garden; Larry O. Calvin, Dallas Zoo and Aquarium; Gary K. Clarke, Topeka Zoological Park; Harold J. Coolidge, National Academy of Sciences; Lee S. Crandall, New York Zoological Park; Donald R. Dietlein, Kansas City Zoological Gardens; James G. Doherty, New York Zoological Park; Don Farst, Columbus (Ohio) Zoological Gardens; J.W. Fitzgerald, Central Park Zoo, New York City; M.H. Gritzmacher, Milwaukee County Zoological Park; Dr. Bernhard Grzimek, Director Emeritus of the Frankfurt (Germany) Zoo; Sybil E. Hamlet, National Zoological Park, Washington, D.C.; Charles J. Hardin, Toledo Zoological Society; Edalee (Mrs. W.B.) Harwell, San Diego Zoological Garden; N.S. Hastings, Zoological Society of Cincinnati; Charles H. Hoessle, St. Louis Zoological Park; Marvin L. Jones, San Francisco Zoological Gardens; Keith K. Kreag, Detroit Zoological Park; Jeremy J.C. Mallinson, Jersey Wildlife Preservation Trust; Edward J. Maruska, Zoological Society of Cincinnati; William E. Meeker, Cheyenne Mountain Zoological Park, Colorado Springs; R. Marlin Perkins, Lincoln Park (Chicago) Zoological Gardens (and, later, St. Louis Zoological Park); Deets Pickett, D.V.M., Kansas City, collector of gorillas and chimpanzees; George Speidel, Milwaukee County Zoological Park; Kenheim Stott, Jr., San Diego Zoological Garden; Dr. Warren D. Thomas, Lincoln Park Zoo, Oklahoma City; Frederick A. Ulmer, Jr., Philadelphia Zoological Garden; Weaver M. Williamson, D.V.M., Chicago Zoological Park.

Introduction

Of all the myriad kinds of animals in which man is interested, it is perhaps safe to say that those that attract his curiosity most are those that most remind him of himself. These are the anthropoid apes—particularly the gorilla. The other apes—chimpanzee, orangutan, and gibbon—share with the gorilla the distinction of being man's nearest living relatives. However, in the matter of "personality," or the ability to interest and impress his human cousins, the giant gorilla—even as a juvenile—is in a class by himself. The casual visitor at the zoo, and the physical anthropologist in his laboratory, each has an intense desire to know the whys and wherefores of this challengingly manlike creature.

In this book I have endeavored to both ask and answer some of the most common questions concerning gorillas. In this procedure I have used terms that should be understandable to the general reader and at the same time possibly of some significance to mammalogists (primatologists) who specialize in anatomical studies of these giant apes.

My interest in "animals" (mammals) in general dates back to the time I was eight years old, when my mother saved up and invested three dollars (not a trivial sum in those days) in a volume entitled *Brehm's Life of Animals,* which I literally had begged her to get for me. That large volume—now bound in three-fourths red morocco leather—after sixty-seven years, I still possess. In it are hundreds of wood engravings by German animal artists of the nineteenth century, mainly the three brothers Specht—Friedrich, August, and Carl Gottlob—who were supreme at this type of illustration. A number of their illustrations of gorillas and other anthropoid apes are herein reproduced.

It was not until about 1940, however, that I was able to work professionally in the fields of zoology and paleontology, which I did while being employed at the California Institute of Technology in Pasadena. About that time I became particularly interested in two huge mountain gorillas, Mbongo and Ngagi, which resided at the San Diego Zoo; and I made a number of trips there to study them. In this I was aided most ably by Mrs. Belle J. Benchley, director of the zoo, who was seeing to the welfare of Mbongo and Ngagi with all the care that would have been bestowed on human children. After Mbongo died in March 1942, I was privileged, at the San Diego Museum of Natural History, to take complete measurements of his disarticulated skull and limb bones.

Also about this time, I entered into correspondence with Dr. Adolph H. Schultz, of the Laboratory of Physical Anthropology at Johns Hopkins University, who was the foremost authority on the skeletal anatomy of the great apes. Dr. Schultz encouraged my studies and writings, one of which was an article entitled "The Gorilla—Largest Living Primate," which appeared in *The Scientific Monthly* for January 1950, pp. 48–57. Some of the many photographs of wild-shot gorillas that appeared in that article are reproduced here in chapter 5.

As the reader will perceive, the information on gorillas presented in this book is largely in the department of physical anthropology rather than in that of behavior, intelligence, physiology, or life history—all of which subjects have been explored (although far from completely) by specialists in these fields. My purpose has been, rather, to correlate the available information on bodily growth in the gorilla, as recorded on living specimens in zoos, with the more precise measurements made on cadavers and skeletal material by such primatologists as Drs. Adolph Schultz, the late Francis Randall, and a few other experts in this highly specialized department. In brief, my aim is to present an account of the gorilla based on all available statistics concerning the growth and physical development of this most interesting near-relative of man. Along with having taken the physical measurements of hundreds of men, women, and children, and of thousands of bones of fossil mammals, mainly horses, zebras, et al, it has been my pleasure to add to these biometric statistics similar data on the bodily structure and conformation of the three higher anthropoid apes.

Quite apart from the universal spectator interest in apes at the zoo is the more academic question of the evolutionary relationship of these higher primates to man. That apes and man *are* related both structurally and zoologically is abundantly evident from the similarity or morphological correspondence of their skeletons in every respect—bone by bone, tooth by tooth. In view of this physical comparability, growth studies of the three higher anthropoid apes are of especial interest in relation to similar statistics on man. This topic is explored in chapter 6.

The scale drawings (Figs. 1, 2, and 3) reproduced herein, which I made of the heads, hands, and feet of primates in general, and (in chapter 5) of the skulls of gorillas in particular, were done only after meticulous measurements of each had been taken. Although many of these drawings were made in the 1940s, it was not until now (after I had "retired" and could devote full time to writing) that I was able to reproduce them in a book about gorillas. Other illustrations reproduced herein from early books and magazine articles on gorillas are the result of a long-standing interest in these most important near-relatives of man.

Often, in adding yet another book to what is already a formidable collection on the same subject, an author will make an apology for his own contribution. My view in this connection is that an author does a favor to his readers if he merely brings together in a single convenient volume a digest of pertinent yet scattered information that has been contributed by previous writers in the field. And if, in addition, the compiler is able to analyze and clarify some of the facts previously stated, yet not properly analyzed, surely he must be making a definite contribution to his subject. Another consideration is that no two authors express themselves in the same way, even when writing on an identical topic. Each will emphasize the points with which he is best acquainted. For this reason it is of advantage—if one really wants to learn about a subject—to get the views of a number of authors and profit by the best that each has to offer.

Obviously, in titling the present book I have taken "artistic (or 'literary?') license." For there is no subject whatever that anyone could know "all about"—especially the subject of gorillas! I can only hope that the reader derives as much enjoyment out of perusing this book as I did in writing it.

—DAVID P. WILLOUGHBY

ALL ABOUT GORILLAS

A lowland (Gaboon) gorilla in his native forest, as pictured by the German animal illustrator Friedrich Specht, about a hundred years ago. Note the exaggerated "sideburn," the hairy (not naked) chest, and the hair on the forearms growing upwards. Possibly Specht had only a museum (stuffed) specimen from which to make his drawing—in which case, he did very well.

1

Evolution and Classification of the Primates

The Order Primates

Before discoursing on the chief subject of this book, the gorilla, it should be opportune to introduce the order of mammals, namely, the Primates, of which this great ape is a member. A brief discussion of the various genera and species included in the order Primates, and of their similarities and differences in relation to the gorilla, should help make clear the specific position of the latter ape within the entire order.

The Primates, as the name implies, is that order within the class Mammalia that comprises the "first," or highest, forms related to man—who himself is placed (by himself!) at the top of the order (see Fig. 6). The other included genera are the apes, monkeys, baboons, marmosets, lemurs, and tarsiers, respectively. While during recent years it has become common practice among zoologists (primatologists) to include also in this order the tree shrews (*Tupaia*), the latter primitive prosimians are still considered by many to be properly classed among the Insectivora (moles, shrews, hedgehogs, etc.).[1] By still other zoologists, or taxonomists, the tree shrews (Figs. 8 and 4) are regarded as being best assigned to a separate order, the Tupaioidea.* The study of the relationships of the various types embraced in the order Primates is known as the science of primatology. This subject deals not only with the living forms of apes, monkeys, lemurs, et al, but particularly also with the living races of mankind and their extinct or fossil forebears.

Reference to Figures 1, 2, and 3, and to the several "family trees" shown in this chapter, should enable the reader to note the great diversity prevailing among the living representatives of the order Primates. Nevertheless, there are certain characteristics common to all the species within the order, as will shortly be described.

First, however, it may be appropriate to define certain terms connected with this order. As previously mentioned, the name *primates*, which is usually pronounced pri-ma'-tez, means "first," or "chief." It applies to man and the apes, monkeys,

---

* The pros and cons of this question are reviewed at length by Stephen I. Rosen in his book *Introduction to the Primates . . . Living and Fossil*—New Jersey: Prentice-Hall, 1974, pp. 33–35. Rosen himself feels that the tree shrews are best regarded as "aberrant insectivores." He adds, however, that the problem of their zoological position should be further examined.

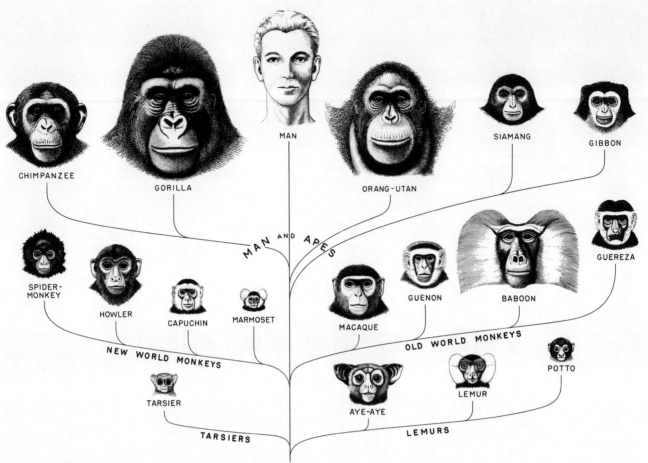

Fig. 1. Heads (drawn to scale) of some representative members of the order Primates. Figures are approximately one-tenth of their natural size.

and lemurs, which collectively are considered the most highly developed of mammals. "Ape" properly refers to one of the four anthropoid apes, which are taillesss and generally approach man most closely in their physical structure. These apes are the gorilla, chimpanzee, orangutan, and gibbon, respectively, although there is a species of macaque called the Barbary ape, which, as in the more highly developed anthropoids, has no tail. To most of the other primates the term *monkey* is popularly applied, although there are vast differences between certain groups of them, such as the baboons, the marmosets, and the lemurs. Shown in Figure 1 are the heads of some of the representatives of the order Primates; and in Figures 2 and 3, the hands and feet, respectively, of the same species (or, more correctly, *generic* types).

Distinguishing Characteristics

Apart from man, an external feature that serves to distinguish most Primates from other mammals is an opposable hallux, or big toe. This feature caused many zoologists of the nineteenth century to group apes and monkeys under the heading Quadrumana, meaning "four-handed." Even where the hallux is relatively small, as in the marmosets, it can be folded inwardly across the sole of the foot. A second distinguishing feature is that both the hand and the foot have five digits, except in the potto, where the second digit is vestigial, and in the spider monkeys and the guereza, in which the thumb is either vestigial or totally absent (see Fig. 2). The fingers and toes of the members of this order are generally equipped with broad, flattened nails, although in some highly specialized types (e.g., the aye-aye, Fig. 9) these may take the form of claws, or clawlike nails. Additional features are two mammae (nipples) located over the pectoral (front chest) muscles; orbits directed forward (permitting binocular vision), encircled by bone and shut off from the temporal fossae (hollows on the sides of the skull); and digits in which the first (thumb or big toe) is opposable to the others, either in the hands, the feet, or both limbs.

Dentition

Concerning the dentition in the Primates, there are never more than two incisors and six cheek teeth above and below on each side, making, with the four canines, at the most a total of thirty-six teeth. In adult higher Primates, with the exception of man and the gibbons, the sexes can readily be determined by the size of the canines, which normally are much larger in males than in females. Even in gibbons, upon close examination, it is found that the canines are somewhat longer in males, although nothing like to the degree existing in the gorilla, chimpanzee, and orangutan, in which giant forms—particularly the gorilla and the orangutan—the males are bodily much larger than the females. The anthropologist Aleš Hrdlička found the average size (linear) of the first and second lower molars in the female as compared with the male (taking the latter to equal 100) to be as follows: man 97.4; gibbon 97.5; chimpanzee 96.8; siamang 92.3; gorilla 89.7; orangutan 89.3.[2] These dental proportions are discussed further in chapter 2.

Nasal Differences

An important point of distinction between the monkeys of the New World (infraorder Platyrrhini) and those of the Old World (infraorder Catarrhini) is in the structure of the nose and nostrils. In the Platyrrhini (meaning "broad-nosed") the nostrils are typically separated by a relatively broad external septum (partitioning cartilage) and point more or less directly forward. In contrast, in the Catarrhini ("narrow-nosed") or Old World monkeys, including the anthropoid apes, the nasal septum is comparatively narrow, and the nostrils are directed more toward each other at the lower ends, giving them a V-shape. These differences may be seen by comparing in Figure 1 the noses of the New World monkeys and the Old World monkeys, the latter including the anthropoid apes.

General Similarities

Nearly a hundred years ago a fairly accurate general description of the anatomical structure of the apes and monkeys was given by the German naturalist Alfred Edmund Brehm, as follows:

> There is greater similarity in the anatomical structure of the different kinds of Apes than would be supposed from their outward appearance. The skeleton has seven cervical vertebrae, from twelve to sixteen dorsal, four to nine lumbar, two to five sacral, and three to thirty-three caudal (going to form the tail); the clavicle is strong; the bones of the forearm are separate and mobile, the wrist [metacarpal, D.P.W.] bones are long, while the finger bones seem stunted in their growth. The feet are supplied with a thumb [opposable big toe, D.P.W.]. The shape of the skull differs very much, according to the greater or lesser prominence of the jaws, and the size of the brain. The arches of the eyebrows are strong and prominent.... Among the muscles, those of the hand attract our attention, as, in comparison with those of a human hand, they seem to be so much simpler. The larynx is not capable of producing sounds that might constitute articulate speech, in our sense of the word; but the well-developed glands of the trachea seem to favor the production of shrill, howling sounds. Special mention must be accorded the cheek pouches, that distinguish certain Apes. These open from the cheeks by a small aperture near the corner of the mouth and serve to store away food. They are most highly developed in

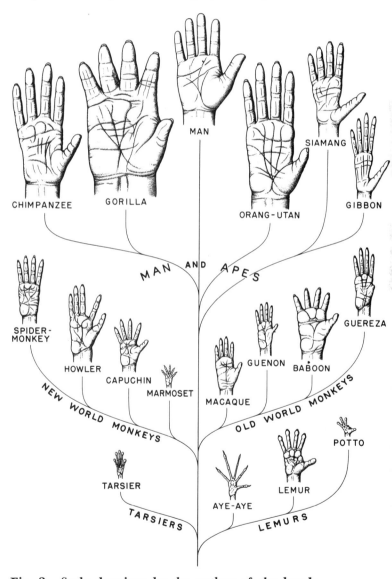

Fig. 2. Scale drawings by the author of the hands of the primates shown in Figure 1. These figures are approximately one-seventh of their natural size.

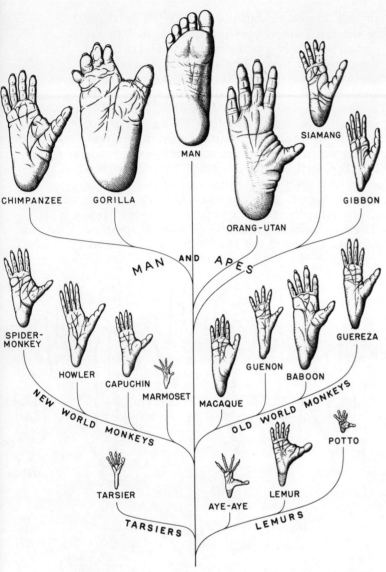

Fig. 3. Scale drawings by the author of the feet of the primates shown in Figure 1. These figures are approximately one-seventh of their natural size.

the Guenons, Macaques and Baboons; they are entirely lacking in the Man-shaped Apes and in all the New World Monkeys.[3]

The brain in Primates, except in the lemurs and tarsiers, presents a relatively high stage of development—the cerebrum (forebrain) being proportionately large and convoluted and covering the cerebellum (hindbrain). Excepting man, the Primates are arboreal (tree-dwelling) forms and are peculiarly well fitted for that mode of existence. The tail is often long, and in various New World monkeys (Cebidae) it is prehensile, being used as a "fifth hand." However, the Anthropomorpha (man and the apes), with which we are here especially concerned, are tailless—the gorilla, among others, having even fewer coccyxgeal vertebrae than

has man. The members of this higher group are distinguished also by having a vermiform appendix and a complex brain—although in the size of the latter organ there is a conspicuous hiatus between the manlike apes and the most primitive men. In all except man, the anthropoids have arms that are longer than the legs, along with a laterally expanded pelvis, a relatively broad (rather than deep, from front to back) chest, and a posture that only on occasion can be fully erect when standing on the feet. Most apes, with the exception of the ponderous adult gorillas, have the ability to swing (brachiate) through the trees by their hands and arms; while the gorilla and the chimpanzee spend much of their time on the ground in quadrupedal walking. As previously noted, man differs markedly from the apes in having a hallux (big toe) that is nonapposable to the other toes—the apposability of digits being restricted to the thumb and fingers. Man's hand and fingers have been likened to an object capable of grasping a sphere; while in the apes the fingers are more hooklike and are adapted for grasping a branch (cylinder).

Diet of Apes and Monkeys

As to the diet of the Primates, they, like bears, are essentially omnivorous, although their environment in many cases practically restricts them to plant food such as fruits, berries, nuts, and whatever edible vegetables are available. It is notable that chimpanzees in the wild will eat insects, eggs, and small birds or mammals whenever these items can be obtained by them. This suggests that hunger, if sufficiently intense, eliminates "natural" habits in the choice of food and was doubtless the chief determinant in the diet of man's early ancestors.*

Among the Primates the usual number of offspring is one, sometimes two (although in man, in rare instances, up to six or even eight; although

---

* J.Z. Young elaborates on the diet question as follows:

> Fruits and seeds provide especially rich sources of nourishment, not requiring the mastication of large quantities of siliceous grasses, upon which the grazing herbivore must spend so much time. The prime need is sharp eyes to find the fruit, hands and skill to pluck it, and taste to sample it. This form of diet may well have had much to do with the development of the quick-witted manipulating primate and its differences from the "slower-brained" ungulate types (though such differences are hard to quantify).
>
> The supply of fruits and seeds often changes and the primate has to resort to bark, shoots, and perhaps small animals, ants, termites, frogs, and lizards. Even the terrestrial baboons who live mainly on grass seeds and roots will eat meat. Most primates are thus catholic in their choice of diet and can digest meat. This is consonant with our own omnivorous diet.... The primate method has been to develop an omnivorous habit and rely upon getting something of everything.[4]

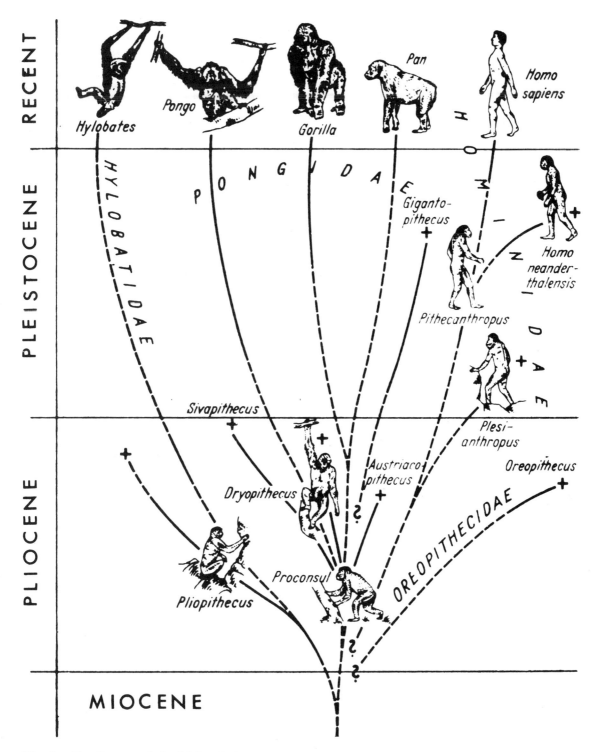

Fig. 4. Family tree of the higher Primates (man and the anthropoid apes). The crosses at the tops of some stems denote the apparent times of extinction of those forms. Where lines are dashed, it indicates uncertain continuity. According to this chart, the *Oreopithecidae*—a group that terminated in the Pliocene—had their origin in the Oligocene, and the *Pongidae* in the Miocene. (After Erich Thenius and Helmut Hofer, 1960.)

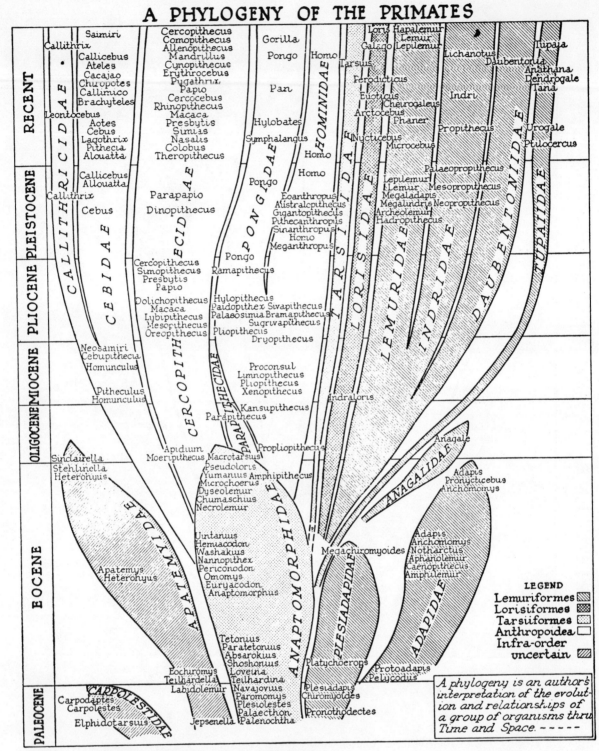

Fig. 5. A detailed phylogeny of the Primates as interpreted by Robert A. Stirton in his paper "Ceboid monkeys from the Miocene of Colombia" (1951). Compare with Figure 4.

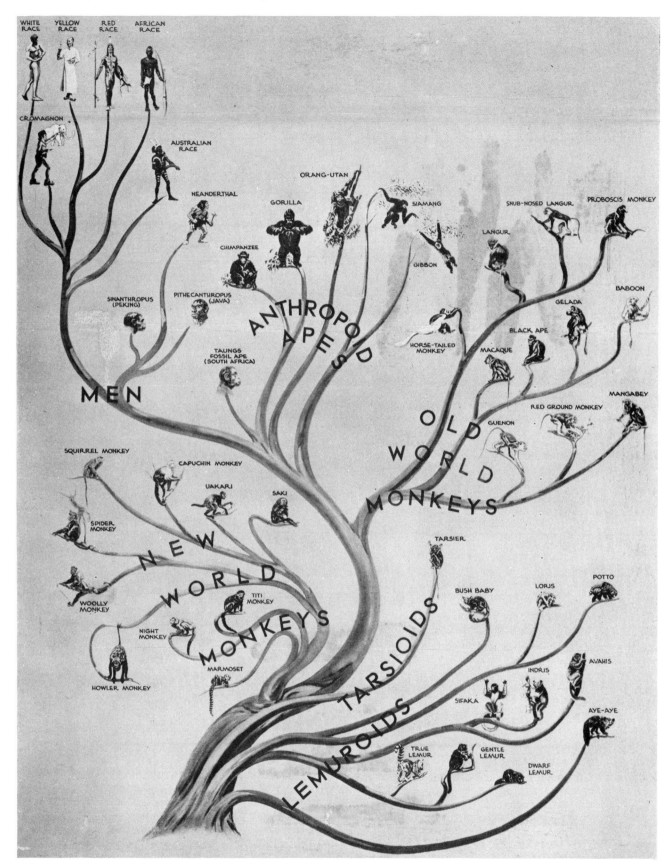

Fig. 6. A pictorial family tree of the Primates, illustrating the great diversity of forms existing within this order of the Mammalia. Note the separation of the Lemuroids and the Tarsioids from Man, the Anthropoid Apes, and the Monkeys (together, the Pithecoids), as tabulated in Figure 7. (Chart by E. S. Lewis; courtesy the American Museum of Natural History.)

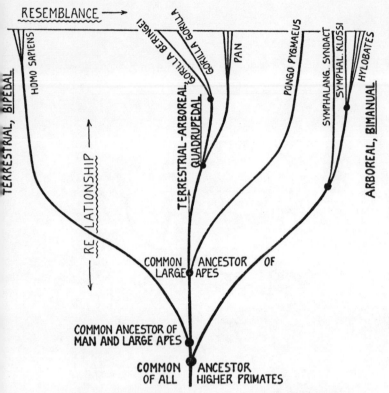

Fig. 7. Hypothetical family tree of the higher primates, indicating their physical resemblances and phylogenetic (evolutionary) relationships. (Courtesy A. H. Schultz. "The Skeleton of the Trunk and Limbs of Higher Primates." *Human Biology* 11, no. 3 (September 1930): 303–438.)

in the latter cases the infants rarely survive). While an offspring of a Primate is born in a fairly advanced state of development, it still must go through a relatively long period of parental care and postnatal growth. This prolonged developmental stage conduces to a more thorough development of the brain, which in turn aids in separating the Primates from various other orders of mammals. Among the latter in particular are the Artiodactyla (oxen, deer, antelopes, and other cloven-hoofed animals) and the Perissodactyla (horses, tapirs, rhinoceroses), in which groups the postnatal growth is rapid, and the development of the brain is correspondingly of shortened duration (see Fig. 47).

Primate Evolution

The evolution of the Primates—which is a special and complex field of study in itself—may for our purpose be summarized conveniently in the hypothetical "family trees" shown in Figures 4, 5, 6, and 7. It may be added that even among specialists in the study of man's ancestral pedigree, there is no universal agreement on his origin. All that can be surmised at the present stage of knowledge on the subject is that—to quote the words of the biologist Raymond Pearl: ". . . at some remote period in the past for which no clear paleontological record has yet been uncovered, man and the other Primates branched off from what had theretofore been a common ancestral stem."[5]

Fig. 9. Some lower forms of the Primates, as illustrated by the old-time German animal artist Friedrich Specht. 1. Aye-aye (*Chiromys madagascariensis*) times one-sixth; 2. Spectral tarsier (*Tarsius spectrum*) times one-fourth; 3. Potto (*Periodicticus potto*) times one-third; 4. Tana, or tree shrew (*Tupaia tana*) times one-fourth. The latter animal, of which there are four existing genera, is now generally regarded as a primitive ancestral type from which evolved the order of Primates as it is constituted today.

ORDER PRIMATES

SUBORDER	INFRAORDER	FAMILY	SUBFAMILY	GENUS	ENGLISH NAME
LEMUROIDEA	LEMURIFORMES	LEMURIDAE	LEMURINAE	LEMUR	Common lemur
				HAPALEMUR	Gentle lemur
				LEPILEMUR	Sportive lemur
			CHEIROGALEINAE	CHEIROGALEUS	Mouse lemur
				MICROCEBUS	Dwarf lemur
		INDRIDAE		INDRI	Indris
				LICHANOTUS	Avahi
				PROPITHECUS	Sifaka
		DAUBENTONIIDAE		DAUBENTONIA	Aye-aye
	LORISIFORMES	LORISIDAE	LORISINAE	LORIS	Slender loris
				NYCTICEBUS	Slow loris
				ARCTOCEBUS	Angwantibo
				PERODICTICUS	Potto
			GALAGINAE	GALAGO	Bush baby
TARSIOIDEA		TARSIIDAE		TARSIUS	Tarsier
PITHECOIDEA	PLATYRRHINAE	HAPALIDAE		HAPALE	Common marmoset
				OEDIPOMIDAS	Marmoset
				LEONTOCEBUS	Tamarin
		CEBIDAE	CALLIMICONINAE	CALLIMICO	Callimico
			AOTINAE	AOTUS	Douroucouli
				CALLICEBUS	Titi
			PITHECIINAE	PITHECIA	Saki
				CHIROPOTES	Saki
				CACAJAO	Uakari
			ALOUATTINAE	ALOUATTA	Howler
			CEBINAE	CEBUS	Capuchin
				SAIMIRI	Squirrel monkey
			ATELINAE	ATELES	Spider monkey
				BRACHYTELES	Woolly spider monkey
				LAGOTHRIX	Woolly monkey
	CATARRHINAE	CERCOPITHECIDAE	CERCOPITHECINAE	MACACA	Macaque
				CYNOPITHECUS	Black ape
				CERCOCEBUS	Mangabey
				PAPIO	Baboon
				THEROPITHECUS	Gelada
				CERCOPITHECUS	Guenon
				ERYTHROCEBUS	Patas monkey
			SEMNOPITHECINAE	SEMNOPITHECUS	Langur
				RHINOPITHECUS	Snub-nosed langur
				NASALIS	Proboscis monkey
				COLOBUS	Guereza
		HYLOBATIDAE		HYLOBATES	Gibbon
				SYMPHALANGUS	Siamang
		PONGIDAE		PONGO	Orang-utan
				PAN	Chimpanzee
				GORILLA	Gorilla
		HOMINIDAE		HOMO	Man

Fig. 8. Classification of the living genera of the order Primates, as adopted by W. L. Strauss, Jr. (see note 1, this chapter). Note that Dr. Strauss regards the ancestral (prosimian) tree shrews, or tupaiids, as being best assigned, neither to the Primates nor to the Insectivores, but to a separate order of their own: the Tupaioidea. (Compare with Figure 6.)

Paleontologists of more recent years may find fault with Pearl's statement, on the grounds that several manlike fossil remains, dating from the lower Miocene, suggest that the human branch may *at that time* have separated from a still-undesignated common ancestor of both man (Hominidae) and the anthropoid apes (Pongidae). This, however, is still conjecture.

Classification

Zoologists (taxonomists) classify animals according to their anatomical structure, embryological development, evolutionary history, food, habitat, and general affinities or similarities. In this system of classification it is necessary to recognize and adopt various categories, or groupings, starting with major groups and gradually descending to minor ones. This method of classification was devised by the Swedish botanist Carl von Linné in 1758. Astutely adopting Latin names to provide a universal and imperishable terminology, he even Latinized his own name, changing it to that by which he is best known: Carolus Linneaus. He applied his system of classification both to plants and animals of all kinds. The system is shown here in its application to our subject, the gorilla, the species of which, as previously pointed out, is a representative of the mammalian order Primates.

KINGDOM Animalia (animals in general, as separated from plants, etc.)
  SUBKINGDOM Vertebrata
    PHYLUM Chordata (animals with a bony spinal column)
      CLASS Mammalia (warm-blooded vertebrates that give milk to their young)
        SUBCLASS Theria (mammals that bring forth living young)
          INFRACLASS Eutheria (placental mammals; excludes monotremes and most marsupials)
            ORDER Primates (man, apes, monkeys, marmosets, lemurs, and tarsiers)
              SUBORDER Anthropoidea (the Primates, exclusive of the lemurs and the tarsiers)
                SUPERFAMILY
                FAMILY } Pongidae (the anthropoid apes: gorilla, chimpanzee, gibbon, siamang, and related fossil types)
                SUBFAMILY

                GENUS } *Gorilla*
                SUBGENUS

                SPECIES The lowland gorilla (*Gorilla gorilla*) and the mountain gorilla (*Gorilla beringei*) *

                SUBSPECIES Two local races, or subspecies, of the typical gorilla have been identified: the western lowland gorilla (*G. gorilla gorilla*) and the eastern lowland gorilla (*G. gorilla graueri*). There appears to be only a single form of the mountain gorilla (*G. beringei*).

* While most zoologists recognize only a single species of this ape, certain marked and constant anatomical or structural differences existing between the lowland and the mountain gorillas would appear to entitle the latter form to be regarded as a distinct species (see chaps. 5, 7).

Notes

1. W. L. Straus, Jr., "The Riddle of Man's Ancestry," *The Quarterly Review of Biology* 24, no. 3 (September 1949): 200–223. Reprinted in *Yearbook of Physical Anthropology* (1949), pp. 134–57.
2. Aleš Hrdlička, *American Journal of Physical Anthropology* 7 (1923):434.
3. A. E. Brehm, *Brehm's Life of Animals* (English edition) (Chicago, 1896), p. 2.
4. J. Z. Young, *An Introduction to the Study of Man* (New York and London: Oxford University Press, 1971), pp. 435–36.
5. Raymond Pearl, *Man the Animal* (Bloomington, Ind.: Principia Press, Inc., 1946), p. 3.

2

Is "Bigfoot" a Descendant of GIGANTOPITHECUS?

"Bigfoot", or the Sasquatch

My main reason for including this chapter is to revise an opinion that I expressed some years ago in an article entitled "The Gorilla—Largest Living Primate,"[1] for that article was published before the hominoid creature popularly known as "Bigfoot" appeared in the news to such an extent that its existence as a living primate could no longer be ignored. And all evidence as to the bodily size of "Bigfoot" indicates that it is both taller and heavier than a gorilla of corresponding age and sex.

"Bigfoot," it should be explained, is the same creature that has long been known to the Indians, as well as to the white residents of the Pacific Northwest, as Sasquatch. While actual sightings of the manlike animal have been comparatively infrequent, its gigantic footprints have been found in scores of locations, and a large number of plaster casts of them have been made. The bipedal posture and gait and the clearly humanoid character of the feet as revealed by the casts show that Bigfoot, or Sasquatch, is manlike rather than apelike, although the latter affinity is evidenced by the inch-long, usually black hair that covers practically the entire body. This hairy covering is clearly shown in a short strip of movie film that the late Roger Patterson, a Yakima, Washington, rancher, made of a female Sasquatch in 1967.* The location was a forest clearing near Bluff Creek in California's Del Norte County. A scale drawing of this photographed specimen, carefully enlarged from the Patterson color film, is shown here in Figure 10, in which I have added the outline of a human female of average size for comparison as to bodily proportions. Detailed measurements taken on a blowup of the Patterson film, along with actual measurements of casts of the feet, show that each foot is 14½ inches in length, and that the erect standing height is about an even seven feet (84 inches). From the latter height and the widths and depths of the body and limbs as measured on the enlarged film, I have estimated that Patterson's

---

* When the film strip was submitted by Patterson to the Walt Disney Studio for examination, the verdict was: "If it is a fake, then it's a masterpiece, and as far as we're concerned the only place in the world where a simulation of that quality could be created would be here, at Disney Studios, and the footage was not made here." (Los Angeles *Herald-Examiner*, October 7, 1975, p. A-10.)

"Abominable Snowlady" had the following girths (in inches): neck, 40; upper arm, 24; forearm, 20; wrist, 12; chest (or bust), 70; waist, 60; hips, 65; thigh, 40; knee, 27; calf, 25; ankle, 16. These girths, together with the standing height of 84 inches, indicate a weight of about 650 pounds. Since presumed male footprints of 17 inches or more in length have often been recorded, it may be deduced that in these individuals the stature is eight feet (96 inches) or over, and the weight between 900 and 1000 pounds. Here, then, is the largest living primate! The height difference between the sexes, if Patterson's female may be taken as an average-sized, female adult, is essentially the same as that which prevails between adult male and female gorillas. The female Sasquatch (Patterson's example), however, is proportionately (to the male) somewhat heavier built than a female gorilla compared with a male.

General Description

The extensive literature presently available on "Bigfoot" gives detailed accounts of almost every aspect of the creature's existence so far as is known; therefore, it does not need to be repeated here. However, some of the conclusions arrived at by John Green, a full-time investigator of the subject, as given in his book *On the Track of the Sasquatch*,[2] are summarized here:

(1) They [the Sasquatch] are not human. Their size, hairiness, and bulkiness go beyond [normal] human limitations, and they show no sign of human mental ability. Plainly, they

Fig. 10. Female Sasquatch, or Bigfoot, seven feet tall, filmed in northern California by Roger Patterson. For a comparison, the figure of a Caucasian woman of average size is shown. For details, see text. (One-eighteenth natural size.)

have achieved survival with their superb physical equipment, not with their brains. Equally plainly, they lack the ability to organize in opposition to their smaller, weaker human cousins, and as a result have been driven to living in the mountain forests, where they do not have to compete with man. That this defeat was at the hands of Stone Age man, who was neither numerous or well-armed, is a clear indication of an extreme difference in mental ability.

(2) They are completely omnivorous, able to make use of vegetable material that man does not eat, but also fond of meat, and well equipped to obtain it.

(3) They can see in the dark. The Yakima youth and several others noted their eyes reflecting car headlights, which the eyes of daytime creatures will not do.

(4) They are strong swimmers. The Indians (and the Saxons) even suggest that they swim well under water. In these last three characteristics they differ markedly from the known apes, which are almost entirely vegetarian, sleep when it is dark, and are helpless in water.

(5) They make a variety of sounds, but the one most associated with them is a very high-pitched and powerful scream. The boys at Tenmile and Mr. Edwards at Mount Ashland both used the term "whistling scream." The Kwakiutl Indians make pouted "whistling" lips the identifying feature of a carving of one of these creatures.

(6) They are very hard to kill. Some of the people who have shot at them may have missed, but not all. Of course, it is often difficult to bring down a large animal on the spot if you do not know just where to hit it, and wounded animals often have to be hunted down. I know of no one who has tried to hunt down a Sasquatch after apparently wounding it.

(7) They are deliberately avoiding contact with humans, and are very successful in doing so. A check of 125 sightings shows only six to have been identifiable as females and five as children. Unless Jacko is the exception, none of the children was alone. In many instances encounter took place because the Sasquatch for some reason was not being elusive, but was actually approaching or investigating humans or human habitation, or even behaving in an aggressive manner.

Mr. Green follows the foregoing observations by citing some instances in which human beings have similarly lived for long periods close to civilized communities without being detected. These instances he relates as follows:

> In 1872 the last dozen or so of the Yahi Indians gave up fighting the white man and disappeared from sight in a small valley on the slopes of Mount Lassen in California. They lived there undetected for 12 years, until the continuous loss of territory to the ranchers made it necessary for them to raid cabins for food. Even then, in a decade of raiding, they were seen only once. In 1894 the last five survivors made their home on two pieces of land no more than half a mile wide and three miles long. Their presence went entirely unknown for 15 years, until some surveyors blundered into their camp. When the last survivor, known to the white world as Ishi, walked into captivity in 1911, he had lived for 39 years in concealment on the fringe of civilization. Yet he and his people had permanent camps and made daily use of fires, even cremating their dead.

Two other publications by John Green, which I am listing at the end of this chapter, report additional and more recent findings on the subject.[3] There are also several papers by Grover S. Krantz, of Washington State University, on the anatomy of the Sasquatch hand and foot, as deduced from plaster casts, and one paper by Wayne Suttles dealing with Sasquatch culture.[4] The book by Ivan Sanderson on "abominable snowmen" should also be perused, as should his article entitled "The Missing Link."[5] As this is being written, a book entitled *The Search for Bigfoot*, by Peter Byrne, has just been published.[6]

Two major questions raised by the presence of the Sasquatch are (1) from what earlier species or genus of primates has it descended; and (2) why have no direct physical remains of the creature as yet been found?

To answer the latter question first, a partial explanation lies in the comments just quoted from John Green; namely, that the Sasquatch is an expert at concealing himself. He does this in heavily forested regions difficult in access on the ground and virtually impossible in detection from the air. And while the creatures of course die, one might just as well ask why skeletons are rarely found of some of the more numerous forest-dwelling animals, such as deer, bear, small carnivores, etc.*

---

* Peter Byrne *The Search for Big Foot* points out: ". . . since I came to the Pacific Northwest, no less than six airplanes have been lost in the tangled wilderness of the coast ranges. . . . Massive searches involving many people, planes and much money have yet to result in one airplane find."

But sooner or later, either one of the Sasquatch will be captured or a skeleton or other remains will be discovered. And until that time, one can only go by what information is turned up.

Early Relationships

As to what earlier species, or genus, of the Hominoidea (man and the anthropoid apes) the Sasquatch has descended from, there is at present no information or opinion. In effect, the question has been sidestepped by reason of the widely held assumption that all likely ancestors of gigantic size became extinct some 500,000 years ago. This assumption probably stems from the latest known remains—such as those of the hominoid form *Gigantopithecus,* of China—having been given that date. When it is considered that fossil remains of manlike apes, or apelike men, are of even rarer occurrence than those of less intelligent and less adaptable animals, it is evident that a scarcity of remains is not necessarily proof of the extinction of a species. One might ask, for example, what became of the thousands of skeletons of African elephants that were left by ivory hunters on the veldt a mere century or so ago, or of the millions of skeletons of the American bison that strewed the Western plains at an equally recent period. Actually, it is remarkable that fossilized remains thousands of years old or older are ever brought to light; and they are, only by reason of having been entombed and preserved under singularly favorable conditions.

Gigantopithecus

Certainly it is reasonable to ask that since innumerable other large mammals—species more vulnerable and less adaptable than man—have succeeded in surviving through the ages, why not also an anthropoid or hominoid of giant size? I venture to suggest that by reason of its huge bodily dimensions, the Sasquatch of today could conceivably be a descendant of the much-publicized fossil primate known as *Gigantopithecus.* The first remains of this giant-sized, presumably manlike creature were several molar teeth that were discovered in a Hong Kong Chinese apothecary shop during the years 1934 to 1939 by the Dutch paleontologist G.H.R. von Koenigswald.[7] Later, other discoveries in the same region were made of additional teeth, along with fragments of jaws. Figure 11 shows the mandible, or lower jaw, of a specimen of *Gigantopithecus* known as the "Kwangsi giant," as restored by the paleontologist H. Weinert. With it, for size com-

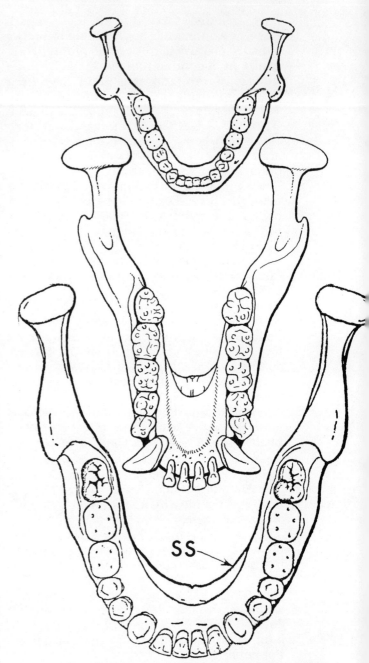

Fig. 11. Mandibles of a modern man (Caucasian), an adult male gorilla, and the "Kwangsi giant" (*Gigantopithecus*), respectively, drawn to the same scale for comparison. Although the mandible of *Gigantopithecus* exhibits the "simian shelf" (SS), it also shows that the teeth form an arch as in man—rather than two parallel rows as in apes. (One-half natural size.)

parison, are shown the mandibles and lower teeth of an adult male gorilla and a modern Caucasian man. My purpose at this point is to demonstrate—contrary to what some other investigators have contended[8]—that there is at least a fairly consistent correlation between the size of the molar teeth

Fig. 12. The prehistoric man-ape or ape-man, known as *Gigantopithecus*, as conceived by artist Neave Parker. The giant primate is shown here with his contemporaries—on the ground, a pair of the fossil subhumans known as Peking man, and in the tree, an orangutan, the species of which manlike ape still exists. For details, see text. (Courtesy *The Illustrated London News*.)

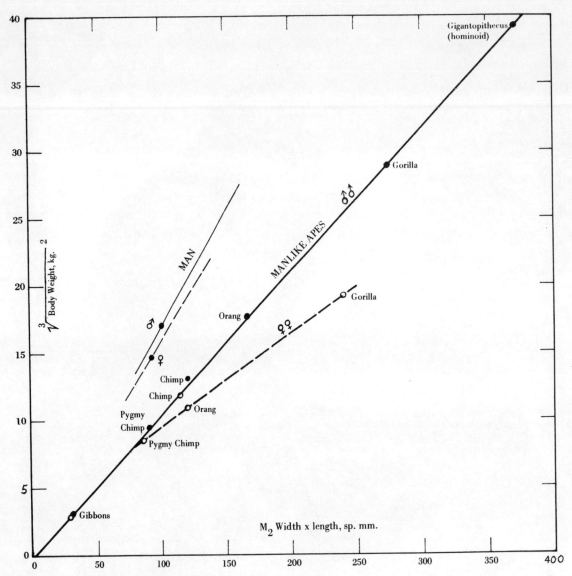

Fig. 13. Body size in relation to M_2 surface area in higher primates. For details, see text.

in anthropoid apes and their weight or body cross-section. This correlation is shown by the figures listed in Table 1, and by the graphic correlation of these measurements as plotted in Figure 12. Reasonably, in animals of the same physical (and physiological) type and feeding habits, there should be a connection between the amount of food required, the overall size of the body (specifically, the average or general cross-section), and the size (crown area) of teeth necessary to masticate the food. I have previously shown that such a correlation between tooth size and body size prevails in horses,[9] and so presumed that it also must exist (although in differing ratios) in apes and man. While, in Figure 12, it is seen that the teeth of the female gorilla, orangutan, and pygmy chimpanzee deviate from the line correlating the teeth of the males of these species, they nevertheless show a correlation between M_2 area and body cross-section.

The known size of the lower second molar (M_2) tooth in von Koenigswald's specimen of *Gigantopithecus*—namely, the dimensions listed in Table 1—indicate that if this fossil primate had the body of a present-day adult male gorilla it would have stood just over six feet (i.e., about 73 inches) in height. The corresponding weight would be about 550 pounds. On the other hand, if *Gigantopithecus* had been of typically human build—that is, in proportion to a man 69 inches in height and weighing 155 pounds, and with teeth proportionately as small as those of a modern Caucasian—he would have stood between ten and eleven feet in

Table 1. Width, Length, and Relative Crown Area in the Lower Second Molar Teeth (M_2) of Higher Primates.* ( ) estimated

Species	Sex	M_2 dimensions, mm			Body Weight, kg	$\sqrt[3]{\text{Weight, kg}^2}$	$\dfrac{M_2 \text{ width} \times \text{length}}{\sqrt[3]{\text{Weight, kg}^2}}$
		width	length	width × length			
Man (white)	♂	9.35	10.69	100.0	70.3	17.03	5.87
″ ″	♀	8.97	10.25	92.0	56.2	14.68	6.27
Gibbon (*H. lar*)	♂	5.0	6.2	31.0	5.7	3.19	9.72
″ ″	♀	4.9	6.1	29.9	5.3	3.04	9.83
Pygmy Chimpanzee	♂	9.0	10.05	90.5	29.5	9.55	9.48
″ ″	♀	8.7	9.7	86.2	25.5	8.66	9.95
Chimpanzee	♂	10.4	11.7	121.7	48.0	13.20	9.22
″	♀	10.1	11.3	114.1	40.5	11.79	9.68
Orangutan	♂	12.5	13.4	167.5	75.0	17.78	9.42
″	♀	10.6	11.3	119.8	37.0	11.10	10.79
Gorilla (lowland)	♂	15.6	17.7	276.1	156.0	28.98	9.53
″ ″	♀	14.5	16.55	240.0	85.0	19.33	12.42
Gigantopithecus	?	18.6	20.0	372.0	(248)	(39.43)	(9.43)

* Man, on the basis of body surface ($\sqrt[3]{\text{weight}^2}$), has molar teeth that have only about sixty percent of the occlusal (crown) area of the teeth of anthropoid apes. This point of difference, while not so frequently emphasized as man's size of brain, is still sufficiently great to completely separate humans (*Homo sapiens*) from all other primates. In baboons, for example, the M_2 area is relatively even larger—approximately thirty percent greater than in anthropoid apes, and accordingly over twice as great, relative to body surface, as in man.

height, just as the Chinese paleontologists who studied the teeth of *Gigantopithecus* had estimated. However, it is exceedingly unlikely that a manlike creature of 500,000 years ago, whatever its height, would have had teeth and jaws proportionately as small as those of a modern, civilized man. Conversely, for a given size of teeth, the stature or standing height of the creature would therefore have been less. If an intermediate height for *Gigantopithecus**—say, 8 feet, or 96 inches—can be assumed, along with the M_2 size listed in Table 1, the weight of the ape-man (that is, the aforementioned 550 pounds) would give him the body build of a modern man six feet in height, weighing 232 pounds, or about the proportions of an average-sized professional wrestler.

Thus, these estimations on the basis of tooth size, while indicating a Primate of giant stature, would be proportionate to a body of "only" 550 pounds or so, and accordingly would fall short of the immense bulk attributed to a full-grown Sasquatch or Bigfoot, even that of the female specimen photographed by Roger Patterson and shown here in Figure 10.

The dimensions of the M_2 teeth listed in Table 1 may be checked by reference to the following books or papers on primate dentition.[11]

Inconclusive as the foregoing comments on Sasquatch or Bigfoot may be, they still indicate—to the present writer, at least—that a race of giant-sized, apelike men could have survived from the mid-Pleistocene (the latest date for their remains) until postglacial times. Then, like many other Old World mammals, these descendants of *Gigantopithecus*, or *Gigantanthropus*, would through the centuries have become, under the present name of Sasquatch or Bigfoot, inhabitants of various parts of today's Canada and the United States, which evidently they have.

Hence, to get back to our starting point, the "largest living primate" can no longer be considered the gorilla, but rather must be recognized as the evidently much larger hominoid species—still known only indirectly—called Sasquatch or Bigfoot.

* Dr. von Koenigswald's associate, the paleontologist Franz Weidenreich, pointed out that since the molar teeth of the Chinese fossil Primates were manlike rather than apelike, the name given to them should be *Gigantanthropus* rather than *Gigantopithecus*.[10]

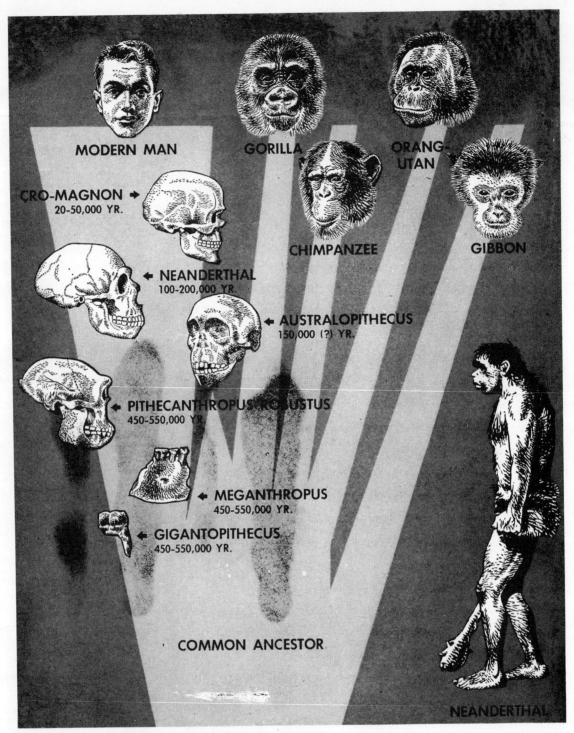

Fig. 14. This simplified family tree, which appeared in a news magazine in 1946, shows how the ancestral branch that led to modern man separated from that of the three higher anthropoid apes at a remote period (while the chart, due to space limitations, shows the divergence taking place only about 500,000 years ago, more than likely the separation occurred as much as twenty million years ago). It also shows that the largest subhuman "ancestor," *Gigantopithecus,* preceded in time the other fossil ape-men whose known remains (only rarely a complete skull) are placed in chronological (geological) sequence on the chart.

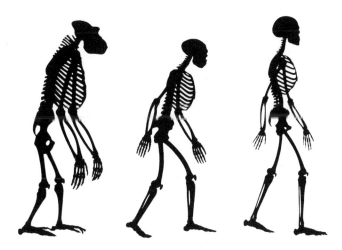

Fig. 15. The "ascent" of man was frequently depicted during the nineteenth century by comparisons such as this, which shows (from left to right) the skeletons of a gorilla, a fossil (Neanderthal) man, and a modern man, respectively. This, however, is misleading, since the separate lines from which man and the anthropoid apes evolved from a common ancestor branched apart millions of years ago. Too, Neanderthal man came to a dead end some 40,000 or more years ago, when his species was overrun and superseded by Cro-Magnon man (Homo sapiens).

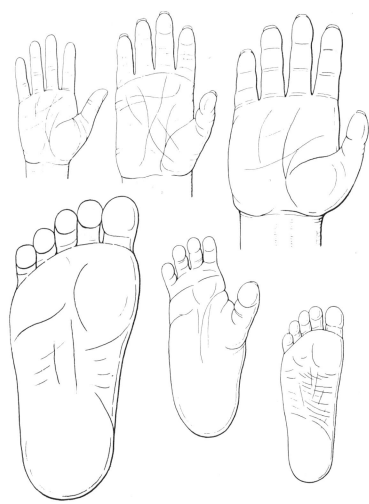

Fig. 16. Right hands and right feet of a man, a male gorilla (*G. gorilla*), and a male Sasquatch, respectively. The feet are shown in reverse order from the hands. The Sasquatch hand (11.5 inches long times 6 inches wide) is a reconstruction from a muddy print left on the side of a house in Fort Bragg, California, 1962. The Sasquatch foot (17.5 inches long times 7.7 inches wide) is the average of seven adult males, of which casts of footprints were secured. (All figures are times 0.2 natural size.)

Notes

1. David P. Willoughby, "The Gorilla—Largest Living Primate," *The Scientific Monthly* 70, no. 1 (January 1950):48–57.
2. John Green, *On the Track of the Sasquatch* (Agassiz, Brit. Col.: Cheam Publishing Co., 1968).
3. John Green, *Year of the Sasquatch* (Agassiz, Brit. Col.: Cheam Publishing Co., 1970); idem, *The Sasquatch File* (Agassiz, Brit. Col.: Cheam Publishing Co., 1973).
4. Grover S. Krantz, "Sasquatch Handprints," *Northwest Anthropological Research Notes* 5, no. 2 (1971):145–51; idem, "Anatomy of the Sasquatch Foot," *Northwest Anthropological Research Notes* 6, no. 1 (1972):91–104; idem, "Additional Notes on Sasquatch Foot Anatomy," *Northwest Anthropological Research Notes* 6, no. 2 (1972):230–41.
5. Ivan T. Sanderson, *Abominable Snowmen; Legend Come to Life* (Philadelphia and New York: Chilton Co., 1961); idem, "The Missing Link," *Argosy* 368, no. 5 (May 1969):23–31. (An inquiry, well illustrated, into the nature and zoological position of a recently killed, then frozen, specimen of an adult male hominid of Neanderthaloid characteristics, examined by Dr. Sanderson and by Dr. Bernard Heuvelmans in Wisconsin, but said to have come originally from an animal exporter in Hong Kong.)
6. Peter Byrne, *The Search for Big Foot* (New York: Acropolis Books, 1975).
7. G. H. R. von Koenigswald, "*Gigantopithecus blacki* von Koenigswald," *Anthropological Papers, American Museum of Natural History* 43 (1952):295–325.
8. S. M. Garn and A. B. Lewis, "Tooth-size, Body-size, and 'Giant' Fossil Man," *American Anthropologist* n.s. 60 (1958):874–80.
9. David P. Willoughby, *Growth and Nutrition in the Horse* (South Brunswick and New York: A. S. Barnes and Co., 1975), p. 135.
10. Franz Weidenreich, *Apes, Giants, and Man* (Chicago: University of Chicago Press, 1947), p. 59.
11. Ales Hrdlicka, *American Journal of Physical Anthropology* 7 (1923):434; Francis Randall, "The Skeletal and Dental Development and Variability of the Gorilla," *Human Biology* 15, no. 4 (December 1943):334–37; D. L. Green, *Gorilla Dental Sexual Dimorphism and Early Hominid Taxonomy*, Symposium Fourth International Congress of Primates, vol. 3, *Craniofacial Biology of Primates* (1973), pp. 82–100; J. Kitahara-Frisch, *Taxonomic and Phylogenetic Uses of the Study of Variability in the Hylobatid Dentition*, ibid., pp. 128–47; Denys H. Goose, *Dental Anthropology* (London: Pergamon Press, 1963), pp. 139–40 ("Modern English Teeth"); F. Twiesselmann and H. Brabant, 1960, Quoted by C. Loring Brace and Paul E. Hahler in "Post-Pleistocene Changes in the Human Dentition," *American Journal of Physical Anthropology* 34 (1971):191–203; David W. Frayer, "Gigantopithecus and its Relationship to Australopithecus," *American Journal of Physical Anthropology* 39 (1973):413–26. See also: Edward W. Cronin, Jr., "The Yeti," *The Atlantic* 236, no. 5 (November 1975):47–53; Michael Grumley, *There are Giants in the Earth* (New York: Doubleday & Co., 1974).

3

Early Accounts of the Gorilla

Hanno's Voyage

By "early" is here meant the period beginning with the first historical mention of the gorilla (c. 500 B.C.), and continuing with later accounts of gorillas by European explorers and zoo keepers until the time (1897) of the first exportation of young captive specimens to the United States.

Practically every historical account of the gorilla begins by mentioning Hanno, a Carthaginian admiral and statesman, who is credited with being the first man to bring to the civilized world knowledge that this huge anthropoid ape existed.[1] However, there has always been confusion and uncertainty as to whether the apes referred to by Hanno were gorillas, chimpanzees, or possibly even baboons. Determination of this question would appear to depend on how far southward along the west coast of Africa Hanno's fleet actually sailed or proceeded.

The story is that about the year 500 B.C., Hanno, in command of some sixty ships or galleys of fifty oars each, along with thousands of men and women and with provisions and other necessaries, set out from Carthage on a voyage of exploration and colonization of the west coast of Africa. During the course of this voyage southward—after the fleet had passed through the straits of Gibraltar—Hanno founded one city and established seven colonies. The terminus of the voyage was a bay called the "Horn of the South," or "Southern Horn"; and it is upon the identification and location of this bay that should determine whether or not the apes encountered by Hanno's sailors were actually gorillas.

In his account of this voyage, Hanno remarked that "one day's journey from a great river," the explorers sighted a lofty mountain. Now the "great river" could well have been the Niger, which flows southward through the central portion of Nigeria and ends in a vast delta between the Bight (bay) of Benin and the Bight of Biafra. These two bights, or bays, shaping the southern coastline of Nigeria into what could be construed as a pair of horns, could accordingly have been the "Horn of the South" mentioned by Hanno. And a "day's journey" in an east-southeast direction from the place where the party sighted a high mountain (Mount Cameroon, a volcanic peak 13,350 feet high, is located near the southern coast) would have brought

35

the explorers close to the mountain as well as to the island (presumably that which is today named Fernando Póo) on which the "wild men," or gorillas, were encountered (see map, Fig. 20).

Hanno's narration follows:

> [on the island] there was a lake, and in this there was another island full of wild men. But much the greater part of them were women with hairy bodies, whom the interpreter called "gorillas." But pursuing them we were not able to take the men; they all escaped, being able to climb the precipices, and to defend themselves with pieces of rock. But three females, who bit and scratched those who led them, were not willing to follow. However, having killed them, we flayed them, and conveyed the skins to Carthage; for we did not sail any farther, as provisions began to fail.*

This review of Hanno's epic voyage of 2,500 years ago should establish that the "wild men" encountered by him and his men were in all probability gorillas and not some other form of hairy primates. The "island" mentioned by Hanno could equally have been a peninsula; but even if it were indeed an island, it was sufficiently close (c. twenty miles) to the southeast corner of Nigeria, bordering the Cameroons (typical gorilla country), for a land bridge to have existed sometime in the past.

About the only factor that would rule out the likelihood of Hanno's "wild men" being either gorillas or chimpanzees was their asserted presence on an island in the middle of a lake. Both species of ape are either nonswimmers or non-river-crossers, and so would not likely have taken residence on ground surrounded by water. In that case, as one author has questioned,[2] could not the hairy apelike creatures that Hanno saw have been some belated survivors of the Australopithecines ("Southern apes") that had existed in south and east Africa a half-million years ago? However, the interpreters clearly referred to the animals as gorillas; so at that identification we must leave them. Least of all could they have been baboons, with which smaller, more widely ranging primates the Carthaginians were thoroughly familiar (and would hardly have referred to dog-sized baboons as "wild women").

---

* Some four hundred years later, when Carthage was taken by the Romans in 146 B.C., two of these skins were found still hanging in the temple of Juno. They were said to have belonged to creatures called "gorgones," which was evidently an interpretation of the name "gorilla" given to them by Hanno's interpreter.

Battell's Account

Many centuries were to pass before the next historical mention of the gorilla was made. This was about the mid-sixteenth century, at which time (1559) an English sailor, Andrew Battell, was for some unexplained reason taken prisoner by the Portuguese at Mayoumba on the coast of the Gaboon. The locale, being only about four degrees south of the equator, was near typical western gorilla territory. Battell's travels and narrations constitute one of the "Pilgrimes" (voyages and adventures), published by Samuel Purchas in 1623.[3] Battell presumably met with both the gorilla and the chimpanzee—former apes he described as follows:

> The woods are so covered with baboones, monkies, apes and parrots that it will feare any man to travaile in them alone. Here are also two kinds of monsters, which are common in these woods and very dangerous. The greatest of these two monsters is called (by the Portuguese) *pongo* in their language, and the lesser is called *engeco*. The pongo [gorilla] is in all proportions like a man, for he is very tall, and hath a man's face, hollow-eyed, with long haires upon his brows. His body is full of haire, but not very thicke, and it is of dunnish color. He differeth not from man, but in his legs, for they have no calfe. He goeth alwaies upon his legs, and carrieth his hands clasped on the nape of his necke when he goeth upon the ground. They sleepe in trees, and build shelters for the raine. They feed upon fruit that they find in the woods, and upon ants, for they eate no kind of flesh. They cannot speake, and have no understanding more than a beast. The people of the countrie, when they travaile in the woods, make fires where they sleepe in the night, and in the morning when they are gone, the pongos will come and seat about the fire till it goeth out, for they have no understanding to lay the wood together. They goe many together, and kill many negroes that travaile in the woods. Many times they fall upon elephants that come to feed where they be, and so beate them with their clubbed fists and pieces of wood that they will runne roaring away from them. The pongos are never taken alive, because they are so strong that ten men cannot hold one of them; but they yet take many of their young ones with poisoned arrowes. The young pongo hangeth on his mother's belly with his hands fast clasped about her, so that, when the country people kill any of the females, they take the young which hangs fast upon the mother. When they die among themselves, they cover the dead with great heapes of boughs and wood, which is commonly found in the forests.

As noted, Battell's "pongo"* is the gorilla, and his "engeco" the chimpanzee. The name *engeco* would appear to be corroborated by the fact that some of the natives of the Gaboon still call the chimpanzee the *enche-eko*. In view of the name *pongo* having been applied by the Portuguese of Battell's time to the gorilla, it is difficult to understand why the same name was adopted, much later, as the generic term for the orangutan, an entirely different ape native to the Asiatic islands of Borneo and Sumatra.

Battell's description of the gorilla, while containing a number of obviously exaggerated statements, may in part have expressed what the natives of the region (who were good at embellishing their experiences!) told him, rather than things that he had actually witnessed. Too, his report, being worded in the phraseology of his day, should take that factor into account. But rather than state, as Battell did, that the only particular in which a gorilla's physique differed from that of a man was in its lack of a "calfe," he should have emphasized instead the ape's much longer arms, shorter legs, long and massive torso, feet that could be used like hands, etc.—all of which would have been immediately evident to any observer of the creature. Although Battell, after his release by the Portuguese, became a soldier in their colonial troops, and in that capacity spent many years in Africa, his account of the gorilla, while presumably first-hand, nevertheless fails to impart the conviction that Battell ever came actually face to face with a living specimen of this retiring, difficult-to-contact "king of the African jungle."

Some Other Reports

After Battell's account, the next mention of the gorilla in European literature would appear to have been that of a French traveler named M. de la Brosse. In a book published in 1738, relating his experiences on the coast of Angola, de la Brosse made the following comments about gorillas:

> Their face is dull, nose snubbed and flat, ears without cushions [?], skin a little lighter [?] than that of a mulatto, hair long and thin [?] on many parts of the body, stomach extremely tight [distended?], heels flat [?]. They walk on two feet, and on all-fours when they have the fancy to do so.

---
* The name *pongo*, as referred to by Battell, may well have been a Portuguese version of the Congo natives' name for the the gorilla, *m'nungu*.

It is clear from this "description" that M. de la Brosse either never saw a gorilla (their range does not now, and probably did not in the early eighteenth century, cross the Congo River and extend southward into Angola) nor was a writer of extravagant fiction. This is additionally noticeable when he says that gorillas attain a height of up to seven feet, live in "huts," use clubs to defend themselves, and carry off native women to live with them. It is small wonder, when yarns such as this were accepted verbatim by credulous European readers, that radically inaccurate conceptions of the gorilla were formed.

In 1774, Lord Monboddo (James Burnet), a learned Scottish jurist, received a letter from a sea captain who had evidently made explorations on the west coast of Africa. This captain, like other observers of his day, was so astonished at the apparent size of the gorilla that, in his account to Lord Monboddo, he described the ape thus: "This wonderful and frightful production of nature walks upright like a man, is from 7 to 9 feet high (!) at maturity, thick in proportion, and amazingly strong."

Evidently the next author to mention the gorilla was Thomas Edward Bowdich (1791–1824), an English traveler and writer, of Bristol. In his book *Mission from Cape Coast Castle to Ashanti*, published in London in 1819, Bowdich gave this account:

> The favourite and most curious subject of our conversation on natural history was the *Ingena*, an animal like the orang-utang, but of much greater size, being five feet in height and four feet [!] across the shoulders. Its paw was said to be still more disproportionate, and one blow of it would cause death. Travellers who go to Kaybe [Kribi?] frequently encounter him. He lies in ambush to kill passers-by, and he principally feeds on wild honey. Among other traits which characterize this animal, and on which all persons agree, it is reported that he builds for himself a hut, in rude imitation of that of the natives, and that he sleeps outside on the roof of this dwelling.

Unfortunately, as is seen, Bowdich repeated the fabrications, or misinterpretations, about the gorilla that were current in his time, and that in the first place may have been told to white men by natives who were eager to put across a sensational story. Note, for instance, Bowdich's use of the words "was said," "all persons agree," "it is reported," etc. Bowdich's early demise was caused by jungle fever, which he contracted at the age of thirty-three.

Savage's and Wilson's Report

The first authoritative and trustworthy report on the gorilla was presented in 1847 by two missionaries to west Africa: Thomas S. Savage, of England, and J. Leighton Wilson, of the United States, who were stationed together in the Gabon. In that year, Dr. Savage wrote to the eminent anatomist Sir Richard Owen, enclosing drawings of the skull of an ape from that region, which was described as being much larger than the chimpanzee, and feared by the natives more than the lion or any other wild beast of the forest. These sketches showed the pronounced bony crests atop the skull, which distinguished the gorilla from the chimpanzee. "At a later date in the same year," writes Owen, "were transmitted to me from Bristol two skulls of the same large species of chimpanzee as that notified in Dr. Savage's letter; they were obtained from the same locality in Africa, and brought clearly to light evidence of the existence in Africa of a second larger and more powerful ape."[4] Later, these and several other ape skulls formed the basis of a description by Owen of the cranial distinctions of the gorilla.[5] However, since Owen regarded this ape as being simply a large species of chimpanzee, he gave it the name *Troglodytes savagei*. It seems incongruous that this famed anatomist should adopt a name (*Troglodytes*) for a *tree*-dwelling ape, which means essentially *cave*-dweller.

Further, about the same time that Dr. Savage supplied Professor Owen with the aforementioned sketches, he also sent a skull of the newly discovered ape, together with a description of the animal itself, to Dr. Jeffries Wyman, an anatomist in Boston. This description was written by Dr. Savage's co-missionary, Dr. Wilson, who, it would appear, was actually the first white man to possess and examine the skull of a gorilla. This skull he had obtained from a native in 1846. But it was Drs. Savage and Wyman who first published a description of this osteological specimen.[6] To it they gave the scientific name *Troglodytes gorilla,* thus establishing the name prior to Owen's use of the specific name *savagei*.

The Genus Gorilla

In 1851, the French naturalist Isidore Geoffroy Saint-Hillaire proposed for the "new" ape a separate genus, *Gorilla,* on the basis of such distinctions as the great cranial crests, the shape of the teeth, the disparity in size between the sexes, etc. According to the rules of nomenclature adopted by zoologists, Saint-Hillaire had no right to supersede the prior generic name of *Troglodytes,* which applied to the chimpanzee and had been extended, both by Wyman and Owen, to include the gorilla. Nevertheless, Saint-Hillaire's proposal received scientific approval.[7] However, his proposed species name, *gena* (a contraction of the native name *ingena, n'gena,* or *engé-ena*), has been replaced by a repetition of the generic name, so that the present full designation (of the western or lowland species) is *Gorilla gorilla*.

Also in 1851, a Captain Harris presented to the Royal College of Surgeons in London the first skeleton of a gorilla that had ever been brought to England; while in the same year, another skeleton was sent to the Academy of Natural Sciences in Philadelphia by H.A. Ford, a medical missionary in equatorial Africa. Ford may have been the first white man to study a living gorilla, for he was given the opportunity to do this over a period of several months with a very young specimen that was kept in captivity by natives. Ford described the young gorilla as being of an intractable nature and that it had once bitten him. Possibly because of its discontent, it lived for only a short while.

Paul Du Chaillu

An epoch-making account of the gorilla in its

Fig. 17. An adult male gorilla in its native habitat, as shown in Du Chaillu's book *Explorations and Adventures in Equatorial Africa,* 1861.

native equatorial jungle was given by the American-born, French-descended explorer Paul Belloni Du Chaillu after his return from the Gabon, to which country he had gone in 1855 for the specific purpose of learning, first-hand, about the giant anthropoid and its actual habits. Alone, and only twenty years of age, Du Chaillu spent nearly four years in the forests of west Africa, living among the natives. Evidently he was the first white man to face a living adult gorilla in its natural haunts. He publicized his observations in a comprehensive book on the subject.[8] Although Du Chaillu's statements about the gorilla met with considerable ridicule and refutation at the time, subsequent findings by other investigators confirmed most of the statements he had made. However, his writings were of a nature that waxed quite spectacular at times, as the following quotation from his book shows:

> Suddenly, as we were yet creeping along in a silence which made even a heavy breath seem loud and distinct, the woods were at once filled with a tremendous barking roar; then the underbush swayed rapidly just ahead, and presently stood before us an immense gorilla. He had gone through the jungle on all fours, but when he saw our party he erected himself and looked us boldly in the face. He stood about a dozen yards from us, and was a sight I think I shall never forget. Nearly six feet high (he proved four inches shorter), with immense body, huge chest, and great muscular arms, with fiercely-glaring, large, deep grey eyes, and a hellish expression of face, which seemed to me some nightmare vision;—there stood before us the king of the African forest. He was not afraid of us; he stood there and beat his breast with his large fists till it resounded like an immense bass drum (which is his mode of bidding defiance), meantime giving vent to roar after roar.
> The roar of the gorilla is the most singular and awful noise heard in these African woods. It begins with a sharp *bark,* like an angry dog, then glides into a deep bass *roll,* which literally and closely resembles the roll of distant thunder along the sky, for which I have sometimes been tempted to take it where I did not see the animal. So deep is it that it seems to proceed less from the mouth and throat than from the deep chest and vast paunch.
> His eyes began to flash fiercer fire as we stood motionless on the defensive, and the crest of short hair which stands on his forehead began to twitch rapidly up and down, while his powerful fangs were shown as he again sent forth a thunderous roar, and now truly he reminded me of nothing but some hellish dream-creature—a being of that hideous order, half man, half beast, which we find pictured by old artists in some representations of the infernal regions. He advanced a few steps, then stopped to utter that hideous roar again, advanced again, and finally stopped when at a distance of about six yards from us. And here, as he began another of his roars and beating his breast in rage, we fired and killed him.
> With a groan which had something terribly human in it, and yet was full of brutishness, he fell forward on his face. The body shook convulsively for a few minutes, the limbs moved about in a struggling way, and then all was quiet—death had done its work, and I had leisure to examine the huge body. It proved to be five feet eight inches high, and the muscular development of the arms and breast showed the immense strength it had possessed.

Winwood Reade

One reader who was particularly upset by what he regarded as a false or exaggerated report on the part of Du Chaillu was a young English writer named William Winwood Reade. So convinced was he on this point that he decided to go to Africa, over the same routes through the Gabon and the Congo that Du Chaillu had taken some six years previously, and to disprove various of the latter explorer's statements regarding the gorilla. Reade spent five months in the gorilla country and upon his return addressed a meeting of the Zoological Society of London as follows:

> The evidence which I now lay before you is composed of statements made to me by men who had killed gorillas. It is collected from three distinct parts of Equatorial Africa, namely, from the Balengi of the Muni river, from the Shekani and Fans of the Gaboon, and from the Commi, Bakeli, etc., of the Fernand Vaz. But from the last river, where gorillas are most plentiful, I obtained the most information.
> The gorilla is found in those thick and solitary places of the forest where animal life is scarce. His food is strictly vegetable. He moves along the ground on all fours; sometimes he goes up into the trees to feed on fruit, and at night he sleeps in a large tree. When the female is pregnant the male builds a nest, where she is confined, and which she abandons as soon as her young one is born.
> The gorilla does not beat its breast like a drum. It utters a kind of short, sharp bark when enraged, and its ordinary cry is of a plaintive nature.

Fig. 18. "Death of My Hunter," another illustration from Du Chaillu's account of his gorilla-hunting trek into the African jungle.

Fig. 19. "Death of the Gorilla," showing Du Chaillu shooting his first specimen of this (then) newly discovered manlike ape.

With respect to its ferocity, the hunters have a proverb, "Leave the *ngina* alone, and it will leave you alone." When it is at bay and wounded, it will attack man; like the stag, the elephant, and other animals naturally timid. But it makes this attack on all fours; the hunters [natives], who are themselves as nimble as apes, often escape from it as men escape from the charge of an elephant. I have seen a man who had been wounded by a gorilla; his wrist was crippled and the marks of the teeth were visible. He told me that the gorilla seized his wrist and dragged it into his mouth; it was contented with having done this, and went off. The nearest approach to an erect posture which the gorilla attains to is by supporting itself by holding on to the branches. When I asked the people of Ngumbi whether a man had ever been killed by a gorilla, they said their fathers had spoken of such a thing, but that nothing of the kind had happened within the memory of anybody living.

I can make one or two positive assertions from my own experience. Although I never succeeded in seeing a gorilla in its wild state, I can assert that it travels on all fours; for I have seen the tracks of its four feet, over and over again. I can assert that it runs away from man, for I have been near enough to hear one running away from me; and I can assert that the young gorilla is as docile as the young chimpanzee in a state of captivity, for I have seen both of them in a state of captivity.

The foregoing account by Reade was supplemented by other details in a book that he had published in 1863.[9] Despite Reade's avowed intention of showing Du Chaillu up as a prevaricator, it does not appear that he succeeded very well. Note, for instance, that he says "the gorilla does not beat its breast." This action is such a characteristic one that it is frequently seen being performed by zoo gorillas. Again, Reade asserts, or indicates, that natives have not been killed by gorillas. Yet a number of such killings have been reported.* Finally, since Reade, during his five-month trek through the jungle, never saw a living gorilla, he concluded that neither Du Chaillu, who had spent nearly four years under the same conditions, nor any other white man had up to that time seen one. If this were so, perhaps all the more credit should be given to Du Chaillu for his known contributions to zoology—such as the gorilla skeletons of which he gives meticulous measurements in his book, and the preserved specimen or specimens that he sent to natural-history museums while he was still in Africa.

An interesting commentary on Du Chaillu was given by William T. Hornaday, who at the time (1915) was director of the New York Zoological Park. He remarked:

> Skulls and science are all very well; but for our knowledge of the gorilla we owe most to Paul Du Chaillu and his popular book, "Equatorial Africa." It was through that hair-raising and altogether masterful presentation that the greatest and most fearsome of the great apes burst upon an astonished world. It is a matter of history that when the doughty explorer landed in America with a priceless collection of gorilla skins and skeletons, expecting and deserving an ovation, he was bitterly disappointed. At that time American zoology was still in its swaddling clothes. There were few museums of any kind, and few persons who cared about ape skins or ape stories from far distant Africa. As a result, Du Chaillu resentfully gathered up his collection and took it over to England, where it was better appreciated.[10]

A few years later, Carl E. Akeley, while feeling that Du Chaillu described probably with accuracy what he saw, deplored his use of emotional embellishment.[11] In 1929, Robert M. Yerkes credited Du Chaillu with a work which is "illuminating, despite its highly colored and exaggerated descriptions." But Yerkes pointed out also the explorer's scholastic shortcomings and unscientific interpretations.[12]

To sum up, Du Chaillu's statements would have been more difficult to have fabricated than to have expressed, as he did, what he probably saw and experienced. His emotions, under the contributing conditions, are understandable.

Dr. Falkenstein's Young Gorilla

Proceeding timewise another few wears, a German "gorilla expedition" disembarked at Loango, in the Middle Congo, in 1873, and carried on explorations until 1876. However, these zoologists found gorillas to be very rare so near the coast (which at Loango was probably their extreme southward range), although they were to be met with "in or near the mountainous region farther inland." The explorers learned that during 1851 and 1852, gorillas had been seen on the coast, where they may have been driven from the interior by want of food; but that since that time none had strayed that far westward. Toward the end of their

---

* For example, an Ituri Forest pygmy who was torn literally limb from limb by an enraged and possibly psychopathic male gorilla. See "The Congo Gorilla," by Wilmon Menard, *Sports Afield*, June 1952, p. 40. See also Figure 97.

sojourn in Africa (i.e., in October 1875), the party's physician, a Dr. Falkenstein, had the good fortune to obtain from a Portuguese trader a young gorilla that had been owned originally by a black man who had shot the mother and then captured the young one. This gorilla was sold by Dr. Falkenstein to the Berlin Aquarium for the then large sum of twenty thousand marks (c. five thousand dollars)—the sum going to the benefit of the expedition funds.

Dr. Falkenstein's young gorilla, which was later given the name of "Pongo," arrived with the members of the German expedition at Liverpool in mid-1876. A Mr. Moore, curator of the Free Public Museum of Liverpool, visited the young gorilla and sent the following account of it to the local *Times* newspaper:

> A veritable young living gorilla was yesterday brought into Liverpool by the German African Society's Expedition, which arrived by the streamship Loanda, from the West Coast. The animal is a young male, in the most perfect health and condition, and measures three feet in height. Its beetling brows, flattened, podgy nose, black muzzle, small ears, and thick fingers, cleft only to the second joint, distinguish it unmistakably from the chimpanzee. Only one other specimen has been brought alive to England. In the winter of 1855–56 a young female gorilla, of much smaller size, was exhibited by the late Mrs. Wombwell in Liverpool and other places.* It died in March, 1856, and was sent to Mr. Waterton, of Walton Hall, who preserved the skin for his own collection, and sent the skeleton to the Leeds Museum. This specimen I saw living in Liverpool and dead in Walton Hall. All subsequent attempts to import the gorilla alive have failed; and unfortunately the British public will have no opportunity of profiting by the present success, as the members of the expedition, with commendable patriotism, are taking the animal on Saturday *vid* Hull to Berlin. Could it have graced our own Zoological Gardens it would have been the lion of the day; for in addition to the great scientific interest of the species, the abounding life, energy, and joyous spirits of the example would have made it a universal favourite. Courteously received at Eberle's Alexandra Hotel by the members of the expedition, I found the creature romping and rolling in full liberty about the private drawing-room, now looking out of the window with all the becoming gravity and sedateness, as though interested, but not disconcerted by the busy multitude and novelty without; then bounding rapidly along on knuckles and feet to examine and poke fun at some newcomer; playfully mumbling at his calves, pulling at his beard (a special delight), clinging to his arms, examining his hat (not at all to its improvement), curiously inquisitive as to his umbrella, and so on with visitor after visitor. If he becomes over-excited by the fun, a gentle box on the ear will bring him to order like a child—like a child, only to be on the romp again immediately. He points with the index finger, claps with his hands, pouts out his tongue, feeds on a mixed diet, decidedly prefers roast meat to boiled, eats strawberries—as I saw—with delicate appreciativeness, is exquisitely clean and mannerly. The palms of his hands are beautifully plump, soft, and black as jet. He has been eight months and a half in the possession of the expedition, has grown some six inches in that time, and is supposed to be between two and three years of age.[13]

The following account of "Pongo," after he had arrived in Germany, was given by Dr. von Hermes, then director of the Berlin Aquarium. The report forms an interesting and informative sequel to that of Mr. Moore, just quoted.

> The Aquarium of Berlin has always set great value on the possession of Anthropomorphous Apes. During the last few years it has been able to procure specimens of all of the four species—the Gibbon, Chimpanzee, Orang-utan, and Gorilla. In this way I had the best of opportunities to study them in captivity and compare them with each other.
> The chief among all the Anthropomorpha is the Gorilla. It seems as if he was born with a patent of nobility among Apes. Our Gorilla [Pongo], about two years old, is nearly twenty-eight inches high. His body is covered with gray, silky hair, the head alone having a reddish color. His thick-set, robust shape, his muscular arms, his smooth, shining black face with well-shaped ears, his large, black, clever eyes—all strike one as exceedingly human. If his nose was not so broad he would look like a Negro boy. What serves to heighten this impression is his awkwardness; all his movements seem those of an ungainly boy rather than an Ape. When he sits there like a Chinese pagoda, his gaze directed upon the spectators, and suddenly with a bright nod claps his hands, he has conquered all hearts at a stroke. He likes company, makes a difference between young and old, male and female. He is kind to little children, likes to kiss them, and allows them liberties, without taking advantage of his superior strength. Older children he does not

---

* In "Wombwell's Traveling Menagerie."

treat so well, although he likes to play with them, to race around tables and chairs which he frequently upsets, playfully slaps their faces sometimes, and also thinks nothing of trying his teeth on their legs. He is fond of ladies, likes to sit in their laps and hug them, or sit still, with his head on their shoulders. He also likes to play in the common cage, but conducts himself there as an unconditional autocrat. Even the Chimpanzee has to obey him, though the Gorilla treats him more as an equal, selecting him as his only playfellow and sometimes bestowing rather rough caresses on him, while he pays no attention to the smaller fry. Sometimes he gets hold of the Chimpanzee and rolls on the floor with him. If the Chimpanzee escapes, the Gorilla falls to the floor, on his hands, like an awkward boy. His gait resembles that of the Chimpanzee: they both walk on the soles of their feet, supporting themselves on the back of the hands. But the Gorilla turns his toes out more, and holds his head higher, producing the impression that he belongs to a better class of society.

His manner of life is just human as his appearance. At about eight o'clock in the morning he wakes up, yawns, scratches himself in various places and remains sleepy and apathetic till his glass of morning milk is brought. That rouses him. He gets up, peers around to see whether he cannot find something to destroy in the room, looks out of the window, claps his hands, and if he has no better company, tries to play with the keeper. The latter must always be with him. The moment he is left alone, he screams. At nine o'clock he is washed, a process in which he finds much pleasure, and expresses his satisfaction in grunts. Living with his keeper, he gets his meals at corresponding times with him. For luncheon he eats a couple of Frankfurt or Vienna sausages, or a sandwich with cheese, or smoked beef. His favorite drink is Weiss beer, and he looks remarkably funny when he tries to hold the large glass with his short, thick fingers and one foot. At one o'clock the keeper's wife brings in dinner. While he was living in my house, last summer [1877], he was evidently yearning for this hour. He always ran to open the door himself, when he heard the bell. As soon as the woman came in, he would investigate the dishes and sometimes help himself to a little of some dainty. She would punish him with a slap, and then he would behave and sit quietly. The first course is a cup of bouillon, which is emptied to the last drop. Then comes a dish of rice or vegetables, preferably potatoes, carrots or parsnips, cooked with meat. The woman insists on his eating properly, and he can handle a spoon quite well, but the minute he thinks he can do so unobserved, he puts his mouth in the dish. He likes a piece of roast fowl best at the end of his meal. At the conclusion of dinner he takes a nap of an hour or an hour and a half, and is then ready for new pranks. In the course of the afternoon he gets some fruit, while his evening meal consists of milk or tea with bread and butter. At nine o'clock he goes to bed. He has a nice mattress and covers himself with a blanket. The keeper stays with him until he is asleep, which does not take long. He likes best to sleep in the same bed with the keeper, hugging him and putting his head on some part of his body. He sleeps all night through and does not awaken until eight in the morning. A glass house in connection with a little conservatory adapted for palms, has been built specially for him, to take the place of the damp atmosphere of his tropical home. In this way I hope that aided by his robust nature, our Gorilla will be spared us and will long be the greatest ornament of our Aquarium, an honor to Germany, a joy to humanity, a glory to science.[14]

Brehm, in his book *Life of Animals* follows the foregoing quotation with these comments:

This Gorilla [Pongo] died on the 13th of November, 1877, after having been watched for nine months in Africa and fifteen months in Berlin, and having successfully made a trip to England.*

Fig. 20. Dr. Falkenstein's young male gorilla, "Pongo" (or M'Pungu), which, in 1876, was the first of his species to be seen in continental Europe, and the second to be seen in England.

* The cause of death was acute inflammation of the bowels, an affliction that in those days carried off many infants and young children.

43

The second gorilla was taken to Europe by Pechuel-Loesche, the former companion of Falkenstein, who arrived with him in 1883, on his return from the Congo, and was also presented to the Berlin Aquarium. He lived under the care of Director Hermes for fourteen months and died of the same disease as the first Gorilla. It is a notable fact that neither of the animals suffered from sea-sickness.

Von Koppenfels

Brehm also mentions the German hunter Hugo von Koppenfels, who in 1874 shot his first gorilla, an adult male, on Christmas of that year. Shortly thereafter he shot a second specimen, a large male, which he judged to weigh about 400 pounds. Von Koppenfels describes his encounter with the second gorilla in much the same terms as were used by Du Chaillu; that is, with the gorilla roaring and beating his chest as he advanced in an erect posture, etc.

The shaggy hair on his head raised itself with a vibrating motion, and it seemed that my terrible opponent was going to attack me. If I had retreated in time, I am fully convinced that the Gorilla would not have approached me, but such was not my intention. Mastering my agitation, I took a steady aim at his heart, and pulled the trigger. The animal jumped high up, and spreading his arms, fell on his face. He had seized in falling, a liana two inches in circumference, and so powerful was his grasp* that he tore it down along with dry and green branches from the tree.[15]

Thus, von Koppenfels must be listed—along with Du Chaillu, Winwood Reade, Dr. Falkenstein, and Pechuel-Loesche—as one of the earliest explorers in modern times to have either personally encountered, or to have contributed knowledge on, the mighty ape of the forests of west Africa known as the engé-ena, or gorilla.

* * *

Some Early Zoo Gorillas

In review, we have seen that up until 1883 only three living gorillas, all immature specimens, had been transported successfully from their native west-African habitat into captivity in Europe. The first was the young female that was exhibited in Liverpool in the winter of 1855–56 by Mrs. Wombwell. This little ape, which was at first thought to be a chimpanzee, was only later recognized for the hitherto unknown species it turned out to be. From then on it was called "The Wombwell gorilla." The second imported gorilla was "Pongo"; and the third, the young male brought by Pechuel-Loesche to the Berlin Aquarium in 1883. None of these young specimens survived long under the then-customary zoo conditions in which it was exposed to human diseases.

Following the foregoing "pioneers" among captive gorillas were seven other young specimens, which were exhibited in the Garden of the Zoological Society of London until 1908. All, in all probability, were individuals belonging to the west-African, or lowland, species.

---

* The explanation of this gorilla's unbreakable grasp is that he evidently fell with his wrists fully extended, in which position the fingers of a gorilla automatically "lock" into a hook shape. This is caused by the tendons of the hand not permitting full extension of the second and third (or distal) phalanges, except when the wrist is flexed.

---

1. Male. Purchased October 1887. Lived 2 months.
2. Female. Purchased March 1896. Lived 5 months.
3. Female infant. Purchased August 1904. Lived 3 weeks.
4. Female c. 6 yrs. Purchased August 1904. Lived 5 weeks.
5. Female. Deposited August 1905. (sent to America after 10 days).
6. Female infant. Deposited March 1906. Lived 2 months.
7. Male. Deposited March 1908. Lived 1 week.

And here are listed some gorillas that were exhibited in Europe following "Pongo" (1876) and Pechuel-Loesche's young male (1883).

1. Male, "Pussi." Received in Liverpool, September 1897. Exhibited in Breslau, Germany, from September 1897 to October 1904: seven years.
2. Young male. Received in Dublin, December 1906. "Lived only a few weeks."
3. Young female, "Empress." Received in Dublin, January 1914, when about twenty months old. Died in May 1917, after being in captivity $3\frac{1}{3}$ yrs.
4. Male, "John Daniel." Received in London in 1918 at the age of about twenty months. Kept in London by Miss Alyse Cunningham from 1918 to March 1921, when he was sold to the Ringling Brothers and Barnum & Bailey Circus in the United States. "John Daniel" was then about $4\frac{1}{2}$ years old. (See also chapter 9.)
5. Male, "Sultan," or "John Daniel II." Received in London in 1923 as an infant. Kept by Miss Cunningham until the time of his death in 1927.
6. Young male. Captured by Ben Burbridge and delivered to the Antwerp Zoo in 1922.
7. Young male. Captured by Ben Burbridge and delivered to the Antwerp Zoo in 1925. Both of these young gorillas, which were of the mountain species, died after being only a few months in captivity.

Detailed accounts of other captive gorillas, specifically those imported into the United States, beginning with the young unnamed male that arrived in Boston in 1897 and lived only five days, are given in chapter 10.

Notes

1. Hanno, *The Voyages of Hanno*, trans. Thomas Falconer (London, 1797).
2. Herbert Wendt, *Out of Noah's Ark*, trans. Michael Bullock (Cambridge, Mass.: The Riverside Press, 1959), p. 14.
3. Samuel Purchas, *Purchas, His Pilgrimes*, 4 vols. (London, 1625). Among this series of accounts, Purchas tells of the adventures of Andrew Battell. A reprint in 20 volumes was published in Glasgow, 1905–7.
4. Richard Lydekker, in *Harmsworth Natural History*, 3 vols. (London, 1910), 1:184.
5. Richard Owen, in *Transactions of the Zoological Society of London* 4, pt. 3 (1851):75–88.
6. Thomas S. Savage and Jeffries Wyman, "Notice of the External Characters and Habits of *Troglodytes gorilla*, a New Species of Orang from the Gaboon River," *Proceedings of the Boston Society of Natural History* 2 (1847):245–47, and 5 (1847):417–43. (The first scientific identification of the gorilla as a new and hitherto unclassified species of manlike ape.)
7. Isidore Geoffroy Saint-Hillaire, "Note sur le Gorille," *Annals of Science and Nature*, 3d ser. 16 (1851):154–58. (The first proposal of a *generic* separation of the gorilla from the chimpanzee.)
8. Paul Belloni Du Chaillu, *Explorations and Adventures in Equatorial Africa* (New York: Harper & Bros., 1861). (See also Du Chaillu's *The Gorilla Country*, 1868.)
9. William Winwood Reade, *Savage Africa* (London, 1863; New York, 1864).
10. William T. Hornaday, "Gorillas, Past and Present," *Bulletin of the New York Zoological Society* 18 (January 1915):1181.
11. Carl E. Akeley, "Is the Gorilla Almost a Man?" *World's Work*, September 1922, p. 527.
12. Robert M. and Ada W. Yerkes, *The Great Apes* (New Haven, Conn.: Yale University Press, 1929), p. 34.
13. In J. Fortune Nott, *Wild Animals Photographed and Described* (London, 1886), pp. 534–35.
14. In A. E. Brehm, *Brehm's Life of Animals* (English edition), "Mammalia" (Chicago, 1896), pp. 14, 16.
15. In ibid., p. 14.

4

Races, Natural Habitats, and Geographical Distribution of the Gorilla

Species and Ranges

Part of the description of any animal should include its natural habitat and the geographical locality in which the animal lives. The habitats of the four existing anthropoid apes are the equatorial forests of west and central Africa and of Malaysia, respectively. The gorilla and the chimpanzee are found only in Africa; the orangutan, in Borneo and Sumatra; and the gibbon, in the warmer regions of Southeast Asia, particularly in and around the Malay Peninsula. The siamang, the largest of the gibbons, is apparently confined to Sumatra.

Closely involved with the geographical distribution of the gorilla is the question of how many species, subspecies, or local races its genus includes. While Elliot, in 1913, listed some fifteen species and subspecies of the gorilla, Coolidge in 1929 reduced the number to one species and two subspecies: (1) the typical or western lowland gorilla (*Gorilla gorilla gorilla*), and (2) the eastern or mountain race (*G. g. beringei*).[1] However, in 1934, Schultz, on the basis of some twenty different anatomical features distinguishing the mountain gorilla from the lowland form, proposed that the mountain gorilla be recognized as a distinct species and not merely a regional variant of the lowland gorilla.[2] Although I am inclined, as a result of my own anatomical studies, to agree with the opinion that Schultz expressed at that time, it would appear that he has since reverted to the conventional point of view (i.e., one species), on the grounds of the since-observed great variability in practically all the physical characteristics of the gorilla. Despite this, I feel that Schultz's earlier point of view has much to commend and so validate it. Accordingly, in this book I shall refer to the lowland gorilla and the mountain gorilla as being distinct species: *G. gorilla* and *G. beringei*, respectively. My reasons for adopting this classification are given in chapter 5.

Coolidge defined the range (as of 1929) of the lowland or coast gorilla as follows:

For the Coast Gorilla, the westernmost boundary approximates the Cross River in the southern

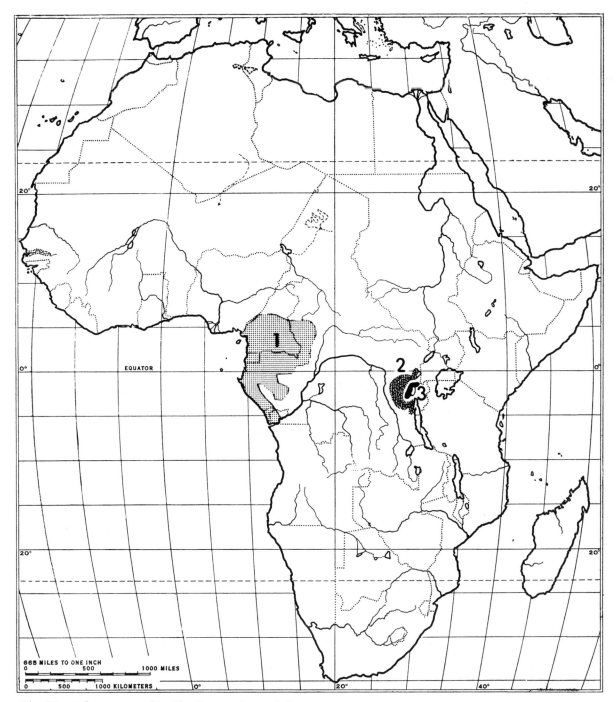

Fig. 21. The present distribution, as far as is known, of the two species (three races) of the gorilla. 1. western lowland gorilla (*G. gorilla gorilla*); 2. eastern lowland gorilla (*G. gorilla graueri*); 3. mountain gorilla (*G. beringei*). The latter species, of which perhaps only one thousand individuals survive, is confined to the area to the west and north of Lake Kivu, shown here by the solid black shape to the left of figure 3.

provinces of Nigeria. The most westerly point actually recorded is Ikom, 8°40′ east and 6° north. The northernmost point is close to Basho, 9°25′ east, 6°7′ north. On the east we have reports from several places such as Wesso and Nola on the Sanaga [Sangha] River. The Sanaga [Sangha] River, about 16°15′, seems to mark the eastern boundary of the range of the Coast Gorilla.* On the southeast the line follows the border of the forest which reaches its southernmost limit at Mayombe [?] on the edge of the Belgian Congo [now Zaire], 5° south, 13° east. Along the Atlantic coast in most places the forest begins a little way inland. Gorillas have been reported actually on the coast, but generally they are found not closer than thirty miles from the sea. They seem especially plentiful along the Gaboon, Ogowe, Camp, and Sanaga [Sangha] Rivers (p. 363).

Of the mountain gorilla, Coolidge gives the distribution as follows:

The Mountain Gorilla is found in a comparatively narrow strip of the eastern Congo [now Zaire]. Its principal habitat is the mountain forest as distinguished from the lowland forest of the Belgian Congo [now western Zaire]. Its northern limit is Mulu, 0°10′ south, 29°10′ east. We find it as far west as Walikale, 1°20′ south, 28°1′ east, where it strays a little into the lowland forest.** The eastern limit seems to be

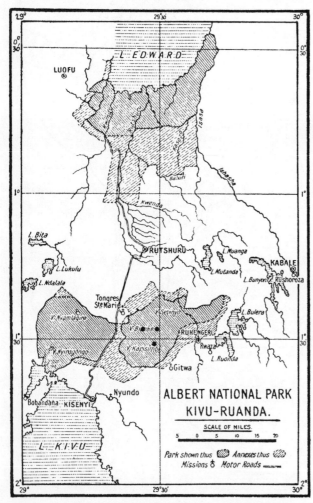

Fig. 22. Albert National Park, on the border of Uganda and Ruanda, which was established on 21 April 1925 by the King of Belgium for the protection of the mountain gorilla. The three round spots indicate Mount Mikeno, Mount Visoke, and Mount Karisimbi, respectively, in the Virunga (volcanic) Range—favorite haunts of this species (*Gorilla beringei*). Note that the three peaks are only about five miles apart.

Fig. 23. Slopes of Mt. Mikeno (14,787 feet), showing the dense vegetation typical of the mountain gorilla's habitat. (After Bingham, 1932.)

* The range, as later established, extends somewhat farther eastward, across the Sangha River to about 17°15′ E.

** This (see Fig. 23) is rather the habitat of the eastern lowland gorilla (*G. g. graueri*). D.P.W.

close to Kigezi in Uganda, 1°15′ south, 29°45′ east. The southern limit is Baraka on Lake Tanganyika, 4°19′ south, 29°2′ east. In this entire region the gorillas that are most known and accessible are the troops that inhabit the volcano regions where Akeley died while studying them. Whether they are entirely isolated from contact with outside gorillas at the present time [1929] is doubtful and has not yet been established. In the mountains back of Baraka, Boko, Uvira, and Katana large troops have been recently found in the upland forests. (p. 383).

In the latter sentence, Coolidge again refers to a locality that probably houses not *G. beringei*, but rather the eastern lowland gorilla (*G. g. graueri*).* This would account also for the "large troops" then reported.

Although the western form of the lowland gorilla is generally referred or alluded to as a single physical type, it has been found that there are significant differences between the skulls of gorillas from the west-coast district of Gabon (or Gabun; Eng., Gaboon) and those from the northernly adjoining Cameroons. These differences are shown in the measurements of gorilla skulls from these two respective localities, recorded by Randall.[3] The study made by the latter author shows that skulls from the Gaboon are predominantly long, while those from the Cameroons are predominantly broad. The respective cranial measurements are given in chapter 5. To quote Randall, p. 323:

> A distribution curve made with the addition of the Coolidge data indicates that a third hyperbrachy [very broad] type is probably present, but this type is also found in both the geographical regions, with a predominance in the Cameroons.

It may be added that prior to this, in 1923, Lord Rothschild had divided the species *gorilla* into three subspecies:

Gorilla gorilla gorilla (Gaboon); skull long with occiput narrow and peaked.
Gorilla gorilla dielhi (Cameroons); skull broad, with occiput broad and flat.
Gorilla gorilla beringei (Mountain)

Thus it is seen that Rothschild regarded the Gaboon gorilla as the typical form of the species.[4] Indeed, it was from the Gaboon that most of the specimens shot by early hunters, such as Du Chaillu, were obtained.

Aschemeier considered the Fernan Vaz [west-central Gaboon] district as the best region for finding both the gorilla and the chimpanzee. It was his experience that a gorilla will occasionally advance to attack an intruder, even when not wounded.[5]

Groves, using the dimensions and proportions of 469 adult male and 278 adult female gorilla skulls in various museum collections, deduced the existence of three races or subspecies:[6]

(1) Western lowland gorilla (*G. g. gorilla*), inhabiting the tropical rain forests of southeast Nigeria, the Cameroons, Gaboon, equatorial Guinea, Cabinda, and the Congo Republic (southern part).
(2) Eastern lowland gorilla (*G. g. graueri*), found in the dense lowland forests of the Mwenga-Fizi region, Utu and Tshiaberimu in eastern Zaire (formerly the Belgian Congo), and the Kayonsa Forest in Uganda (?). Groves gives to this subspecies the scientific name *G. g. manyema*, after a skull in Lord Rothschild's collection that was listed in 1908 by that name. However, this subspecies had already been named by the systematist Paul Matschie in 1914. The name Matschie gave to it was *G. gorilla graueri*, after the German hunter Grauer, who in 1908 and 1910 had collected sixteen specimens in the Ugoma Mountains west of Lake Tanganyika.
(3) Mountain gorilla (*G. g. beringei*), from the Virunga volcanic range including Mounts Mikeno, Karisimbi, Visoke, Mahavura and Sabinio, and Kahuzi, west of Lake Kivu.

Grauer's gorilla, it should be added, while presumably a variety or subspecies of the western lowland gorilla, exhibits a number of physical proportions that place it between those of the lowland (western) and the mountain forms. These bodily differences are commented upon in chapter 5. The geographical range of *G. g. graueri*, as is shown in Fig. 23, comprises an area of at least 20,000 square miles. In comparison, the habitat of the true mountain gorilla (*G. beringei*) would appear to be an area of not more than a thousand square miles. In contrast, the vast area occupied by the western lowland gorilla contains some 250,000 square miles. Between the areas inhabited by the western and the eastern lowland gorillas is a stretch of broad-

---

* Cuthbert Christy, in his book *Big Game and Pygmies* (London, 1924), points out (pp. xxvii–xxix) that "the range of the gorilla was at first, in the nineteenth century, thought to be confined to northwest central Africa, to the Gaboon and the Cameroons. But George Grenfell reported its existence in North Congoland; and shot an example near the River Mutima on the north bank of the Northern Congo in 1904." To this he adds: ". . . a fact persistently ignored by English zoologists."

leaf evergreen forest over 600 miles long in which no living gorillas, and only a few gorilla skulls, have ever been found. The probability, however, is that in Pliocene or Pleistocene times, when apes and monkeys were distributed more widely than they are at present, gorillas occupied this entire east-west equatorial area. This would account for the marked similarity that exists between the present-day western and eastern lowland forms or races.

Cousins,[7] presumably following Groves's (1967) classification, has applied it to zoo gorillas in an endeavor to ascertain more definitely to which race each specimen belongs. However, in Cousins's survey, there is no mention of quantitative cranial and skeletal differences—such as the relative lengths of the limb bones, the vertebral segments, or the trunk length, along with such diagnostic features as the relative widths of the pelvis (bi-iliac) and the shoulders (bi-acromial), etc.—measurable criteria that identify the work of the primatologists Adolph Schultz and Francis Randall. Rather, Cousins depends upon such features as the shape of the nose and nostrils, which he asserts are sufficiently different in the three races to distinguish or separate them. Such qualitative criteria, which were extensively employed by the Italian criminologist Lombroso (1836–1903) to "separate" convicts from law-abiding citizens, are notoriously subjective and so are of decidedly less value than careful measurements of various parts of the body, whether of men or of apes. In any case, Cousins recognizes as mountain gorillas (*G. g. beringei*) only a few of the many zoo gorillas that were previously labeled as such, and identifies these specimens rather as *G. g. graueri*, even in cases where the specimens are known to have come from the Lake Kivu (Mt. Kahuzi) and other typical mountain-gorilla locales.

That many of the external characteristics and even some of the skeletal proportions "overlap" in the ranges of these criteria in *G. g. graueri* and *G. beringei* is a matter of observation and measurement. The point is that if a single one of these characteristics can be shown to be nonoverlapping, or completely separated, between the types (here, gorillas) being compared, even among 1000 specimens of each, that difference should in the present writer's opinion be sufficient to indicate a species distinction between the two forms. Such a degree of difference prevails between the western lowland gorilla (*G. g. gorilla*) and the mountain gorilla (*G. beringei*)—the respective body and limb measurements of which are listed in chapter 5. Of Grauer's, or the eastern lowland, gorilla, only five sets of limb-bone measurements (3 adult male and 2 adult female) are known to the writer, and these meager statistics are insufficient to establish many of the differentiating body measurements, such as trunk length, neck length, arm length, leg length, shoulder and hip breadth, etc.

To sum up, it should be added that not all primatologists accept either Groves's designation of *G. g. graueri* as a subspecies of the western lowland gorilla, or his premise that the gorillas inhabiting zones 4, 5, and 6 on his map (Fig. 24) must therefore all be *graueri* (rather than possibly *beringei*). Evidently the problem of gorilla classification cannot be resolved until more measurements of *G. g. graueri*—particularly of the skull, the cervical vertebrae, and the pelvis—become available. Possibly through the collection of skeletons of *graueri* from the most westerly areas of their observed range, specimens may be had that will not be regarded as being either *G. beringei* or *graueri-beringei* hybrids, if the latter actually occur.

* * *

Natural Habitat

The immediate surroundings or habitat of gorillas, while consisting essentially of dense jungle, varies according to altitude and also somewhat to longitude. In the equatorial forests of west and central Africa, from the Gulf of Guinea eastward across the Belgian Congo (now Zaire) some 1200 miles, the vegetation is lush and profuse, due to a high mean temperature accompanied by an annual rainful of from 60 to as much as 160 inches. Van Oertzen pictures the primeval forest of the lowland gorilla as a sad, melancholy region that only rarely permits a beam of sunlight to break through.[8] The general elevation is from about 2000 to 3000 feet, with only a few peaks above the latter height. Rivers and streams flow through the forest and during the wet season, which lasts for about eight months, overflow into the surrounding areas. Thus, the ground is alluvial, soft, and swampy. While this vast jungle is the home of numerous forest-dwelling natives, the hot and humid climate is not particularly healthful to most Europeans, who frequently contract malaria or some other tropical disease. Even the gorillas, while accustomed to the frequent and often heavy rainfall, seem greatly relieved when the sun comes out and they can stretch and dry themselves.

Jones and Sabater Pi list some forty species of plants indigenous to the rain forest in Rio Muni, the most common of which is *Pycnanthus angolensis*.[9] A peculiarity of tropical forests is that they

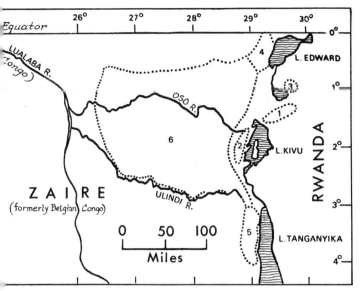

Fig. 24. Ranges of the mountain gorilla (*G. beringei*), localities 1 and 2; and of the eastern lowland gorilla (*G. g. graueri*), localities 3, 4, 5, and 6—as recognized by C. P. Groves, 1970. 1. Virunga volcanoes; 2, Mt. Kahuzi district; 3. Kayonsa forest (a portion of); 4. Mt. Tshiaberimu; 5. Mwenga-Fizi district; 6. Utu district, lowland forest.

grow in infertile soil. As leaves and fruit fall to the ground and decay, the minerals in them are quickly reabsorbed by the roots of the growing plants and trees, which thus flower and produce fruit throughout the year. The tops of the broadleafed trees, which grow to a height of 150 to 200 feet, form a veritable roof of foliage, through which sunlight rarely penetrates. Beneath this canopy it is shady, and the trunks of the towering trees are often smooth and unbranched. Due to the lack of sunshine, there are no flowers. The heavy shade also prevents grass from growing, with the result that no grazing animals occur. Ropelike lianas stretch and loop from tree to tree, while various parasitic plants, creepers, and vines hang from or wrap around the thick tree trunks. On the ground and the lower levels of the trees is the natural habitat of the gorilla and the chimpanzee, while at higher levels monkeys, an occasional leopard, snakes, birds, and insects each adapt to the most favorable environment. So dense is the vegetation in some places that gorilla hunters may go for days, weeks, or even months without seeing one of the apes, which are adept at concealing themselves and in these surroundings have the utmost opportunity to do so.

Despite all this, there are occasional clearings in the forest, and here and there a native farm or plantation. When, as sometimes occurs, a band of gorillas in search of food raids such a cultivated area, it gives the owners ample excuse for spearing or shooting the apes. Too, as more and more forest areas become occupied by lumber companies and other concerns interested in products that the forest produces, so will the habitat of the lowland gorillas be cut down.

The foods consumed by lowland gorillas in the wild consist almost entirely of vegetables, herbaceous plants, and fruits. And the apes appear to be choosy eaters. Of some plants they consume only the leaves, and of others only the tenderest parts of the roots, bark, or pith. Of plants bearing fruit, such as the batuna, which grows throughout the lowland forest, they often prefer the stalk to the fruit itself, twisting and breaking open the stalk with their powerful teeth and jaws, to get at the succulent interior of the plant. The wild mangrove, which forms a staple article of food for the chimpanzee, is rarely if ever touched by the lowland gorilla of the same habitat. Naturally, whenever a band of these gorillas wanders into a deserted *shamba*, or once-cultivated field, the apes eagerly consume as much bananas, plantains, and sugar cane as they can find. And it follows that if the native farmers are still about, they either drive the gorillas away or shoot them, notwithstanding that the apes are under rigid legal protection.

Elaborating on this point, the Reverend A.I. Good (1947), who for thirty-seven years was a Presbyterian missionary in the Cameroons, remarks:

> Where the population is sparse and the gorillas numerous, they may become quite a scourge to the people, as they are very destructive to banana gardens. Even a large banana garden may be almost ruined if a band of gorillas elects to spend several days in it. Since firearms are prohibited to all natives except a favored few, their only recourse is to noise. And so to save a garden, they must be continually on the watch. From time to time they must spend several days beating on tin pans, drumming and yelling, to discourage some old gorilla who has a craving for banana shoots.
>
> The gorilla is on the list of animals protected absolutely by the government, no one being allowed to kill it at any time unless furnished with a scientific permit. Theoretically the natives are also under this rule, but practically no one pays attention if they succeed in killing one with their primitive weapons. Occasionally a French official, upon special petition by the people, will even send a native soldier with a rifle to kill one or two gorillas that have become unusually destructive to gardens in some particular locality.

However, comparatively few are killed in all these ways, and the gorilla enjoys practical immunity to danger from man and seems fully as numerous as ever. In my judgment, these strict conservation measures are good, as the gorilla deserves protection. On the other hand, considering how very few are taken for scientific purposes, some of the present strictness might well be relaxed for such collectors as can furnish proof that the specimens are to be used for scientific purposes, as the effect on the total gorilla population would be negligible. More are killed and eaten annually by the natives, and completely wasted for the scientific world, than scientific collectors would ever request.[10]

The type of locality (terrain) inhabited by the mountain gorilla is, as the name implies, of higher elevation than that of the western or lowland species, along with exhibiting differing kinds of vegetation. Apparently, too, the vegetation is different to a certain extent in various sectors of the general habitat. Of the gorillas inhabiting the Virunga Mountains (see Fig. 23), Barns (1923) writes as follows:

> Gorillas are usually found in the thickest bamboo forest, but on the occasion of which I write I had driven them out of their usual haunts below camp, and they had, as I foresaw, taken to the open forest above me, taking refuge in the *kloofs* that run down from the peak of the Karisimbi Volcano. These are covered with an extraordinary tangle of succulent herbage, thigh deep with nettles, docks, sorrels, hemlocks, and blackberries, as well as larger growths of vernonias, balsams, lobelias, and senecios. Above this wonderland the magnificent pink-boled hagenia trees spread their fairy foliage, their low-hanging branches thick with green moss-pads resembling great velvet cushions—veritable seats for the mighty, for young gorillas may sometimes be seen squatting on the lowest of them, or the older ones making use of the often hollow and bent trunks for their sleeping quarters.[11]

Also referring to his personal experiences in the home of the mountain gorillas, Burbridge (1928) remarks:

> These foothills and mountain slopes, at chill altitudes of eleven to twelve thousand feet, are clothed with dense vegetation because of the continuous [continual] tropical rains. The limbs of the great trees are festooned with flowering orchids, pale green mosses, and many blossoming vines and shrubs. And beneath is large-leafed succulent vegetation growing head high, of so soft and juicy a texture that a handful, if squeezed, would fill with liquid a small goblet. [Located] in a veritable labyrinth of jungle-sown ravines and netted forests, it seems nature had endowed these fastnesses with every care and forethought for the preservation of the gigantic ape. Half a dozen varieties of edible vegetables grow around beds of wild celery; the tender shoots from bamboo clusters prod upward from leafy floors; upon tree and shrub, berries are scattered through the endless miles of jungle stronghold. Mother Nature, while kindly to the gorilla, did not fail to lift a protecting hand to throw every obstacle of entanglement and barricade against man's intrusion into these solitudes. For some reason, ominous it may be, elephants, lions and other killers rarely invade the forests of the gorilla. The crafty leopard alone left here and there in the gorilla's trail the occasional traces of his spoor.[12]

Bingham also, in describing the environment of the Virunga Mountain gorillas, as observed by him in 1930, says:

> The "spotty" vegetation is only one aspect of a varied environment which has in compact scope fascinating suggestions of gorilla ecology. There are abrupt changes in climate, weather and temperature. Precipitation, altitude, air currents, seasonal exposures and plant life appear to be interrelated in the natural habitat of the gorillas. There is ample rainfall in the upper triangle and on the adjacent slopes of the central massif which brings forth from the fertile soil (lava ash) a luxuriant vegetation of trees and undergrowth. Gorillas appeared on Mount Mikeno at the lower edge of the forest of the upper triangle, and in the diminishing tree growth of the higher slopes. Some [of the gorillas] at least reach the frostline of the central massif where there are occasional flurries of snow, while others are found in a tropical or bamboo zone.[13]

Evidently the "wild celery" (*Umbelliferae*) plant, which has the general appearance of the wild parsnip, is one of the favorite foods of the Virunga Mountain gorillas. Dian Fossey (1971)[14] found that a favored delicacy was the fruit of *Pygeum africanum,* which resembles an oversized cherry. Whenever the trees bore fruit—which was only for two or three months during each year—the gorillas, seemingly knowing this, would quickly consume everything in sight, even the largest males climbing to the tops of the trees!

The extensive range (over 20,000 square miles) of the eastern lowland gorilla (*G. g. graueri*) located west (no. 6 on the map in Fig. 23) of the narrow strip occupied by the mountain species (*G. beringei*) is supplemented by the smaller areas numbered 3, 4, and 5 on the map. Area no. 3, the Kayonsa forest, is described by Akroyd as covering about 150 square miles. Through the forest there are (or were, in 1934) only two regular trails, one of which, if not used frequently, has to be kept cut. The area contains a number of impenetrable forests, which extend up the mountain slopes from 6000 to 7900 feet. The gorillas in this forest live under different conditions from the mountain species on the Virunga volcanoes; for instance, they build their "nests," or sleeping platforms, not on the ground or just above it, but in trees from ten to twenty feet above the ground. The mountain gorillas are said also to be more "migratory" than the eastern lowland gorillas. This may be because the Virunga gorillas are frequently disturbed by the natives cutting bamboo for charcoal, and also because of seasonal variations in the amounts of edible vegetation, caused by differences in the respective amounts of rainfall occurring on the northern and the southern slopes of the volcanoes.[15]

Pitman, who explored the same region shortly after Akroyd, gives a more extensive account of the lowland gorillas that live in the Kayonsa forest. The following (nonconsecutive) quotations are from various pages of his report:

> The occurrence of Gorillas in the Kayonsa region of Uganda [about midway between the Virunga volcanoes and Lake Edward] has been known for many years.
>
> There is in the Kayonsa a complete absence of bamboo, wild celery, dock, and similar juicy-stemmed plants such as abound in the humid, high altitudes, forcing the Gorilla to confine its diet to a mixture of leaves, berries, ferns, the tender fronds of tree-ferns, parts of the wild banana stems and leaves, and fibrous bark peeled off a variety of shrubs in the undergrowth.
>
> Owing to a lack of what apparently are normal food constituents the Gorilla has become more enterprising in search of food, and in consequence climbs trees freely to a known height of at least 50 feet.
>
> Normally the troops vary in size from five to eight or nine, [but one troop was said to include nearly two dozen].
>
> The Wambutte [Pygmies] are extremely tolerant of the Gorillas, but not so the other local natives, who would readily endeavour to exterminate the lot, were it not for the fact, of which they are well aware, that these splendid animals are absolutely protected.[16]

* * *

Populations

As to the present, or recent, numbers of gorillas in their native habitats, there have been many estimates. Of the mountain gorilla (*G. beringei*), which appears to have the most accessible environment, and which definitely exhibits the smallest population, Akeley in 1923 estimated that on the three Virunga volcanoes—Mikeno, Karisimbi, and Visoke—there were altogether between fifty and one hundred gorillas.[17] Burbridge made an estimate of between one thousand and two thousand individuals for the entire mountain-gorilla population.[18]

Dr. Jean M. Derscheid, a Belgian zoologist who was a member of the second Carl Akeley African Expedition (1926), made an estimate of the number of gorillas then in the Lake Kivu (Virunga volcanoes) region. His conclusion was that there were, all told, probably from 450 to 650 individuals. Derscheid found the gorillas most numerous at elevations between 2700 and 3500 meters (c. 8800 to 11,500 feet), with extreme occurrences at 1900 and 3900 meters (c. 6200 and 12,800 feet). He encountered a few solitary old males, but more usually families or troops composed of anywhere from seven to forty-three individuals. He remarked on the surprisingly small proportion of young gorillas in the groups.[19]

Capt. Pitman, in 1935, estimated the number of eastern lowland gorillas in the Kayonsa-Niwashenya habitat, an area of about sixty square miles near the usual mountain gorilla confines, to be at least eighty. This would be a population density of four gorillas to three square miles.[20]

Martin Johnson hazarded the guess that in the Alumbongo Mountains alone, west of Lake Kivu, there must have been at that time (1930) about 20,000 gorillas (judging by the numerousness of their tracks?).[21] Commander Attilio Gatti also felt that the total number of gorillas in the Kivu region, due to obscure, uncounted bands, was "very considerable."[22]

Emlin and Schaller (1960) estimated that the total population of mountain gorillas was between 3000 and 15,000 individuals.[23]

In marked contrast to the latter opinions is that of the conservateur Adrien Deschryver, who as recently as 1973 estimated that there were approximately 250 mountain gorillas in the Kahuzi-Biega National Park (west of Lake Kivu) and no more than 300 to 350 inhabiting the Virunga volcanoes, making a total "world" population of this species of less than 600 specimens.[24]

Most estimates of the numbers of lowland go-

Fig. 25. A typical habitat group of the mountain gorilla (*Gorilla beringei*), as displayed in the Akeley Hall of African Mammals in the American Museum of Natural History, New York. The gorillas were obtained on Carl Akeley's 1923 African Expedition and were mounted by him in lifelike poses. The background was painted by William R. Leigh, while the flora and the foreground were prepared by Albert E. Butler and assistants. In the background are shown the three volcanic peaks—Visoke, Karisimbi, and Mikeno, the slopes of which provide living space for perhaps five hundred individuals of this diminishing species of manlike ape. (Photo courtesy the American Museum of Natural History.)

rillas, especially those in the dense western forests of the Gaboon and the Cameroons, where anything like an accurate count would be impossible, must accordingly be speculative; for example, the British anthropologist Arthur Keith in 1896, on the basis of accounts furnished by travellers and hunters, inferred that the total population of these apes was "well under 10,000"; yet in 1914 he raised this estimate to 20,000 to 30,000.[25]

In 1967 the zoologists Clyde Jones and J. Sabater Pi estimated that the population of the lowland gorillas in Rio Muni, a west African state having an area of approximately 10,000 square miles, was no more than "a few hundred animals."[26] On the basis of the areas actually inhabited by the gorillas, these authors arrived at population-density figures ranging from 0.58 gorillas per square kilometer in the Amuminzok-Aninzok district to 0.86 in the Mt. Alen district. This is equivalent to 1.5–2.2 gorillas per square mile. Taking the round figure of two gorillas per square mile, this would indicate an inhabited area of only about 150 square miles out of Rio Muni's total of 10,000 square miles. This seems an exceedingly low ratio for a region supposed to be well-populated gorilla country. In contrast, Sabater Pi in 1964 had estimated the population of gorillas in Rio Muni at approximately 5,000 individuals.[27]

Blancou (1961) estimated that the total population of western lowland gorillas was somewhere between 10,000 and 20,000.[28] However, on the basis of 2 gorillas per square mile, and the ratio of inhabited area to total area being no higher than that found in Rio Muni, the expected population of gorillas would only be about 7,500.

Actually, rigid protection on the one hand and continued shooting on the other make impossible any really reliable estimation of either the lowland- or the mountain-gorilla populations. It is generally agreed, however, that at present only the mountain gorilla is truly in the position of being an "endangered species" of this ape.

* * *

Other Animals of the Rain Forests

In the foregoing discussion, the gorilla has been so exclusively the subject that to an uninitiated reader it might seem to be the sole wild animal inhabiting the areas attributed to it! This, of course, is not the case. Actually, the number of chimpanzees inhabiting the same areas far exceeds that of gorillas. In addition, in the typical western lowland forest—and often in the eastern gorilla's range as well—there are numerous species of other mammals both small and large. Hippopotami, for example, frequent the rivers and larger streams throughout most of Africa south of the Sahara. Another large and widely distributed animal is the Cape buffalo, which in one or another of its varieties ranges from the densely forested areas to mountain grassland as high as 13,000 feet. Even more widespread than either of the foregoing herbivores is the leopard, which, except in most of Cape Colony, is practically omnipresent below the Sahara.

In the excellent field guide to African mammals authored by Dorst and Dandelot,[29] the numerous distribution maps show more than forty species of animals, the ranges of which either coincide with or overlap those of the gorilla. These forest-sharing animals include various species of monkeys and lemurs, numerous small carnivores (as well as in some places the lion), elephants, warthogs, bush pigs, perhaps a dozen forms of antelope, and such odd animals (anteaters) as the aardvark and the pangolin. Thus, gorillas have plenty of "company," none of which, with the exception of an occasional attack by a leopard or a python (usually upon young animals), causes them any trouble. Rarely also does one hear of any encounter with a crocodile, possibly because gorillas, like the other manlike apes, cannot swim, and so avoid wherever possible any contact with large bodies of water.

To sum up, there are one or two species of gorillas (depending upon what criteria one uses to determine the question), and at least three varieties or local races—all of which are confined to the equatorial forests of Africa. The diet of gorillas is essentially herbivorous and frugivorous and is normally supplied in adequate amounts by the ever-growing plants and fruits of their lush tropical forests. Population-wise, both the western and the eastern lowland gorillas appear to be thriving; but the numbers of the mountain gorilla today, and evidently for some years past, are so small that it definitely places the species on the "endangered" list of animals for which continued, and perhaps even more rigid, legal protection is needed.

Notes

1. Daniel Giraud Elliot, *A Review of the Primates*, 3 vols. (New York, 1913); Harold Jefferson Coolidge, Jr., "A Review of the Genus Gorilla, *Memoirs, Museum of Comparative Zoology* 50, no. 4 (1929):291–381.
2. Adolph Hans Schultz, "Some Distinguishing Characters of the Mountain Gorilla," *Journal of Mammalogy* 15, no. 1 (February 1934):51–61.
3. Francis E. Randall, "The Skeletal and Dental Development and Variability of the Gorilla," *Human Biology* 15, no. 3 (September 1943):236–51; no. 4 (December 1943):307–37.
4. Walter Rothschild, "Notes on Anthropoid Apes," *Proceedings of the Zoological Society of London* (1923), pp. 176–77.
5. C. R. Aschemeier, "On the Gorilla and the Chimpanzee," *Journal of Mammalogy* 2, no. 2 (1921):90–92.
6. Colin P. Groves, "Ecology and Taxonomy of the Gorilla," *Nature* (London) 213 (1967):890–93.
7. Don Cousins, "Classification of Captive Gorillas," *International Zoo Yearbook*, vol. 14 (1974), pp. 155–59.
8. In Jasper von Oertzen, *Wildnis und Gefangenschaft* (Berlin: Kameruner Tierstudien, 1913), p. 3.
9. Clyde Jones and Jorge Sabater Pi, "Comparative Ecology of *Gorilla gorilla* and *Pan troglodytes*," *Bibliotheca Primatologica* (Basel), no. 13 (1971), p. 30.
10. I. A. Good, "Gorilla-land," *Natural History* 56, no. 1 (January 1947):37.
11. T. Alexander Barns, *Across the Great Craterland to the Congo* (London, 1923), pp. 139–40.
12. Ben Burbridge, *Gorilla* (New York and London, 1928).
13. Harold C. Bingham, "Gorillas in a Native Habitat," *Carnegie Institute of Washington* 426 (1932):59.
14. Dian Fossey, "More Years with Mountain Gorillas," *National Geographic* 140, no. 4 (October 1971):582.
15. R. Akroyd, "The British Museum Expedition to the Birunga Volcanoes, 1933-4," *Proceedings of the Linnean Society, London*, pt. 1 (1934–35), pp. 17–21.
16. C. R. S. Pitman, "The Gorillas of the Kayonsa Region, Western Kigezi, S. W. Uganda," *Proceedings of the Zoological Society of London*, pt. 3 (1935), pp. 477–94.
17. Carl E. Akeley, *In Brightest Africa* (New York, 1923), p. 248.
18. Burbridge, *Gorilla*, pp. 215–16.

19. J. M. Derscheid, "Notes sur les Gorilles des Volcans du Kivu (Parc National Albert)," *Annales Societie Royale du Zoologie Belgique* 58 (1928):149–59.
20. Pitman, "The Gorillas of the Kayonsa Region," p. 480.
21. Martin Johnson, *Congorilla* (New York, 1931).
22. Attilio Gatti, "Gorilla," *Popular Mechanics*, September 1932, p. 418. (See also *Field and Stream*, October 1932, p. 73.)
23. J. Emlen and G. Schaller, "Distribution and Status of the Mountain Gorilla," *Zoologica* 45 (1960):41–52.
24. Adrien Deschryver, *Eleventh Annual Report, The Jersey Wildlife Preservation Trust* (1974), p. 20.
25. Arthur Keith, quoted in Robert M. and Ada W. Yerkes, *The Great Apes* (New Haven, Conn.: Yale University Press, 1929).
26. Jones and Sabater Pi, "Comparative Ecology," p. 13.
27. Jorge Sabater Pi, "Distribution actual de los gorilas de llanura en Rio Muni," *Publicaciones del Servicio Municipal del Parque Zoologico de Barcelona* (1964).
28. L. Blancou, "Destruction and Protection of the Wild Life in French Equatorial and French West Africa. Part V, Primates," *African Wild Life* 15 (1961):29–34.
29. Jean Dorst and Pierre Dandelot, *A Field Guide to the Larger Mammals of Africa* (Boston: Houghton Mifflin Co., 1970).

5

Size, Physique and Growth, and Development of the Gorilla

Size of the Gorilla

When, in 1847, news was brought to Europeans of the existence of a hitherto unknown (to them) "giant" African ape called the gorilla, naturally one of the first questions people asked was, "How big is it?" And for many years thereafter, a flood of reports came from travellers and hunters in which, apparently, each claimed to having heard of, or actually collected personally, a larger specimen of the gorilla than any previously recorded. A review of this particular question may help throw light on what size an adult male gorilla actually attains. Many years ago I contributed an article just on this subject;[1] and it is from that article, with a few additions and updatings, that the following particulars are taken. Unless otherwise stated, all references are to the western lowland species (*Gorilla gorilla typicus*).

Before protective restrictions came into force, and before the gorilla was the well-publicized animal to be seen in most major zoos today, the hunting of these giant primates was conducted along lines similar to the hunting of any other imposing and desired museum specimen. In these early gorilla hunts, the object of the hunter naturally was to obtain the largest and finest individual possible. And, as the adult male gorilla is of much greater stature and bulk than the female, the biggest male in a troop was usually the specimen sought. As a result of such wild-killed "specimens" being measured and recorded, gradually a literature on the size of the gorilla, particularly of adult males thought to be of "record" dimensions, came into existence. From these early sportsmen's records, most popular works on natural history have drawn their information as to the size of the gorilla.

To state, as is commonly done, merely the size of a large or "record" specimen, however, is to provide only a vague idea of the average or typical size of the species in question; and it is always in relation to the typical (rather than the exceptional) specimen that comparisons should be made.

Perhaps the first specific reference to the size of the gorilla was that reported by the English traveller and writer Thomas Henry Bowdich (1791–

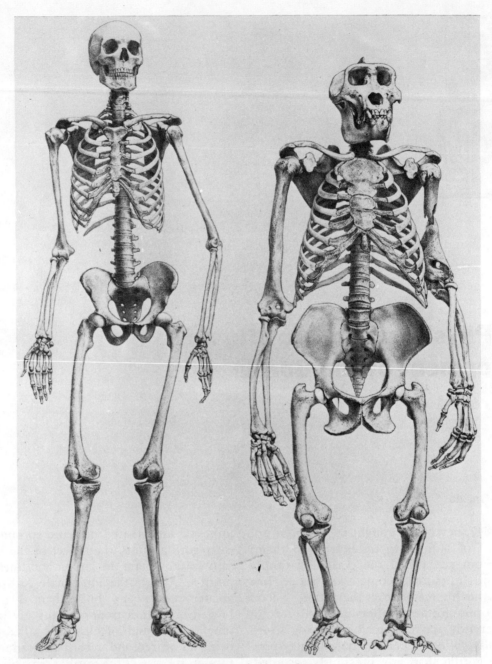

Fig. 26. Skeletons of a man (Caucasian) and an adult male gorilla, for comparison. The bones are the same in each example, except for size and proportions. The skeleton of the gorilla, which in life stood about 5 feet 2 inches, is (or was) in the collection of the Royal College of Surgeons, London. It was the first specimen of a gorilla brought to England (1859). The figures are each about one-tenth natural size. About two inches must be added to the height of the skeleton to obtain the living height. Compare with Figure 27.

1824).[2] He stated, "on the authority of the natives of the Gaboon," that this ape was generally about 5 feet in height. This may have referred to the height of the gorilla as it stood nearly erect but with its knees slightly bent. In 1859, the first embalmed gorilla specimen to reach England was described by the eminent anatomist Richard Owen as being 5 feet 6 inches in standing height, 42 inches in sitting height, and with arms each 40 inches long.[3] However, from the lengths of the limb bones of the skeleton of this specimen, which were carefully measured and recorded, it developed that the probable height of this gorilla in life was about 1575 mm (62 inches). The skeleton of this historic specimen of a wild-killed gorilla, which is (or was) in the collection of the Royal College of Surgeons, London, compared with that of a man of ordinary size, is shown in Figure 27. Note the severe break and shortening of the left upper-arm bone, an injury that had perhaps been healed for some time before the ape was shot.

The first person really to publicize the wild, free-living gorilla was, as is mentioned here in chapter 3, the famous hunter Paul Du Chaillu, who in 1861 wrote a book, *Explorations and Adventures in Equatorial Africa,* in which he recounted his experiences with this and other jungle animals. The largest of nine gorillas, an adult male shot by Du Chaillu, was stated by him to have stood 5 feet 8 inches in height.[4] Again, however, the limb-bone measurements of this specimen, which were taken by Dr. J.E. Gray, show the ape to have been not more than 63 inches in living standing height.[5]

Some Wild-Killed Gorillas

One of the most publicized photographs ever taken of a wild-killed gorilla was that obtained by the German trader H. Paschen. It shows the ape placed in a sitting position on the ground with natives standing behind it (Fig. 28). The specimen, an adult male, was killed near Tonsu, in the Cameroons, in 1900. Its skin and skeleton were prepared by J.F.G. Umlauff, of Hamburg, Germany, and were exhibited in the Museum Umlauff of that city. Later, the stuffed specimen was purchased for 20,000 marks (then 4,760 dollars) by Lord Walter Rothschild, and placed in the Rothschild Museum at Tring, Hertfordshire, England. Particular interest attaches to this specimen in that it was stated to be of gigantic size—"The Largest Gorilla Ever Shot," as one published account put it. Its length, lying, was given as 2.07

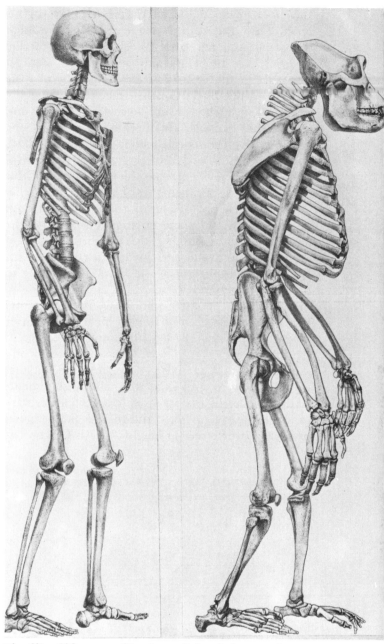

Fig. 27. Side or profile views of the skeletons of man and gorilla shown in front views in Figure 26. The man in life stood about 5 feet 11 inches (1803 mm), and the gorilla, about 5 feet 2 inches (1575 mm). The skeleton of the gorilla was brought to England in 1859 and was the first complete skeleton of that then-new species of ape to be seen there.

meters (6 feet 9½ inches), and its span (spread of arms) was 2.80 meters (9 feet 2¼ inches). Upon investigation, however, it was found that the "length, lying" was taken from the crown of the head to the end of the middle toe, not to the heel. Again, if the armspread of this gorilla were actually 9 feet 2¼ inches, its standing height would have been about 6 feet 2 inches, not 6 feet 9½ inches.

But, to confuse matters, W.T. Hornaday, in his book *The American Natural History*, says that this same specimen, shot by H. Paschen, stood 5 feet 6 inches in height and weighed (by estimation) 500 pounds.[6] In *The Living Animals of the World*, edited by J. Cornish, the height is given as 5 feet 5 inches, span "over eight feet," and weight 400 pounds.[7] In Brehm's *Tierleben* the height is given as "over 2 meters" (6 feet 7 inches), and weight, estimated, 250 kilograms (551 pounds).[8]

It should be evident from the photograph of this oft-mentioned specimen that if it were seated immediately in front of one of the natives, the top of its head would come no higher than, say, the native's navel. Assuming the standing height of the natives to average 66 inches (a generous estimate), and the height of the navel accordingly to be between 40 and 41 inches, a sitting height in the gorilla of the latter amount would correspond to a standing height of 64 or 65 inches. The latter height would also check with Cornish's statement that the gorilla's span was "over eight feet" and the weight 400 pounds. A height of 65 inches would also fit in with Lord Rothschild's statement: "Adult males of the three well-defined races vary in height from 5 to 6 feet, and there is no specimen preserved over 6 feet in height."[9] This Cameroons specimen, being in Rothschild's possession at the time, presumably was included in this dictum.

Rothschild, however, went on to say that in the French periodical *L'Lllustration,* for 14 February 1920, page 129, there was a photograph of a gorilla 9 feet 4 inches (!) in height, according to M. Villars-Darasse, "and the photograph certainly shows a gigantic animal" (Fig. 29). A comparison of the size of the gorilla with that of the native sitting alongside it (assuming the man to have been 6 feet in height and proportioned accordingly) reveals that the sitting height of the gorilla could hardly have been more than 43 inches. This would denote a standing height of 5 feet 8½ inches, which is that of a good-sized but not very large gorilla.

Still another lowland gorilla of asserted gigantic size is mentioned in the French periodical *La Nature,* for 29 July 1905, page 129. This specimen, it is stated, weighed 770 pounds (!) and measured 7 feet 6½ inches in height and 3 feet 7 inches across the shoulders. Fortunately, the illustrations, which show the dead ape in a sitting posture, also show two adult natives, one standing and the other sitting, which makes it possible to check the foregoing alleged measurements. Assuming the natives to have been of average stature, and making due allowance for perspective (since the gorilla is con-

Fig. 28. Adult male gorilla shot by H. Paschen near Yaounde, Cameroons, in 1900. The mounted specimen is now in the Rothschild Museum, Tring, England. (Photo by H. Paschen.)

Fig. 29. A fine specimen of the western lowland gorilla, shot by M. Villars-Darasse in the forests of Bambio, Haute-Lobaze. While stated to have been 9 feet 4 inches (!) in height, this male gorilla probably stood about 5 feet 8 or 9 inches. (Photo from *L'Illustration,* February 14, 1920.)

siderably nearer the camera than are the natives), it can be seen that the specimen pictured is little, if any, larger than an average-sized lowland gorilla. Therefore, why such exaggerated dimensions should have been claimed for it is difficult to understand. If the publishing of such unconfirmed figures appeared only in the original contribution it would be bad enough, but in this instance the stated measurements were quoted in such standard and justly popular works as Lydekker's *Wild Life of the World* and Brehm's *Tierleben*.[10] Precisely in this way do mere claims or assertions come in time to be assumed as established facts; that is, unless someone goes to considerable length to disprove them.

How Apes Stand Upright

To digress, it should be pointed out that none of the higher manlike apes (gorilla, chimpanzee, orangutan) can (or at least does during life) stand fully erect as does man. Thus, the so-called standing height in apes is really the total length from crown to heel in the lying position. This, in large and heavy specimens, such as adult male gorillas, may be as much as two or three inches greater than if the height were actually measured in a standing posture. In a strictly scientific comparison of the proportions of the body in man and apes, specialists in this field of study generally use as a basic measurement the anterior trunk height (from sternal notch to pubis). However, since the standing height is such a familiar measure of size in man, and moreover is a standard basic measurement in the fields of both anthropometry and artistic anatomy, it is here adopted as the "height" of the gorilla.

In a fair number of gorilla specimens obtained for museums, the skeletons as well as the skins were carefully prepared for mounting. In a few of these specimens, measurements fortunately were recorded both in the field (of the external body) and in the laboratory (either of the bones of the skeleton or of the embalmed body). In such cases I have checked the dependability of the field measurements by reconstructions of the probable living dimensions of the individual from its own bones, making due allowance for shrinkage, etc. By thus using the actual skeletal elements of a specimen, more accurate estimations of its probable size in life are possible than if field measurements alone (which can be highly erroneous, as has been shown) were relied upon.

Akeley's Mountain Gorilla

A well-publicized museum specimen of the mountain gorilla is the adult male (standing figure) in the gorilla group in the Akeley Memorial African Hall at the American Museum of Natural History (Fig. 26). This specimen (AMNH no. 54089), known as "the lone male of Karisimbi," was shot by Mr. H.E. Bradley on the Carl Akeley African Expedition of 1921 (Fig. 30).

In publishing the body measurements of this gorilla as taken in the field, Akeley gave the height as 5 feet 7½ inches and the span, or "reach," as 97 inches.[11] Let us see how these figures compare with a reconstruction of the living dimensions derived from the arm and leg bones of this specimen. The combined lengths of the upper arm and forearm bones and the skeleton of the hand indicate a living arm length (from shoulder to fingertip) of 1004 millimeters, or 39½ inches. The length of the clavicles indicates a living shoulder breadth (skeletal) of 446 mm, or 17½ inches. The living span, or horizontal arm stretch, may be derived by taking twice the length of the arm, plus the skeletal shoulder breadth, minus the "loss" in shoulder breadth occasioned by raising the arms to shoulder level (in the gorilla usually about 6 percent). In Bradley's Karisimbi specimen the span thus derived is 2307 mm, or a trifle less than 91 inches. This—even making allowance for individual variation—is quite irreconcilable with the

Fig. 30. "The Lone Male of Karisimbi," which was shot by Mr. H. E. Bradley on the Carl Akeley African Expedition of 1921. From left to right are Mrs. Bradley, Mr. Akeley, and Mr. Bradley. (Photo courtesy the American Museum of Natural History, New York.)

97-inch "reach" given by Akeley. A clue to the discrepancy may be found in the measurements of the same specimen published later by Mary Hastings Bradley, wife of H.E. Bradley, and herself one of the members of the Akeley expedition. In this later account Mrs. Bradley gives the height as 5 feet 7½ inches (the same as Akeley gave it), but the span as 92½ inches and the "reach," from the ground to the tip of the upraised hand, as 98 inches.[12] If we work backward from the latter figure for the height of the upraised hand (middle fingertip), we obtain an implied stature of 5 feet 5½ inches. The combined lengths of the bones of the arms and legs also imply a stature of 5 feet 5½ inches. The span (92½ inches) implies a stature of 5 feet 6 inches. Thus it would appear that the true standing height of this "lone male of Karisimbi" in life was possibly as little as 5 feet 5 inches or as much as 5 feet 6 inches, but probably about midway between these estimations, or 5 feet 5½ inches.

It should be mentioned that in taking the measurements of standing or sitting height in gorillas the "crown pad" on top of the skull should not be included. This pad of connective tissue, which is probably present in the majority of fully adult gorillas—particularly in males of the mountain species—is often as much as two inches in thickness. If it were included in the height of the Karisimbi male, as measured in the field, then it alone would suffice to account for the discrepancy in height here noted.

Another seeming disagreement in this specimen is the chest girth in relation to the weight, which was given by Akeley as 360 pounds. As will be shown later, the weight of a gorilla can be closely estimated where the height (or sitting height) and the chest girth are known. Conversely, the chest girth can be deduced where the weight and the height (or sitting height) are known; and the chest girth commensurate with a standing height of 65.5 inches and a weight of 360 pounds, as in the Karsimbi male shot by Bradley, comes out as 54.7 inches, not 62. How such an enormous error could have been made is difficult to fathom; but since both the arm girth (18 inches) and the calf girth (15¾ inches), as well as the height and weight of this specimen all fall into line, it is evident that the chest girth is in error. A 62-inch chest, even on a gorilla only 5 feet in height, would result in a body weight of about 425 pounds. Thus, the Karisimbi gorilla was not a "large" specimen, as was at first thought, but rather was closer in weight to an average-sized lowland gorilla than to one of its own (mountain) species.

Gatti's, Raven's, and Barns's Specimens

More recently, the well-known African explorer Commander Attilio Gatti shot an adult male mountain gorilla, which was asserted to measure 8 feet 9 inches from the soles of its feet to the tips of its upraised hands, and to weigh 530 pounds.[13] These measurements, if correctly taken, indicate a standing height of about 5 feet 10 inches, which is that of a really large gorilla (Fig. 31). (In his article in *Field & Stream*, Gatti gives the weight of this gorilla as 482 pounds.[14] The weight units, however, were evidently metric pounds or half-kilograms, 482 of which equal approximately 531 English pounds.)

In July 1929, a primate-collecting expedition, under the joint auspices of Columbia University and the American Museum of Natural History, left America for equatorial Africa under the leadership of Henry C. Raven.[15] This expedition returned with three adult male western lowland gorillas and two adult male mountain gorillas. These five specimens, which were embalmed, were later studied and measured by Dr. Adolph H. Schultz, then of the Department of Physical Anthropology of The John Hopkins University, Baltimore. Among the many measurements taken by Dr. Schultz, the standing heights (converted from millimeters) of the three lowland gorillas were 5 feet 5 inches, 5 feet 5 inches, and 5 feet 7½ inches, respectively.

Fig. 32. Adult male mountain (Kivu) gorilla obtained by Henry Raven of the Columbia University-American Museum Expedition of 1929. From left to right: Dr. J. H. McGregor, Dr. W. K. Gregory, and Henry Raven. (Photo courtesy The American Museum of Natural History.)

Fig. 31. A very large (531-pound) mountain gorilla shot by Commander Attilio Gatti (left) in 1930 for the Museum of the Royal University, Florence, Italy. Next to the gorilla are two pygmies of the Mambute tribe, who located the huge specimen in the forest of Tchibinda, near Lake Kivu. (Photo courtesy Gatti-Hallicrafters Expedition.)

Fig. 33. One of the largest lowland gorillas on record, the mounted male specimen (No. 16981 in Table 3) on display at the Academy of Natural Sciences, Philadelphia. In life he stood 5 feet 10 inches and weighed about 530 pounds. The circular insert shows the head of the adult female gorilla in the same museum group. This expertly prepared "gorillarama" is the work of Jonas Brothers Studios. (Photo courtesy Academy of Natural Sciences, Philadelphia.)

Fig. 34. A very large male lowland gorilla secured by the Vanderbilt African Expedition of 1934. This specimen was found in the most easterly locality yet reported for a lowland gorilla, namely, on the east bank of the Sangha River in French equatorial Africa (now the Congo Republic). (Photo from H. J. Coolidge, 1936.)

Of the two mountain gorillas, one stood 5 feet 7½ inches; the other stood 5 feet 8 inches and weighed 467 pounds (Fig 32).

In 1934 the Vanderbilt Expedition of the Academy of Natural Sciences of Philadelphia obtained in the French Cameroons two adult male gorillas and one immature male.[16] The larger of the two adults is the standing figure in the gorilla group in the Museum of the Philadelphia Academy of Sciences (Fig. 33). This very large specimen stood in life about 5 feet 10 inches and weighed (by estimation from its height and girth measurements) about 530 pounds. Another adult male, which was not so tall (namely, about 5 feet 7¾ inches), had even larger girths and weighed possibly as much as 550 pounds. This gorilla was obtained in the most easterly location yet recorded for a living example of the western lowland species; namely, in the forest about 15 miles east of the Sangha River and 22 miles northeast of Nola (Fig. 34).

The English explorer-hunter T. Alexander Barns, in an African expedition he headed in 1920, trekked through the homelands of both the mountain (Kivu) and the eastern lowland gorillas, the latter being encountered in what was then the eastern part of the Belgian Congo. Evidently he collected specimens of both gorilla species, as the accompanying illustrations reproduced from his book show[17] (Figs. 35, 36, 37).

As a final example of a wild-killed "giant" gorilla that turned out to be of quite ordinary dimensions, we show Figure 38. It is interesting to note that while this specimen was said to have been obtained in the Cameroons of west Africa, its pelage (hair-coat) was as long, thick, and shaggy as that of any individual of the mountain species, which normally live at considerably higher (and colder) altitudes. This specimen, then, should provide a good example for those zoologists who prefer to "lump" rather than "split"; that is, to regard all existing gorillas as being of a single, rather than two or more, species.

As will be indicated as this chapter continues, the actual body proportions of gorillas and other apes may be closely estimated by using (1) accurate dimensions of the bones of the skeleton, particularly the lengths of the limb bones; (2) periodically taken measurements of living zoo gorillas, from newborn up to the age of, say, four or five years; and (3) measurements of zoo (or other captive) gorillas of adult or subadult age that have either been tranquilized or are measurable by reason of being exceptionally tame and manageable.

Fig. 35. An especially long-haired adult male mountain gorilla obtained in the Lake Kivu region by T. Alexander Barns on his African expedition of 1920. This photograph is from Barns's book *The Wonderland of the Eastern Congo*, 1922.

Fig. 36. An adult male eastern lowland gorilla shot by Alfred C. Collins for the Field Museum of Natural History, Chicago. Collins accompanied T. Alexander Barns on one of the latter's collecting expeditions in the eastern Congo about 1921 to 1922. While the gorilla here appears gigantic, due to the closeness of the camera, it was actually only of average size, weighing 350 pounds. (Photo by Barns.)

Fig. 37. The huge hulk of a full-grown male mountain gorilla, in comparison with a Watussi (Rwanda) native 5 feet 10 inches in height. However, the ground may be sloping downward from the gorilla to the man. Reproduced from *The Wonderland of the Eastern Congo*, by T. A. Barns, 1923.

Fig. 38. An adult male lowland gorilla killed some years ago in the Cameroons, west Africa. While a fine-looking specimen, it nevertheless appears to be far short of the "7 feet in height" announced in news dispatches at the time. (Photo courtesy Press Association, Inc.)

Using all available data that appeared reliable, the writer has established the following heights and weights as being normal or typical for adult gorillas living in the wild.*

Carl Akeley, as well as various other hunters of gorillas, expressed doubt that any individual specimen had ever reached a standing height of more than 6 feet. In the above series of estimations, the ranges in height and weight are based on a population of 1000 individuals for each sex of the lowland gorilla, and 600 for each sex of the mountain species. The factor governing the ranges in height is what is known in statistical terminology as the Coefficient of Variation. In gorillas this figure averages about 4.00. the same as in man; and it implies that in a population of 1000, the range will be plus or minus 13 percent, or times 1.130. For 600 cases, using the same C.V. of 4.00, the

* The average heights listed here are the same as those I presented in my 1950 article in *The Scientific Monthly* (see Note no. 1). Likewise, the average weights of the lowland gorillas are the same. The weights of the mountain gorillas, however, have been reduced somewhat because of additional cases modifying the average figures.

Table 2. Typical Heights and Weights of Adult Gorillas.

Species	Height, in.	Range	Weight, lbs.	Range
Western Lowland males	63.0	55–71	344	230–520
" " females	52.8	46–60	188	126–280
Mountain (Kivu) males	66.2	59–74	430	300–620
" " females	55.0	49–61	220	160–310

plus-or-minus range is times 1.113. In turn, the ranges in weight are based on the cube of the height, along with making a small allowance for individuals of extreme weight carrying a little extra fat.

T. Alexander Barns, incidentally, claimed that his tallest mountain gorilla measured 6 feet 2 inches—the same as the maximum height to be expected among 1600 specimens. However, in Barns's specimen this dimension would have been less if the animal had been measured in an upright position rather than lying flat on the ground. Unfortunately, there appears no way of checking the height of this gorilla from its bones, since no one seems to know to which museum, if any, the skeleton, or carcass, was sent by Barns.

It is interesting to note that one of the most knowledgeable statements as to the typical height of the gorilla (western lowland male) was given as long ago as 1886 by J. Fortune Nott, who in his book *Wild Animals Photographed and Described* says (p. 526): "The average height of this beast is a few inches over five feet, and its muscular strength is prodigious." That Nott was in a sense prophetic in this statement is shown by the heights of the museum specimens listed in Table 3, following.

Additional information on the bodily measurements of various well-known captive gorillas is given both later in this chapter and in chapter 10.

* * *

Having now shown what the general size (height and weight) of mature gorillas is, it may be opportune, before giving figures on the growth of these apes from birth to maturity, to present a general description of the typical western lowland species. It may be remarked that before the mountain species was discovered, in 1902, it was a relatively simple matter to discuss the "gorilla," since there was apparently only one form of the animal, and an author did not have to continually qualify his statements by referring to species, subspecies, races, or geographic variants of this manlike ape.

Even so, the overwhelming majority of gorillas on exhibition in zoos and menageries today, as in the past, are of the familiar western lowland species, namely, individuals which have either come, or descended, from the west-African countries of the Gaboon, the Cameroons, or French equatorial Africa (now the Republic of the Congo). It is to this "western" species of the gorilla in particular that the following description applies.

It will be noted from Table 2 that adult male gorillas greatly exceed adult females in both height and weight, the superiority in height being about 20 percent and in weight almost double that of the females. In both of these respects the greater size of the male develops only gradually during the juvenile period but accelerates enormously from puberty toward the adult stage. Too, the head and face in infant gorillas is more humanlike than later on, when the features, especially in the males, often take on a fierce, brutal appearance.

Compared with the proportions of a man, a gorilla has a trunk (sitting height) about 20 percent longer, and consequently, legs about 20 percent shorter; that is, the man's trunk is only about 10 percent longer than his legs, whereas the gorilla's trunk is no less than 70 percent longer. An almost equally vast difference between man and ape prevails in the proportions of the arms. For while a man's "span," or horizontal stretch of arms, is usually only from 2 to 4 percent greater than his stature, in an adult lowland gorilla the span is nearly 50 percent greater than the stature (or standing height). Moreover, this proportion is so typical or characteristic that early hunters of the gorilla, who often took "field" measurements of the specimens they shot, would have derived more accurate estimations of the heights of their trophies by multiplying the arm stretch by 0.67 than by trying to take the "height" of the apes properly in a lying position (often measuring to the end of the toes instead of to the heels!).

To express the main body and limb proportions of the gorilla in another way: in order for this ape to be of humanlike build, its legs, to match its trunk, would have to be nearly a foot (or al-

Table 3. Standing Heights of Adult Western Lowland Gorillas (*Gorilla gorilla gorilla*), as Calculated from the Lengths of their Limb Bones.   ( ) estimated

	* Collection and Catalogue No.	Measured by	Humerus	Radius	Femur	Tibia	Total Length, 4 bones, mm	Estimated Living Standing Height mm	Estimated Living Standing Height inches
	— — —	Mollison, 1911	403	318	340	266	1327	1459	57.45
	Nat. M. 174722	Schultz, 1930	403	314	352	275	1344	1477	58.15
	Amer. Mus. 69398	″ ″	407	332	344	278	1361	1494	58.82
	— — —	Mollison, 1911	412	321	353	279	1365	1498	58.98
	Ph. Acad. 5530	Schultz, 1930	406	344	341	278	1369	1502	59.15
	Nat. M. 174723	″ ″	402	335	358	287	1382	1516	59.67
	Wist. I. 4491	″ ″	412	341	357	275	1385	1519	59.79
	P.A.L. II	″ ″	411	337	356	288	1392	1526	60.08
	P.A.L. I	″ ″	424	335	362	296	1417	1552	61.10
	Brit. Mus. (?)	J. E. Gray, 1861	432	330	368	292	1422	1557	61.29
	Ph. Acad. 11805	Schultz, 1930	431	346	361	285	1423	1558	61.33
	— — —	Mollison, 1911	428	342	365	288	1423	1558	61.33
	Amer. Mus. 54356	Schultz, 1930	421	338	359	306	1424	1559	61.37
	— — —	Mollison, 1911	428	355	363	284	1430	1565	61.61
MALES	Du Chaillu, 1861	J. E. Gray, 1861	(443)	346	362	289	(1440)	(1575)	(62.01)
	— — —	Duvernoy, 1856	440	332	371	303	1446	1581	62.24
	Nat. M. 220325	Schultz, 1930	433	352	374	292	1451	1586	62.44
	Zo. Col.	″ ″	432	348	375	302	1457	1592	62.68
	W.R.U. B624	Straus	425	362	378	293	1458	1593	62.72
	Ph. Acad. 16982	Schultz, 1935?	426	357	372	303	1458	1593	62.72
	— — —	Lorenz, 1917	440	352	367	315	1474	1610	63.38
	— — —	″ ″	427	356	379	312	1474	1610	63.38
	P.A.L. 13	Schultz, 1930	447	349	379	300	1475	1611	63.42
	— — —	Mollison, 1911	441	357	380	300	1478	1614	63.54
	— — —	″ ″	427	354	394	303	1478	1614	63.54
	— — —	″ ″	435	370	382	294	1481	1617	63.66
	Roy. Coll. Surgeons	R. Owen, 1859	437	366	368	312	1483	1619	63.74
	— — —	Mollison, 1911	437	358	382	309	1486	1622	63.36
	— — —	″ ″	434	360	378	318	1490	1626	64.02
	P.A.L. 12	Schultz, 1930	442	354	383	313	1492	1628	64.10
	W.R.U. B1020	Straus	460	348	380	315	1503	1640	64.57
	"Phil" St. Louis N.H. Mus.	C. Hoessle, 1975	429	375	391	308	1503	1664	65.50
	P.A.L. 8	Schultz, 1930	480	367	396	316	1559	1697	66.81
	Ph. Acad. 16981	Schultz, 1935?	497	393	423	324	1637	1777	69.95
	Average (34)	— — —	430.94	348.35	370.38	297.00	1446.67	1581.9	62.28
	** ″ (93)	Schultz, 1937	436.8	351.2	376.0	300.4	1464.4	1600.0	63.00
	W.R.U. 626	Straus	344	291	296	233	1164	1281	50.42
	— — —	Mollison, 1911	356	285	294	236	1171	1288	50.71
	— — —	Lorenz, 1917	354	281	298	240	1173	1290	50.79
	Nat. M. 220060	Schultz, 1930	357	283	303	238	1181	1298	51.10
	P.A.L. 2	″ ″	357	279	300	247	1183	1300	51.18
	Dealer, Berlin	″ ″	362	290	300	236	1188	1305	51.39
	— — —	Mollison, 1911	353	286	299	252	1190	1307	51.47
	W.R.U. B824	Straus	357	294	301	245	1197	1315	51.77
	— — —	Mollison, 1911	359	298	302	241	1200	1318	51.88
FEMALES	W.R.U. B627	Straus	360	303	298	249	1210	1328	52.28
	P.A.L. 14	Schultz, 1930	375	290	301	249	1215	1333	52.48
	P.A.L. 6	″ ″	365	291	309	256	1221	1339	52.72
	Ph. Acad. 3319	″ ″	358	295	313	263	1229	1347	53.03
	— — —	Mollison, 1911	365	299	318	270	1252	1371	53.98
	— — —	Bolk, 1926	380	305	323	254	1261	1380	54.33
	— — —	Mollison, 1911	382	302	323	262	1268	1387	54.63
	— — —	Deniker, 1885	370	302	340	266	1278	1397	55.00
	— — —	Mollison, 1911	378	306	331	274	1289	1409	55.48
	— — —	Duvernoy, 1856	340	312	345	277	1324	1445	56.89
	Nat. M. 154553	Schultz, 1930	398	325	345	269	1337	1458	57.40
	Amer. Mus. 54355	″ ″	403	312	352	275	1342	1463	57.60
	Average (21)	— — —	367.73	296.62	313.86	253.90	1232.11	1350.7	53.18
	** ″ (60)	Schultz, 1937	368.0	294.0	311.0	251.0	1224.0	1342.4	52.85

* Nat. M.   = U. S. National Museum, Washington, D.C.
  Amer. Mus. = American Museum of Natural History, New York City.
  Ph. Acad. = Academy of Natural Sciences, Philadelphia.
  Wist. I.   = Wistar Institute of Anatomy, Philadelphia.
  P.A.L.   = Physical Anthropology Laboratory, Johns Hopkins University, Baltimore.
  W.R.U.   = Western Reserve University, Cleveland.
  Zo. Col.   = Department of Zoology, Columbia University.
** Schultz' data on limb bone lengths only. Height estimations by author.

Table 4. Standing Heights of Adult Mountain Gorillas (*Gorilla beringei*),
as Calculated from the Lengths of their Limb Bones.

	* Collection and Catalogue No.	Measured by	Humerus	Radius	Femur	Tibia	Total Length, 4 bones, mm	Estimated Living Standing Height mm	inches
MALES	Amer. M. 54090	Schultz, 1930	390	345	372	295	1402	1587	62.50
	Amer. M. 54089	" "	412	360	378	314	1464	1653	65.08
	Nat. M. 239883	" "	435	362	380	302	1479	1669	65.70
	San Diego N.H. Mus. ("Mbongo")	Willoughby, 1942	448	350	397	322	1517	1714	67.50
	M.C.Z. 23182	Schultz, ?	451	369	404	307	1531	1724	67.87
	Amer. M. 115609 ("Ngagi")	Schultz, 1944	460	355	395	325	1535	1728	68.03
	Average (6)	---	431.83	356.83	387.67	310.83	1487.16	1677.5	66.04
	Assumed Typical	---	433.3	358.1	389.0	311.9	1492.3	1683.0	66.25
FEMALES	P.A.L. 7 (sub-adult "Congo")	Schultz, 1930	326	274	296	242	1138	1296	51.03
	Amer. M. 54091	" "	344	292	311	248	1195	1357	53.42
	Amer. M. 54092	" "	362	302	326	280	1270	1437	56.56
	Average (3)	---	344.00	289.33	311.00	256.67	1201.0	1363.3	53.68
	Assumed Typical	---	353.0	297.0	319.2	263.4	1232.6	1397.0	55.00

* M.C.Z. = Museum of Comparative Zoology, Harvard University.
(For other abbreviations, see Table 3.)

Table 5. Standing Heights of Adult Eastern Lowland Gorillas (*Gorilla gorilla graueri*),
as Calculated from the Lengths of their Limb Bones.

	Collection and Catalogue No.	Measured by	Humerus	Radius	Femur	Tibia	Total Length, 4 bones, mm	Estimated Living Standing Height mm	inches
MALES	---	L. von Liburnau, 1917	428	339	366	281	1414	1578	62.12
	---	" " " "	447	341	383	300	1471	1637	64.46
	---	" " " "	465	351	390	315	1521	1689	66.50
	Average (3)	---	446.67	343.67	379.67	298.67	1468.67	1634.8	64.36
FEMALES	---	" " " "	372	281	317	246	1216	1351	53.19
	---	" " " "	385	287	330	251	1253	1389	54.69
	Average (2)	---	378.50	284.00	323.50	248.50	1234.50	1370.0	53.94

most 50 percent) longer than they are. On the other hand, each of the gorilla's arms would have to be about a foot (or nearly 30 percent) shorter than they are. Even then the ape's span would be about 4 inches greater than the man's, by reason of its broader shoulders. In short, the differing proportions of a gorilla's trunk, arms, and legs, compared with those of a human being, reflect the essentially differing environments and modes of living in which each species exists: the ape, a vegetarian, dwelling largely in trees; while the man, of omnivorous diet and endless adaptability, is basically a creature of open country who walks bipedally on the ground.

Physical Distinctions

The body and limbs of the gorilla are covered with coarse, brownish black hair, the length and precise coloration of which varies to a marked degree individually and with age. In males at about the age of nine to ten years, or puberty, the lower back normally becomes grey. This has caused such individuals to be referred to as "silverbacked males." In some cases the typical "saddle" of grey hair may extend so as to cover the buttocks and even the thighs. On the forearms of mature males the hair is often exceptionally long and shaggy. Normally there is a wooly undergrowth of hair

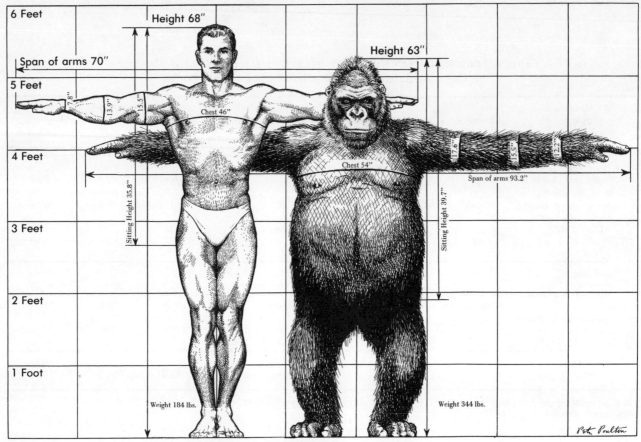

Fig. 39. A muscular, well-proportioned, weight-trained athlete compared with an adult male lowland gorilla of average size. This carefully executed scale drawing is by Peter Poulton. (Times one-sixteenth natural size.)

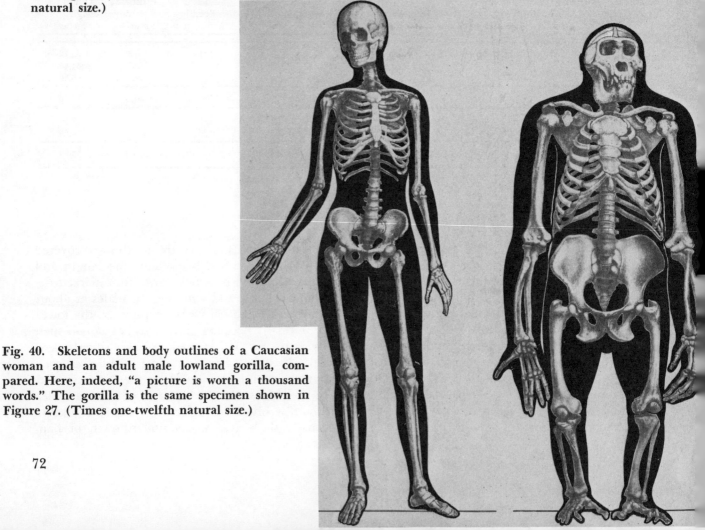

Fig. 40. Skeletons and body outlines of a Caucasian woman and an adult male lowland gorilla, compared. Here, indeed, "a picture is worth a thousand words." The gorilla is the same specimen shown in Figure 27. (Times one-twelfth natural size.)

over the entire body, with the exception of the chest, which in old males is practically bare, as is also the upper abdomen. In individuals of all ages the palms of the hands and soles of the feet are free of hair. While the backs of the hands are hairy, most segments of the fingers are bare, as are also the toes. On the crown of the head in many individuals there is a patch of reddish or rusty brown hair. According to the anatomist Duckworth:

> The hairs [in the gorilla] are said to be implanted in groups of five on the head and torso, and herein the Gorilla is contrasted with the Chimpanzee. The color of the hair is of a "dusty" grey tint, for each hair is "banded", the root and tip being grey with an intervening zone of black. Thus again the Gorilla differs from the Chimpanzee (with its distinctly black hair relieved by local patches of white).[18]

The skin in mature gorillas is black—in some cases intensely so; and the sweat glands emit a pungent, rubberlike odor. At birth, however, the skin is generally pinkish, the coloration of which becomes darker with age. In infant gorillas also there are often unpigmented, almost white spots or blotches on the hands and feet, which likewise disappear with age. In all adult manlike apes the skin is incredibly tough and resistive, and apparently impervious to pain. Thus, these animals are enabled to rip, tear, push, and pull objects with their hands that would be impossible to a human being (see chapter 7).

The head of the gorilla, particularly in adult or old males, is massive and rugged. Viewed from the side, the projecting muzzle, chinless jaw, beetling brows, and flattened nose are in striking contrast to the almost perpendicular facial outline of man, with his prominent nasal appendage and well-developed chin. The eyes of the gorilla are only slightly larger than those of man, notwithstanding that the bony orbits housing the eyeballs have a volume about 60 percent greater.[19] Thus, there is more "space" surrounding the eyeball of a gorilla than that of a man. The iris in the gorilla is generally of a beige brown color and is surrounded by a narrow ring of black. The sclera, or outer covering of the eyeball, is white, as in man; but the "whites" of the eyes are not often seen except when the ape looks sidewise. As the photographs in this book abundantly show, both the nose and the mouth in the gorilla are exceedingly broad, while the ears—in contrast to those of the chimpanzee—are relatively small and lie close to the head. The nose is distinguished from that of the chimpanzee by the greater length of the nasal bones, which descend below the level of the lower border of the eye sockets. It has also characteristic deep folds running from the angles of the nose to the corners of the upper lips.

Particularly massive are the teeth and jaws of the gorilla, especially those of adult males, in which the canines of the upper jaw range from an inch to nearly two inches in length, thus being almost equal in size to those of a lion or a tiger. The other teeth are also very large. Why these manlike creatures should require such powerful means of mastication is somewhat of a mystery. In the mountain gorilla it can be appreciated that powerful jaws and teeth are practically a necessity in cutting and crushing the tough bamboo stalks on which this species mainly feeds; but in the lowland gorilla, which subsists largely on soft herbs, vegetables, leaves, and fruits, there appears little need for such enormous masticatory development. But it is there; and the powerful muscles operating the jaws (which work in a straight up-and-down manner rather than also sidewise, as in man) require for their attachments correspondingly sturdy zygomatic arches (cheek bones), along with a keellike formation of the midsagittal top of the skull. And to counterbalance the heavy, forward-jutting head, great size and strength in the extensor (back) muscles of the neck is needed. Additional power or leverage is furnished by the exceptionally long processes (backward-protruding spines) on the neck vertebrae (Fig. 57).

The long and broad trunk of the gorilla corresponds with its high and wide pelvis, which has to be large to accommodate the vast quantities of relatively unnutritious vegetable matter that form the animal's diet. Again, the size and muscular development of the arms corresponds with that of the chest. Indeed, the ratio of arm girth to chest girth in the gorilla is surprisingly similar to that in man, with the exception of the enormous wrists and hands in the ape. The gorilla's hands are further noteworthy in that the fingers are "webbed" nearly to the first joints and are markedly tapered toward the ends. The tendons attached to the bones of the fingers are so arranged that the fingers can be extended fully only when the wrist is flexed; that is, when the wrist is extended, as in bringing the hand into line with the forearm, the fingers automatically draw into a clutching position that is maintained by powerful tendons and ligaments. Through this anatomical adaptation, gorillas—but to an even greater extent the more tree-living

Fig. 41. Upright and quadrupedal walking gaits as shown in a half-grown lowland gorilla at the Frankfurt, Germany, Zoo. (Photos by G. Budich.)

Table 6. Ratios between Various Limb-bone Lengths in Gorillas of the Three Recognized Types. Sexes Combined. (Based on Data in Tables 3, 4, 5). ( ) estimated

Species or Race	Number	Radius / Humerus	Tibia / Femur	Tibia / Humerus	Humerus + Radius / Femur + Tibia	♂ Humerus / Trunk Height (anterior)	Ilium Length / Humerus
Mountain	9	83.13	80.96	72.89	112.46	64.86	63.00
Western Lowland	153	80.20	79.99	68.61	117.00	76.23	52.50
Eastern Lowland	5	76.18	77.93	66.38	116.25	(75.70)	(54.00)
		Brachial Index	Crural Index	Tibiohumeral Index	Intermembral Index		

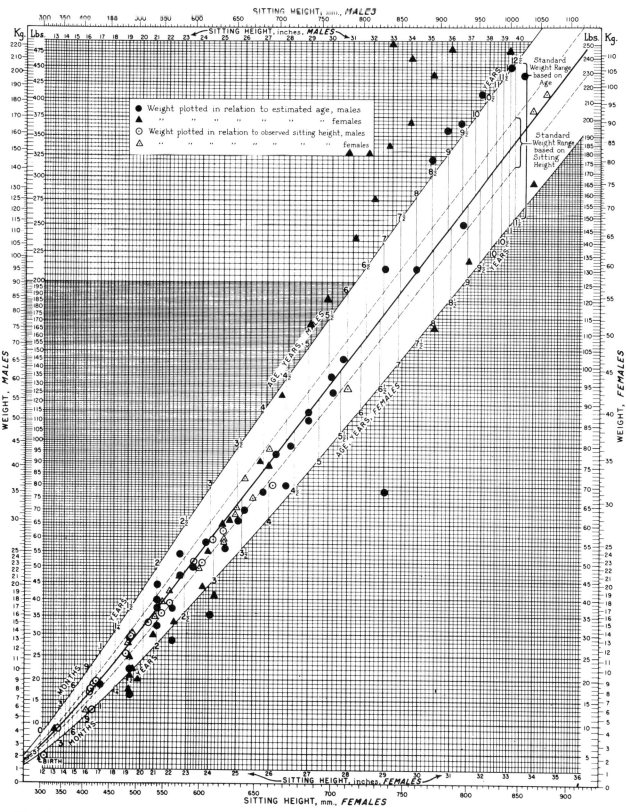

Fig. 42.

(brachiating) smaller apes—are able to suspend themselves by their arms with a minimum of stress on the muscles of the hands and fingers. Actually, in the sense of being gripping (rather than hook-like or suspensory) organs, the gorilla's feet, with their well-separated and powerfully muscled big toes, which act as thumbs, are more efficient than their hands.

How A Gorilla Walks

The type of ground locomotion normally employed by the gorilla, which is called "knuckle-walking," has come about through the circumstance that this heavy-built ape does not comfortably stand fully erect, as does man; that is, it is difficult, especially for an older gorilla, to straighten its hip joints and its knee joints at the same time—or, if they are even nearly straightened, to maintain them in that position for any length of time. Consequently, the gorilla's natural mode of locomotion on the ground is quadrupedal, or on all-fours, rather than bipedal, or on its feet alone, as is normal for man (Fig. 41). While the "knuckle-walking" is usually done on the middle joints (second phalanges) of the fingers, sometimes the clenched fist is used, at least for a resting position, in which case it is the proximal joints (first phalanges) on which the weight is supported. The knuckle-walking habit of the gorilla, which is adopted also by the chimpanzee, gives these anthropoid apes a unique "four-footed" stance, in which, due to their long arms, the head and shoulders are carried considerably higher than the hindquarters.*

Additional comments on the physical makeup of the gorilla are given later in this chapter in connection with the topics of skeletal structure, muscular development, dentition, etc.

* The characteristic inclination of the back, or spinal column, of an adult male gorilla in the quadrupedal posture appears to be about twenty-five degrees from the horizontal. This inclination, with the head held upright, makes the height to the crown of the head about 1.21 times the arm length, and the standing height about 1.27 times the quadrupedal height.

* * *

Some Differentiating Characters in the Three Races of Gorillas

Ever since Colin Groves, in 1967, on the basis of numerous skull measurements, classified the gorilla as comprising three races or subspecies, adhering to Coolidge's 1929 conclusion that only one species was represented, most zoologists appear to have accepted Groves' dictum. I say "most" because there are many who do not so think. The taxonomist Ernest Schwarz, for example, believed that the existing forms of gorillas were best divided into two species: western or lowland, and eastern or mountain. The German anatomist C. Vogel, on the basis of differences in the size of the lower jaw and in the position of the *foramen mentale,* also believes the mountain gorilla (*Gorilla beringei*) to be specifically distinct from the lowland form.[20] Adolph Schultz, who established no fewer than twenty differentiating characteristics, regarded the mountain gorilla as representing a distinct species. Although Schultz later evidently adopted the "single species" concept, on the basis of gorillas being highly variable individually, my opinion is that his first exposition of the numerous dimensions and proportions in which the mountain gorilla differs to a species degree from the lowland form, is a more valid conclusion than to lump the two forms together and regard the differences in the mountain gorilla as being—to quote one casual writer—"mostly minor ones." Schultz, while aware that some of his twenty listed differences might prove inconsistent, in his original investigation declared:

> It seems unquestionable, for instance, that the mountain gorilla is characterized by a comparatively long palate, by a variety of peculiar conditions of the foot [including a greater adaptation for walking on the *ground*], and by proportionately shorter arms[21] [the latter being reflected in the ape's *shorter reach* as compared with its height].

Table 7. Living Lowland Gorillas for which various measurements in addition to Weight were Recorded. ( ) estimated. All linear measurements are in millimeters.

No.	Zoo or Collection	Age	Sex	Weight Kg.	Height Standing	Height Sitting	Span, or Horizontal Reach	Girth Chest	Girth Belly	Length Hand	Length Foot	Name	Index of Body Build*
1	Basel	Birth	♀	1.81	(401)	280	(521)	(280)	(292)	(69)	(84)	Goma	4.36
2	Washington	"	♂	2.27	450	(312)	(592)	(296)	(307)	(76)	(96)	Tomoka	(4.21)
3	Jersey	1 mo.	♂	3.31	(452)	312	(592)	361	(373)	86	102	Mamfe	4.78
4	San Diego	3 mos.	♀	4.08	508	(353)	686	368	(381)	(94)	(119)	Alvila	(4.53)

5	" "	6 "	♀	6.30	584	(401)	808	419	(434)	(109)	(135)	"	(4.61)
6	H. Bartel	10 " 1	♂	7.70	(600)	414	(820)	480	(495)	(112)	(137)	Bimbo	4.77
7	Philadelphia	10 " 2	♂	7.90	(603)	416	(823)	483	(498)	(112)	(137)	Bamboo	4.79
8	San Diego	12 "	♀	11.07	711	(485)	(991)	516	(538)	(132)	(157)	Alvila	(4.60)
9	H. Bartel	14 " 3	♂	8.80	(615)	426	(843)	493	(511)	(114)	(140)	Bimbo	4.85
10	San Diego	15.5 "	♂	11.45	714	480	940	531	559	124	157	Albert	4.70
11	" "	18.1 "	♀	12.61	721	485	978	559	584	130	160	Bouba	4.80
12	" "	19 "	♀	13.70	726	493	1016	571	610	140	178	Bata	4.85
13	J. L. Buck	20 " 4	♂	13.60	724	490	(1003)	574	(594)	(132)	(160)	Ngi	4.87
14	San Diego	20.5 "	♂	14.97	744	521	1011	591	626	135	170	Albert	4.73
15	" "	23.1 "	♀	15.65	775	533	1054	610	640	140	172	Bouba	4.69
16	" "	24 "	♀	17.69	807	546	1156	622	661	152	193	Bata	4.77
17	New York	24 " 5	♂	16.10	(810)	546	(1133)	598	(620)	(140)	(170)	Jimmie	4.62
18	San Diego	24.5 "	♂	17.35	787	541	1118	616	635	140	178	Albert	4.79
19	A. H. Schultz	27 "	♀	16.10	(807)	547	(1133)	598	(635)	150	180	Janet	4.62
20	San Diego	27.1 "	♀	19.10	831	559	1219	641	680	147	178	Bouba	4.78
21	" "	28 "	♀	21.05	851	571	1270	654	692	163	198	Bata	4.83
22	" "	28.5 "	♂	20.41	826	565	1168	635	660	142	190	Albert	4.84
23	" "	31.1 "	♀	22.00	889	597	1232	660	724	140	190	Bouba	4.69
24	" "	32 "	♀	25.17	930	622	1270	673	737	170	208	Bata	4.71
25	" "	32.5 "	♂	23.24	864	597	1251	648	635	152	198	Albert	4.78
26	" "	35.1 "	♀	25.40	927	622	1308	711	718	165	203	Bouba	4.73
27	" "	36 "	♀	28.35	940	632	1403	749	718	178	216	Bata	4.82
28	" "	36.5 "	♂	26.76	927	622	1308	711	686	165	208	Albert	4.81
29	" "	38.8 "	♀	29.26	970	635	1372	749	762	178	216	Bouba	4.85
30	" "	39.7 "	♀	32.52	972	641	1486	762	762	190	229	Bata	4.98
31	" "	40.5 "	♂	28.12	904	635	1372	762	737	165	221	Albert	4.79
32	" "	43.1 "	♀	30.39	991	648	1440	818	813	190	224	Bouba	4.82
33	" "	44 "	♀	35.83	1016	660	1524	824	797	190	254	Bata	5.00
34	" "	48.4 "	♀	42.64	1092	711	1575	864	838	216	261	"	4.91
35	Cincinnati	9.5 yrs.	♀	93.0	1372	864	2032	1143	1219	(208)	(251)	Susie	5.24
36	"	10.5 "	♀	97.5	1410	889	2032	1168	1245	(213)	(257)	"	5.18
37	"	12.5 "	♀	136.5	1499	927	2134	1270	1422	(221)	(267)	"	5.55
38	"	14.5 "	♀	152.0	1537	952	2184	1321	1575	(224)	(269)	"	5.61
39	"	15.5 "	♀	165.0	1549	953	2235	1346	1600	(226)	(272)	"	5.76
40	San Diego	15.5 " 6	♂	263.3	1714	1096	2476	1753	1829	268	323	Mbongo (Mountain Sp.)	5.85
Typical Newborn		Birth	♂	2.15	439	305	571	292	305	74	97	———	4.23
"	"	"	♀	2.00	427	297	551	284	302	71	94	———	4.27
Typical Adult		12½ yrs.	♂	156	1600	1008	2367	1372	1424	241	291	———	5.34
"	"	11½ yrs.	♀	85	1342	846	1968	1107	1172	206	248	———	5.20

1 Author's estimation on basis of body weight; age reported as 12 months.
2 "   "   "   "   "   "   "   "   "   " 14 "
3 "   "   "   "   "   "   "   "   "   " 16 "
4 "   "   "   "   "   "   "   "   "   " 24 "
5 "   "   "   "   "   "   "   "   "   " 29 "
6 Measurements were taken shortly after death (see text).

* 1000 x $\sqrt[3]{\text{Weight}}$ ÷ Sitting Height

These individuals may all have been malnourished, as may be seen by comparison with those in the San Diego Zoo (cf. also with Fig. 42).

Here is a copy of Schultz's twenty characteristics differentiating the mountain gorilla from the lowland gorilla, along with five additional characteristics listed by Coolidge (1929) and four added by the present writer.

1. Greater length of trunk
2. Higher-situated nipples
3. Narrower hips
4. Lesser length of neck*
5. Shorter lower limbs in relation to height of trunk
5a. Longer lower limbs in relation to length of upper limbs
6. Much shorter upper limbs
7. Broader and shorter hand
8. Great toe reaching farther distally
9. Great toe branching from sole more distally
10. Less convex joint at base of hallux
11. Relatively shorter outer lateral toes
12. Usually webbed toes
13. Higher face
14. Narrower width between eyes
15. Smaller average number of thoracolumbar vertebrae
16. Absolutely and relatively shorter humerus**
17. Absolutely and relatively longer clavicle
18. Absolutely and relatively longer ilium
19. Relatively longer radius
20. Peculiarly curved vertebral border of scapula

And according to Coolidge (1929)

21. Longer palate
22. Generally narrower skull
23. Thicker pelage
24. Large amount of black hair
25. Fleshy callosity on the crest (of the head)

* In Schultz's paper, no. 4 is incorrectly worded "*Greater length of neck.*"

** Later specimens added to Schultz's list by the present author make the term *absolutely* in no. 16 no longer applicable.

And according to Willoughby (1950)

26. Relatively shorter arm-spread or span
27. Broader shoulders (due to longer clavicles)
28. Thicker neck
29. Greater height and weight

Of the foregoing and other differential characteristics, perhaps the most fundamental and significant are the proportionate lengths of the limb bones, the proportions of the pelvis, and the relative lengths of the several segments of the vertebral column; for example, Schultz has shown that the length of neck in relation to the length of trunk in two specimens of the mountain gorilla averages 28.2 percent (27.0–29.4), while in three specimens of the western lowland gorilla the average ratio is 38.7 (37.1–40.0). If the ranges in this ratio as shown in these small series are expanded statistically so as to embrace 1,000 examples of each type of gorilla, the range in the mountain type becomes 22.2 to 34.2, and in the lowland type 33.1 to 44.3.* The same statistical relationships show that among lowland gorillas only one in about 350 would be expected to have as small a neck-trunk ratio as the largest ratio present among mountain gorillas, and, conversely, that among the latter only one in 350 would have as large a ratio as the smallest ratio among lowland gorillas.

Tables 3, 4, and 5 list various museum skeletal specimens of gorillas, giving in each case the lengths of the humerus (upper arm), radius (forearm), femur (thigh), and tibia (lower leg). Most of these statistics have been drawn from Schultz;[23] and from the total of the bone lengths in each example the probable living standing height has been estimated by the writer from these equations:

* The procedure for converting smaller ranges to the Standard Range (= 1000 cases) proposed by George Gaylord Simpson is explained by him in his paper *Range as a Zoological Character*.[22]

Adult male western lowland gorillas, height = 1.025 (Humerus+Radius+Femur+Tibia) +100.5 mm.
Adult female western lowland gorillas, height = 1.028 (H+R+F+T) + 84.1 mm.**
Adult male eastern lowland gorillas, height = 1.043 (H+R+F+T) +103.0 mm.
 " female " " " = 1.043 (" " ") + 82.4 mm.
Adult male mountain gorillas, height = 1.057 (H+R+F+T) +105.6 mm.
 " female " " " = 1.064 (H+R+F+T) + 85.4 mm.

** For a given total length of the four limb bones combined, the living standing height in adult male gorillas averages 12 mm, or about a half-inch more than in adult female gorillas. Within a population of several dozen gorillas the largest females are of about the same size (height and weight) as the smallest males (*i.e.*, about 57.5 inches and 240 pounds).

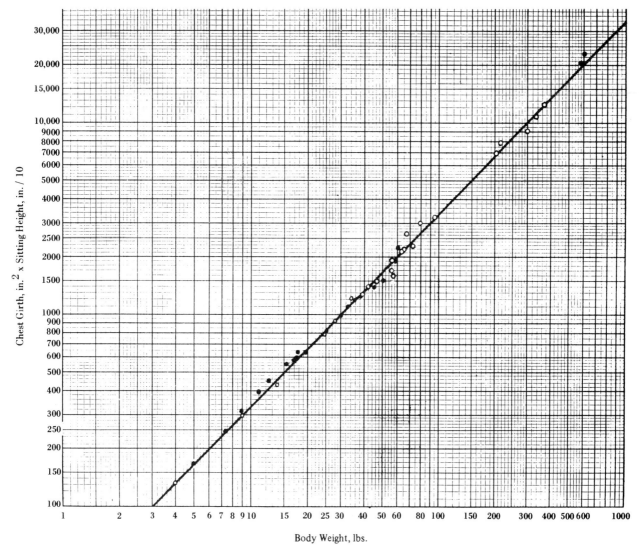

Fig. 43. Although this graph is laid out on a double logarithmic grid, the correlation formula is a simple, linear one, namely: body weight equals 0.00297 (chest girth, inches, squared times sitting height). This single formula applies to both sexes at all stages of growth, although strictly to *G. gorilla* and not *G. beringei*. However, I have shown one adult male mountain gorilla next to the topmost entry on the graph. More correctly, for mountain gorillas, the multiplier is 0.00288 because of their relatively (to weight) larger chests and sitting height. On this graph the solid spots indicate males, and the open circles, females. The observations are those listed in Table 7.

At first it might seem that from only the two leg bones (femur and tibia) the total height could be estimated as well as from all four bones (since the arm bones do not contribute directly to the height), but this is not the case. The humerus and radius, no less than the femur and tibia, express the general size of the body, particularly the trunk length or sitting height, which comprises a major portion of the total height.

Table 5 presents, from the meager data available, the limb-bone lengths and the estimated standing heights in the eastern lowland or Grauer's gorilla, which is seen to be intermediate in size between the western lowland and the mountain gorillas, although closer to the former than to the latter, at least in the males. The proportionate average sizes of Grauer's gorillas may be estimated as about 64 inches and 370 pounds in adult males, and 54 inches and 200 pounds in adult females.

Although some authors consider the eastern lowland gorilla to be, rather, a variant of the mountain gorilla, this belief is not borne out by a comparison of the proportions of the limb bones, such as are listed in Table 6. The ratios given there show, on the contrary, that the eastern lowland gorilla de-

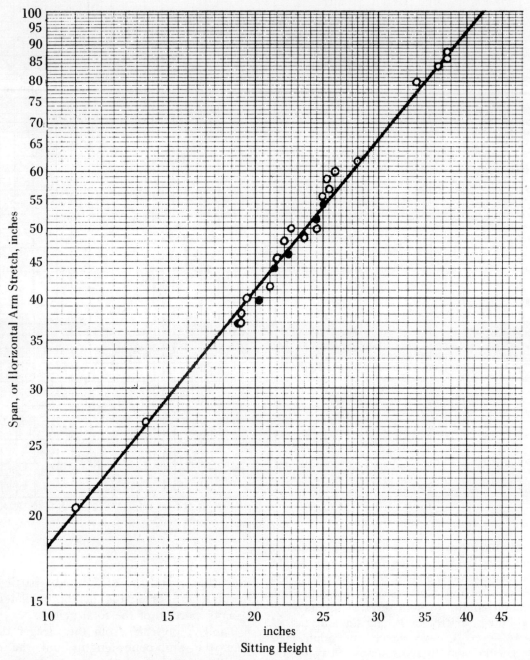

Fig. 44. This graph reveals a fairly close linear correlation between span and sitting height in gorillas, although a more direct correlation exists between arm length plus shoulder (biacromial) width and span. Here the solid spots indicate males, and the open circles, females. The observations are those listed in Table 7, the topmost circles being the measurements of "Susie," formerly of the Cincinnati Zoo, and the other observations mostly those of Albert, Bouba, and Bata, of the San Diego Zoo. Albert was measured periodically until the age of about forty months, and the females Bouba and Bata, until about forty-eight months.

Growth of Gorillas in Height, Weight, and Other Body Measurements

Before presenting typical external body measurements of adult gorillas, it may be appropriate to list various measurements that have been taken of living zoo specimens of younger apes, in which it was possible to secure such data before the apes, especially the males, grew too large and strong to work with in safety.

Figure 42, "Growth Grid for Gorillas," is a chart or graph that I drew up in the 1940s but which is as useful today as it was then, when the various major zoos in the United States featured such fine male specimens as Bushman, Makoko, Bamboo, and Albert, and such females as Oka, Bouba, Bata, and Susie. A careful examination of the graph should make clear the procedure by which the various gorillas were charted from birth to the adult stage. While in most of the specimens only the weight was known, in a few others, such as those measured by Dr. Schultz and at the San

Fig. 45. "Mbongo," a mountain gorilla of the San Diego Zoo (1931–1942), showing the high headcrest and shaggy hair-coat characteristic of this species. The measurements of Mbongo's skeleton, which were taken by the author, also show marked differences in proportions from those of lowland gorillas. (Photo by E. H. Boldrick, courtesy San Diego Zoo.)

viates in the first three ratios even farther from the mountain species than does the western lowland gorilla; and that, with the western gorilla, it is sufficiently deviate in the last three ratios to be clearly distinct from the mountain form. In the humerus-trunk height ratio, for instance, the difference is sufficiently great—even between two series of 10,000 individuals each—to separate completely the mountain gorilla from the western lowland form. For the eastern lowland gorilla, much more skeletal material is needed for better identification. Meanwhile, however, it is seen that this subspecies is even farther removed physically from the mountain gorilla than is the western or typical form. Therefore, along with the latter, *G. g. graueri* should, I feel, be considered specifically distinct from *G. beringei*, the mountain gorilla. (See also chapter 7 and Figure 82.)

* * *

Fig. 46. "Ngagi" (1931–1944), a long-time companion of Mbongo at the San Diego Zoo. Both gorillas, which were captured by Martin Johnson in December 1939 in the Alumbongo Mountains northwest of the Virunga Volcanoes in Uganda, exhibit some characteristics of the eastern lowland gorilla (*G. g. graueri*). Possibly they are hybrid individuals, although predominantly of the mountain species (*Gorilla beringei*).

Diego Zoo, the sitting height was also recorded, which provided an additional means for estimating age. It may be added that in practically every case, age had to be estimated, since at the time the graph was drawn, no gorillas had as yet been born in captivity. It will be noted on the graph that the body weights in relation to age vary widely in both sexes. While some of this variation may be accounted for by differences in body build—that is, skeletal thickness in relation to height, just as in slender, medium, and stocky-built human beings—in most of the gorillas charted it is more likely due to variations in nutrition, some of the imported younger apes being afflicted with "homesickness" and so not eating well, while most of those that survived to older ages apparently were overfed. It may be assumed that those gorillas whose placements on the graph conform most closely with the centermost inclined line were the healthiest specimens and exhibited the most typical trend in growth.

Table 7 lists a greater variety of body measurements than Figure 42 and so is more useful for correlating each of these measurements with age and sex. While the measurements of the San Diego gorillas—Albert, Bouba, and Bata—and of Susie of the Cincinnati Zoo were all originally recorded in inches and fractions, they have been converted here into millimeters, which today is the preferred (metric) measurement among primatologists. Using the metric system also greatly simplifies the job of typesetting, by eliminating fractions and/or decimal points. At the bottom of Table 7 are given the typical body measurements of both newborn and young adult lowland gorillas of both sexes, consistent with well-established limb-bone lengths, standing and sitting heights, and chest and belly girths proportionate to height and weight from birth to maturity.

In the right-hand column of Table 7 are figures that show at each age the width or girth of each gorilla relative to its body length or sitting height.

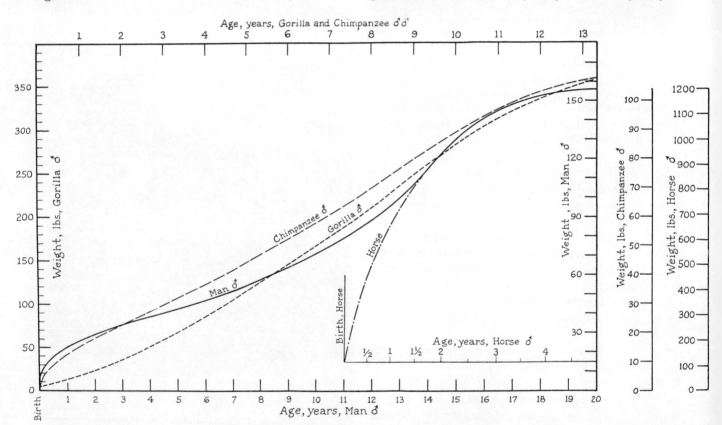

Fig. 47. Comparison of growth curves for man, gorilla, chimpanzee, and horse (Thoroughbred or saddle), respectively, with age and weight scales adjusted for equivalence. The weight scale for the chimpanzee also represents approximately the percentage of mature weight. Note the greatly extended early period of growth in the primates, as compared with that in the horse. Note also that the gorilla does not exhibit the initial postnatal spurt of growth shown by man and the chimpanzee but increases in weight at an almost uniform rate during the first few years. Man's weight curve differs from that of the apes by having a "pubertal inflection" (rate change) at thirteen to fourteen years. For other details, see text.

Table 8. Typical Body Measurements of Male Lowland Gorillas in Relation to Age (Metric Units)
All measurements except weight are in millimeters.

Age	Weight, Kg.	Standing Height	Sitting Height	Anterior Trunk Height	Chest Girth	Belly Girth	Span, or Horiz. Reach	Arm Length	Leg Length	Hand Length	Foot Length	Index of * Body Build
Birth	2.18	439	305	154	292	305	571	251	188	74	97	4.25
1 mo.	2.95	465	323	165	336	348	612	269	197	81	107	4.45
3 mos.	3.95	503	348	180	373	389	666	295	211	91	117	4.54
6 mos.	5.35	551	381	199	419	434	739	328	232	102	127	4.59
12 mos.	9.1	627	434	231	516	533	864	378	264	117	142	4.80
18 mos.	12.5	711	485	261	571	592	996	437	301	132	160	4.79
2 yrs.	16.8	795	538	293	627	650	1123	498	335	147	180	4.76
3 yrs.	26.9	937	620	342	734	762	1344	589	406	165	201	4.83
5 yrs.	52.2	1130	734	410	937	970	1641	724	508	190	231	5.09
7 yrs.	79.8	1290	825	464	1090	1130	1887	828	582	211	257	5.21
9 yrs.	111.6	1433	909	514	1224	1270	2108	927	662	224	272	5.29
11 yrs.	142	1544	975	553	1328	1379	2281	1003	724	234	282	5.34
12½ yrs.	156	1600	1008	573	1372	1423	2367	1040	752	241	291	5.34
Adult/Birth	72.6	3.64	3.30	3.72	4.70	4.67	4.14	4.14	4.00	3.28	3.01	1.26

Same as above, in English Units (pounds and inches)

Age	Weight	Standing Height	Sitting Height	Anterior Trunk Height	Chest Girth	Belly Girth	Span, or Horiz. Reach	Arm Length	Leg Length	Hand Length	Foot Length	**
Birth	4.8	17.3	12.0	6.1	11.5	12.0	22.5	9.9	7.4	2.9	3.8	4.25
1 mo.	6.5	18.3	12.7	6.5	13.2	13.7	24.1	10.6	7.8	3.2	4.2	4.45
3 mos.	8.7	19.8	13.7	7.1	14.7	15.3	26.2	11.6	8.3	3.6	4.6	4.54
6 mos.	11.8	21.7	15.0	7.8	16.5	17.1	29.1	12.9	9.1	4.0	5.0	4.59
12 mos.	20.0	24.7	17.1	9.1	20.3	21.0	34.0	14.9	10.4	4.6	5.6	4.80
18 mos.	27.6	28.0	19.1	10.3	22.5	23.3	39.2	17.2	11.8	5.2	6.3	4.79
2 yrs.	37.0	31.3	21.2	11.5	24.7	25.6	44.2	19.4	13.2	5.8	7.1	4.76
3 yrs.	59.3	36.9	24.4	13.4	28.9	30.0	52.9	23.2	16.0	6.5	7.9	4.83
5 yrs.	115.0	44.5	28.9	16.1	36.9	38.2	64.6	28.5	20.0	7.5	9.1	5.09
7 yrs.	176	50.8	32.5	18.3	42.9	44.5	74.3	32.6	22.9	8.3	10.1	5.21
9 yrs.	246	56.4	35.8	20.2	48.2	50.0	83.0	36.5	26.0	8.8	10.7	5.29
11 yrs.	313	60.8	38.4	21.8	52.3	54.3	89.8	39.5	28.5	9.2	11.1	5.34
12½ yrs.	344	63.0	39.7	22.6	54.0	56.0	93.2	41.0	29.6	9.5	11.4	5.34

Adult/Birth = same as above
*Index of Body Build (Metric) = 1000 x $\sqrt[3]{\text{Weight, kg.}}$ ÷ Sitting Height, mm.
** " " " (English) = 30.256 x $\sqrt[3]{\text{Weight, lbs.}}$ ÷ Sitting Height, in.

Table 9. Typical Body Measurements of Female Lowland Gorillas in Relation to Age (Metric Units)
All measurements except weight are in millimeters.

Age	Weight, Kg.	Standing Height	Sitting Height	Anterior Trunk Height	Chest Girth	Belly Girth	Span, or Horiz. Reach	Arm Length	Leg Length	Hand Length	Foot Length	Index of Body Build *
Birth	2.00	427	297	152	285	302	551	244	183	71	94	4.27
1 mo.	2.81	452	315	163	330	348	592	262	192	79	104	4.48
3 mos.	3.81	493	343	180	373	394	653	290	206	89	114	4.55
6 mos.	5.26	544	376	200	419	444	736	328	228	99	127	4.62
12 mos.	9.1	627	432	234	516	546	864	383	264	117	142	4.83
18 mos.	12.5	711	485	266	571	605	996	442	304	132	159	4.79
2 yrs.	16.8	795	538	299	627	665	1123	498	341	147	178	4.76
3 yrs.	26.1	927	612	344	726	770	1333	589	413	160	198	4.86
5 yrs.	41.5	1069	691	392	859	909	1547	686	490	178	218	5.01
7 yrs.	55.3	1166	747	426	952	1008	1699	757	540	188	229	5.10
9 yrs.	70.3	1262	800	458	1034	1095	1844	818	580	198	241	5.16
11 yrs.	83	1331	838	481	1095	1158	1951	866	614	206	246	5.20
11½ yrs.	85	1342	846	486	1102	1168	1968	874	621	207	248	5.20
Adult / Birth	42.6	3.15	2.85	3.20	3.86	3.87	3.57	3.57	3.39	2.89	2.63	1.22

Same as above, in English Units (pounds and inches)

Age	Weight	Standing Height	Sitting Height	Anterior Trunk Height	Chest Girth	Belly Girth	Span, or Horiz. Reach	Arm Length	Leg Length	Hand Length	Foot Length	Index **
Birth	4.4	16.8	11.7	6.0	11.2	11.9	21.7	9.6	7.2	2.8	3.7	4.27
1 mo.	6.2	17.8	12.4	6.4	13.0	13.7	23.3	10.3	7.6	3.1	4.1	4.48
3 mos.	8.4	19.4	13.5	7.1	14.7	15.5	25.7	11.4	8.1	3.5	4.5	4.55
6 mos.	11.6	21.4	14.8	7.9	16.5	17.5	29.0	12.9	9.0	3.9	5.0	4.62
12 mos.	20.0	24.7	17.0	9.2	20.3	21.5	34.0	15.1	10.4	4.6	5.6	4.83
18 mos.	27.6	28.0	19.1	10.5	22.5	23.8	39.2	17.4	12.0	5.2	6.2	4.79
2 yrs.	37.0	31.3	21.2	11.8	24.7	26.2	44.2	19.6	13.4	5.8	7.0	4.76
3 yrs.	57.5	36.5	24.1	13.5	28.6	30.3	52.5	23.2	16.3	6.3	7.8	4.86
5 yrs.	91.5	42.1	27.2	15.4	33.8	35.8	60.9	27.0	19.3	7.0	8.6	5.01
7 yrs.	122	45.9	29.4	16.8	37.5	39.7	66.9	29.8	21.3	7.4	9.0	5.10
9 yrs.	155	49.7	31.5	18.0	40.7	43.1	72.6	32.2	22.8	7.8	9.5	5.16
11 yrs.	183	52.4	33.0	18.9	43.1	45.6	76.8	34.1	24.2	8.1	9.7	5.20
11½ yrs.	188	52.8	33.2	19.1	43.4	46.0	77.5	34.4	24.5	8.1	9.7	5.20

Adult/Birth = same as above
*Index of Body Build (Metric) = 1000 x $\sqrt[3]{\text{Weight, kg.}}$ ÷ Sitting Height, mm.
** " " " " (English) = 30.256 x $\sqrt[3]{\text{Weight, lbs.}}$ ÷ Sitting Height, in.

By comparing the adult indices of body build thus established (see the bottom-most two rows) with those at birth (that is, in males, 5.34/4.23), it is seen that the relative body and limb girth in mature gorillas is about 25 percent greater than at birth. A similar ratio exists in man; but in humans there is a vast increase from birth to maturity in the length of the lower limbs, whereas in apes the increase is much less, with the result that the general proportions of the body are quite similar throughout the lifetime of the animal.

All linear measurements among gorillas are closely correlated both with one another and with the weight per unit of height, the square root of the latter measurement providing a ratio directly comparable with linear measurements, such as those of length and girth (circumference). Many such correlations, as made by the writer and then plotted on graph paper, make possible the establishment of reliable measurements of all parts of the body and limbs in both sexes from birth to the adult stage.

Figure 43—in which inches and pounds are used—illustrates the close connection between weight and relative body volume as derived from chest girth squared then multiplied by sitting height. From the formula given under the chart, chest girth, sitting height, and body weight may each be derived provided that two of the three measurements are known. The formula adopted assumes a belly girth in males of 1.038 times chest girth, and in females 1.059 times chest girth. Where the chest and belly girths differ by more than 3 inches, a better estimation of weight is obtained by taking into account belly girth as well as chest girth; thus:

Body weight, pounds, males = chest girth $\times$ belly girth $\times$ sitting height, inches, $\times$ .002860
Body weight, pounds, females = chest girth $\times$ belly girth $\times$ sitting height, inches, $\times$ .002805

Incidentally, the term *waist* (rather than *belly*) can hardly be used in connection with gorillas, since normally at all stages, from birth to maturity, the abdomen is larger in girth than either the chest or the hips, and so presents no "waist" or smallest part.

Figure 44 shows that two seemingly dissociated length measurements—in this case sitting height and span—are correlated sufficiently well to enable one to be predicted with fair accuracy from the other. The correlating formulae are:

Span, inches, 2 years to adult, males = 2.65 $\times$ sitting height — 12.0
Span, inches, 2 years to adult, females = 2.78 $\times$ sitting height — 14.5
Span, inches, birth to 2 years, males = 2.41 $\times$ sitting height — 6.4
Span, inches, birth to 2 years, females = 2.42 $\times$ sitting height — 6.6

An almost endless series of correlation formulas for body and limb measurements in gorillas could be given if all parts of the body were to be interrelated. Such extensive information, however, is best presented in technical papers rather than in a book intended for general readers. Nevertheless, it should be appropriate here to show at least the growth from birth to maturity in the basic measurements of the body and limbs. This, for lowland (western) gorillas of both sexes, is done in Tables 8 and 9. While based largely on the measurements of living zoo gorillas as listed in Table 7, all other available sources of information on the subject have been drawn upon and are listed under the notes.

Figure 47 is a graphic presentation of three slow-growing primates—man, gorilla, and chimpanzee—as compared with the horse, a rapidly growing ungulate that is as far advanced physically at birth as a boy is at 11 years. The relatively slow rate of bodily growth in the primates is one

Fig. 48. Carl Stemmler, head keeper at the Basel, Switzerland, Zoological Gardens, weighs reluctant two-year-old "Jambo" by having his mother, "Achilla," hold him, while he then deducts Achilla's 160 pounds from the total of 186 pounds (1963). (Photo by Paul Steinemann.)

Table 10. *Proportions* of the Body in Male Lowland Gorillas at Various Ages in Relation to Standing Height (= 100.0)

Age	Standing Height	Sitting Height	Anterior Trunk Height	Chest Girth	Belly Girth	Span, or Horiz. Reach	Arm Length	Leg Length	Hand Length	Foot Length
Birth	100.0	69.4	35.0	66.5	69.4	130.0	57.2	42.9	16.8	22.1
1 mo.	100.0	69.4	35.4	72.1	74.9	131.7	57.9	42.4	17.5	23.0
3 mos.	100.0	69.2	35.7	74.2	77.0	132.3	58.6	42.0	18.2	23.3
6 mos.	100.0	69.1	36.2	76.0	78.8	134.1	59.5	42.1	18.4	23.0
12 mos.	100.0	69.1	36.8	82.2	85.0	137.7	60.3	42.1	18.6	22.6
18 mos.	100.0	68.2	36.8	80.4	83.2	140.0	61.4	42.3	18.6	22.5
2 yrs.	100.0	67.7	36.8	78.9	81.8	141.2	62.0	42.2	18.5	22.5
3 yrs.	100.0	66.1	36.6	78.3	81.2	143.4	62.9	43.3	17.6	21.5
5 yrs.	100.0	65.0	36.2	82.9	86.0	145.2	64.0	45.0	16.8	20.4
7 yrs.	100.0	64.0	36.0	84.4	87.6	146.3	64.2	45.1	16.3	19.9
9 yrs.	100.0	63.5	35.9	85.5	88.7	147.2	64.7	46.2	15.7	19.0
11 yrs.	100.0	63.2	35.8	86.0	89.2	147.7	65.0	46.9	15.2	18.3
12½ yrs.	100.0	63.0	35.8	85.8	89.0	147.9	65.0	47.0	15.1	18.2
Adult / Birth	1.000 (Reciprocal) →	0.908 (1.101)	1.023	1.290	1.282	1.138	1.136	1.097	0.900 (1.111)	0.824 (1.214)

Same as above, in Female Lowland Gorillas

Age	Standing Height	Sitting Height	Anterior Trunk Height	Chest Girth	Belly Girth	Span, or Horiz. Reach	Arm Length	Leg Length	Hand Length	Foot Length
Birth	100.0	69.6	35.6	66.8	70.8	129.2	57.2	43.0	16.6	22.0
1 mo.	100.0	69.7	36.0	73.0	77.0	130.9	58.0	42.5	17.5	23.0
3 mos.	100.0	69.6	36.5	75.7	78.9	132.5	58.8	41.8	18.1	23.1
6 mos.	100.0	69.3	36.8	77.1	79.9	135.5	60.3	41.9	18.2	23.3
12 mos.	100.0	68.8	37.3	82.2	87.0	137.7	61.1	42.1	18.7	22.7
18 mos.	100.0	68.2	37.5	80.3	85.0	140.0	62.2	42.7	18.6	22.4
2 yrs.	100.0	67.7	37.6	78.9	83.7	141.2	62.6	42.9	18.5	22.4
3 yrs.	100.0	66.5	37.1	78.3	83.0	143.8	63.5	44.5	17.3	21.4
5 yrs.	100.0	64.6	36.6	80.3	85.0	144.7	64.2	45.8	16.7	20.4
7 yrs.	100.0	64.0	36.5	81.7	86.5	145.8	64.9	46.3	16.1	19.6
9 yrs.	100.0	63.4	36.3	81.9	86.7	146.1	64.9	46.0	15.7	19.1
11 yrs.	100.0	63.0	36.2	82.3	87.0	146.6	65.1	46.1	15.5	18.5
11½ yrs.	100.0	63.0	36.2	82.1	87.1	146.7	65.1	46.3	15.4	18.5
Adult / Birth	1.000 (Reciprocal) →	0.905 (1.105)	1.017	1.229	1.230	1.135	1.139	1.077	0.927 (1.079)	0.838 (1.193)

of the factors contributing to their higher brain development and intelligence (see chapter 7).

It is now opportune to comment briefly on each of the measurements listed in Table 8 and to indicate the significance of certain measurements in the gorilla, as compared with those in man (and in some instances the chimpanzee).

Age

It is impossible to state a specific age at which the higher anthropoid apes attain "maturity," anymore than to do this in the case of man. Body weight, for instance, in many kinds of mammals commonly increases, although usually at a diminishing rate, throughout the lifetime of the individual. This appears to be the case in the higher primates, although man, with his superior intelligence and foresight, is able to thwart the process by restricting his food intake. The cessation of growth in length of the long bones of the limbs is perhaps the best criterion of essential "mature" size, and this arrival at full limb-bone length is reflected in the virtual cessation of growth in stature or standing height. In man, according to recent statistics on growth in height and weight of school children in the United States, full stature is reached by 18 or 19 years of age in boys and as early as 16 or 17 years of age in girls.[24] In the present study I have assumed, on this basis, that growth in stature or long-bone length in the gorilla is completed, on the average, by the age of 12½ years in males and 11½ years in females.* This assumption makes the rate of growth in the gorilla (also the chimpanzee—and probably the orangutan) just 1½ times as rapid as in man, as is shown in Figure 47. This ratio is substantiated by the eruption time of the permanent dentitions in man and the gorilla (Table 14). But, as previously mentioned, there is a wide range of individual variation in this respect —some gorillas being essentially full-grown by the age of 10 years, while others may take until 15 years or even older. "Mbongo," the male mountain gorilla of the San Diego Zoo, was shown by his limb bones, the epiphyses of which were still slightly "open," to have been growing still at the age of 15½ years. However, it would appear that the assumption of an "adult" age of 12½ years for male gorillas and 11½ years for female gorillas is satisfactory as a basis for figures expressing average or typical physical growth in these anthropoids. The gestation period in the gorilla, or the age at birth, is nearly as long as that in man. It averages 265 days (c. 8¾ months), ranging from about 236 to 290 days (c. 7¾ to 9½ months).

Weight

This measurement ordinarily presents no problem with young gorillas but may be a considerable one with adult or subadult specimens. And sometimes even a very small gorilla will balk at being put on the scales. This happened with "Jambo," the first male gorilla born in Europe (Basel, Switzerland), when he was about 2 years of age. Jambo's keeper, Carl Stemmler, solved the difficulty by first getting Jambo's well-trained mother, "Achilla," onto the scales, then having her hold Jambo while he weighed both together (Fig. 48). Achilla, who was very light for an adult female gorilla, weighed 160 pounds and Jambo 26 pounds, which was also very light for a two-year-old male. Some gorilla keepers have used a similar procedure, by weighing themselves while supporting a half-grown (and still tractable) gorilla on their back. Full-grown gorillas may be weighed by equipping their living quarters with a platform scale that can be read outside the enclosure. "Guy," the adult male at the London Zoo, has been weighed regularly by this means, as were "Mbongo" and "Ngagi" of the San Diego Zoo, each of which came eventually to weigh over 600 pounds. Outside the glass-walled "living room" of "Samson," of the Milwaukee Zoo, is a spring scale that registers weights up to 2000 pounds and which shows the current weight of Samson (who has weighed as much as 610 pounds) whenever he gets onto the scale's platform on the other side of the ¾-inch-thick laminated glass wall, through which visitors may observe him.

Stature, or Standing Height

While this measurement is a time-honored basis for the establishment of body proportions, both in scientific anthropometry and in artistic anatomy, in apes it is not a similarly satisfactory basis because apes cannot readily assume a posture in which both the body and the lower limbs are fully straightened at the same time. Although the joints in apes, such as the gorilla, are remarkably supple, it would appear that the habitual quadrupedal posture in these animals, particularly the gorilla, has resulted in corresponding alterations in the muscles involved, such as a lesser development (in comparison with man) of the extensors of the hip (gluteus maximus) and possibly some shortening of the hamstrings—the latter condition making it an effort to straighten the knees when the body is carried semierect. For this reason, the "standing

---

* Gavan and Swindler assume an adult age of 11 years for each of the three great apes—orangutan, chimpanzee, and gorilla.[25]

Fig. 49. "Albert," of the San Diego Zoo, who weighed eight pounds on arrival on 10 August 1949. He is being held by Mrs. Belle Benchley, then director of the zoo. (Photo by G. E. Kilpatrick, courtesy San Diego Zoo. Approximately 0.45 natural size.)

percent. The wide divergence of this ratio in adult gorillas and man is the result of the marked lengthening of the lower limbs in man from birth to maturity; while in apes the ratio of body length to leg length during the same period changes comparatively little. In mountain gorillas the sitting-height/standing-height index averages 63.8 percent in adults and about 70 percent in newborns. Sitting height is useful also, in connection with chest girth, as a means of estimating body weight (see Fig. 43).

Trunk Height

A measurement commonly used by primatologists as a basis from which to calculate the proportions of the body and limbs in apes and monkeys is the height or length of the front of the torso from the symphysis pubis (pubic arch) to the suprasternale (top of the breast-bone or sternum). This measurement is properly known as the anterior trunk height (or length, since in apes and human cadavers it is taken with the body in a lying or supine position). It may be obtained on

Fig. 50. "Albert," the first of the San Diego Zoo's male lowland gorillas, shown here at the age of twenty years, is one of the few zoo-raised gorillas to have attained full mature size without becoming fat and overweight. (Photo by Ron Garrison, courtesy San Diego Zoo. Compare with Figure 49.)

height" as listed in Tables 8 and 9, in adult and subadult gorillas, represents what the *expected* measurement would be, as compared with the sitting height, leg length, and span. In young, more manageable specimens, the standing height is closely approximated by measuring the ape in the lying (supine) position, using boards or sliding calipers held against the crown of the head and the heels, the feet being at right angles to the lower legs.

Sitting Height

This measurement is a useful and practicable one in apes, since it can be taken with a fair degree of accuracy and can be used to estimate erect standing height. Sitting height in adult lowland gorillas of both sexes averages 63 percent of the standing height. In man, averaging the two sexes, the corresponding ratio is only about 53 percent. At birth this comparison is much closer, the sitting height in gorillas averaging about 69.5 percent of the standing height and in human babies 67.4

cadavers with a high degree of accuracy, provided the upper border of the symphysis pubis can be located properly beneath the overlying soft tissues. However, this body measurement is practically impossible to secure with accuracy on a living gorilla—particularly one beyond infantile age—and is therefore never used by zoo keepers. The measurements of anterior trunk height given in Table 8, therefore, are estimations based on the relation of this measurement to that of sitting height as determined on cadavers of gorillas by Dr. Adolph Schultz.

Chest Girth

This, one of the least difficult measurements to secure on an ape, is simply the greatest circumference of the torso as taken immediately under the arms and across the biggest bulge of the chest muscles in front and back. The tape, however, should be pulled sufficiently taut to minimize the effect of hair and so express essentially the chest girth directly next to the skin. Too, the arms should be hanging at the sides and not raised to shoulder level. As listed in Tables 8 and 9, chest girth varies from about 6 inches less than standing height at birth to 9 or 10 inches less than height in adults (of both sexes). Instances in which the chest girth surpasses the standing height occur only in adult male gorillas and indicate specimens of relatively enormous bulk and weight. Again, trophy hunters who wish to be credited with specimens of "giant" size are apt to exaggerate the measurement of chest girth as much as that of standing height or arm spread. Overstated measurements of girth can readily be checked by comparing them with the probable height and weight of the specimen.

Belly Girth

Usually the largest girth of the body in a gorilla is that taken around the most protruding part of the abdomen—the "barrel," as it is called by hunters. This girth ranges from an inch or less larger than the chest girth at birth to 2 or 3 inches larger at maturity, being proportionately (to the chest) somewhat larger in females than in males. If the belly measures 6 inches or more larger than the chest—as it came to do in "Susie," the Cincinnati Zoo's adult female (see Table 7)—an obese condition may be assumed, regardless of what the ape may weigh. Belly girth squared may be used in conjunction with chest girth squared and sitting height (the three measurements being multiplied together) to provide a remarkably reliable "formula" for estimating body weight where it (weight) is not known.

Span, or Horizontal Arm Stretch

This measurement, which is almost invariably included among those taken by trophy hunters, is the distance between the middle fingertips with the arms fully straightened at shoulder-level. The tapeline may have to be curved a bit at each end of the span in order to conform with the not fully straightened fingers. While spans of up to 10 feet have been claimed by some publicity-seeking gorilla hunters, it is probable that a correctly measured span of 9 feet, or even slightly less, represents the maximum, since it would be proportionate to a standing height of 6 feet 1 inch in a lowland gorilla, or at least six feet four inches in a mountain gorilla—neither height of which has been authenically recorded. In adult lowland gorillas the span averages $1.479 \times$ standing height in males and $1.467 \times$ standing height in females. At birth these ratios are about 1.30 and 1.29, respectively. In adult mountain gorillas, because of their greater standing height and relatively shorter arms, the span-height ratio averages about 1.40 in males and 1.39 in females. For subadult mountain gorillas the span-height ratio has been recorded only in one specimen, as far as I know. This was the female "Congo," which was measured at the age of $8\frac{1}{3}$ years by Professor Yerkes and was found to have a standing height of 50 inches and a horizontal span of 71 inches—a ratio of 1.42. Another "span" measurement is that which is taken from the soles of the feet to the tips of the middle fingers in a lying position, with the arms stretched vertically overhead. In adult lowland gorillas this upward reach averages about $1\frac{1}{2}$ inches more than the horizontal span in males and about $2\frac{1}{2}$ inches more in females. In mountain gorillas (of both sexes) the difference is about 6 inches greater overhead than horizontally. In Tables 8 and 9 the regular (horizontal) span measurement has been derived from arm length plus shoulder (biacromial) breadth, the formula for both sexes at all ages being: SPAN $= 2 \times$ arm length $+$ biacromial width $\times 0.94$. Omitting biacromial width, the span is equal to $2.275 \times$ arm length in lowland males (all ages), and $\times 2.254$ in lowland females (all ages). Arm length, in turn, is derived from the combined lengths of upper arm, forearm, and hand. Finally, upper-arm length (as in life) is equal to humerus length (dry bone) $\times 1.015$; while forearm length is equal to radius length (dry bone) $\times 1.015$,

Arm Length

This length, which is measured with a sliding calipers, is the distance between the tip of the acromial process (acromion) and the end of the middle finger (dactylion), with the arm held straight. In adult lowland gorillas of both sexes each arm measures about 65 percent of the standing height, while in newborns the ratio is only a trifle over 57 percent. The ratio increases rapidly with age, and at 4 years averages 63.5 in males and 63.8 in females. There is an interesting relationship also of arm length to sitting height—this ratio at birth being 82.3, at 6 years practically 100.0, and at 12 years 102.3; that is, arm length increases from birth to maturity about 12 percent faster than sitting height. How arm length is derived from the combined lengths of upper arm, forearm, and hand is explained under the measurement of SPAN, above.

Leg Length

What anthropologists commonly adopt as "leg length" is the height of the greater trochanter of the femur (trochanterion) from the ground; or, with the body lying, from the heels. This overall length of the lower limb is essentially the combined length of the thigh (femur), lower leg (tibia), and foot (from heel to ankle joint). It is equal to femur length plus tibia length $\times$ 1.114 in males, and $\times$ 1.117 in females. Relative to standing height, leg length (or trochanter height) equals 47.0 percent in adult male lowland gorillas and 46.3 percent in adult females. At birth this index in males is 42.8 and in females about 43.0. When relative leg length and relative sitting height are added, there is an "overlapping" of from 9 to 12 percent, which represents the percentage of the standing height that the top of the femur (trochanter) is above the level of the seat.

Hand Length

Figures for this measurement have been derived both from the listings in Table 7 and from measurements taken on gorilla cadavers by Dr. Schultz. The index of hand length-standing height in lowland gorillas averages 16.7 percent at birth; 18.6 at one year; and 15.1 male, 15.4 female in adults. The hands in mountain gorillas are shorter and relatively (to their length) broader than in the lowland species.

Foot Length

In adult lowland gorillas (both sexes), foot length is commonly about 20 percent longer than hand length. In newborns the ratio averages over 1.30 $\times$ hand length and diminishes to the 1.20 ratio at about 4 years. The foot in the mountain gorilla has less space between the big toe and the other toes and is less adapted for grasping and more for walking on the ground than in the lowland species.

Table 8 gives the typical body and limb measurements of male lowland gorillas from newborn to adult (the upper half of the table is in metric units and the lower half in English inches and pounds). Table 9 does the same for female gorillas. In the bottom row of each of the metric tables is given the ratio of the measurement at maturity to that at birth. These ratios show the relative rates of increase in absolute size. In Table 10, the standing height at each age and in both sexes is made the basis (100.0) to which each of the other linear measurements is related. These comparisons show the changes in bodily proportion that occur during growth. The sitting height, for example, at birth is about 69.5 percent of the standing height and, during the period of growth, gradually decreases to 63 percent of the height in adults. This relative decrease occurs because the absolute leg length increases faster than absolute trunk length or sitting height; the latter measurement therefore becomes a lesser percentage of the total standing height. Some of the other proportions show no such steady change. Relative hand length and foot length, for instance, increase until the age of 1 or 2 years, then decrease to maturity. Conversely, relative leg length at first decreases until 1 or 2 years, then increases. The lower row in the table for each sex shows the degree to which the body and limb proportions change from newborn to adult. The most variable measurements are naturally the girths of chest and belly, since these circumference measurements express changes in muscular or soft-tissue development, which is more subject to differences generally in growth, nutrition, and bodily activity (or inactivity) than is bone length and/or thickness.

The two sexes in gorillas are so similar throughout the whole period of infancy and well into the juvenile stage that, so far as body measurements are concerned, it is often difficult to tell one sex from the other. Only after the age of 3 or 4 years do males begin to outstrip females in height, weight, and girth, which they do at an accelerating rate. If any of the absolute measurements listed in

Tables 8 and 9 are plotted on graph paper with respect to age, this increasing separation of the sexes will be seen.

Although no attempt is made herein to establish other (mostly girth) measurements in gorillas throughout the entire period of growth, enough data have been available to do this at least for newborns and adults of both sexes in western lowland gorillas and for adults (only) of both sexes in mountain gorillas. Practically no information on these measurements is available for eastern lowland gorillas, which in any case are probably very similar to the western race, except for being, on the average, slightly taller and heavier. Table 11 gives the estimations for both adult and newborn gorillas.

Bodily Proportions in the Gorilla as Compared With Man

Compared with man, the girth measurements in gorillas of the upper arms and forearms, in relation to chest girth, are very similar, indicating that the force of the movements of the arms in both man and ape directly governs the muscular size of the chest. On the other hand, girth of the neck in adult male* gorillas averages 47 percent of chest girth, while in a symmetrically proportioned human male athlete (e.g., Fig. 39) the ratio is only 38 to 39 percent. This extra size of the neck in the gorilla (and other apes) is present from birth onward. In the adult male gorilla, in particular, the size results largely from the great length of the spinous processes of the cervical vertebrae (see Figs. 51 and 62)—this causes the neck to be deeper from front to back than from side to side, and therefore oval in cross-section, rather than round as in man.

The wrists, too, in gorillas are even larger, relative to chest girth, than is the neck, the wrist-chest ratio in the adult male gorilla averaging 22.7, while in man (Caucasian) the ratio is about 17.0, or only 75 percent as large as in the gorilla. The massive development of the wrists in apes, which makes the forearms look cylindrical rather than oval and tapering as in man, is accompanied by a pronounced development of the hands and fingers, especially in gorillas, where these parts are relatively broad and thick as well as long. This immense development is well shown in the skeletons of man and gorilla in Figures 27 and 51, where the hands of the man, with their relatively slender bones, appear almost feeble in comparison with the hamlike "paws" of the gorilla. At this point an important distinction should be mentioned: in man, the gorilla, and the chimpanzee, there is one less bone in the wrist (carpus) than in all the other primates.[26] In Figures 27 and 51 it will be observed that the humerus (upper arm bone) in the gorilla is fully as long as the femur (thigh bone) in the man, besides being of greater thickness. This should give some idea of the immense strength possessed by gorillas in their upper limbs, which are possibly stronger than a human athlete's legs. Again, the strength of the arms in the gorilla is indicated by the size (girth) of the wrists, which is practically the same as that of the ankles. In adult man, in contrast, the wrists are typically only 80 percent of the girth of the ankles.

Figures 27, 40, and 51 show also many of the other skeletal differences that exist between man and gorilla. The least differences, morphologically, are in the long bones of the upper and lower limbs, which in the gorilla diverge only by reason of their size (thickness and length). Between the two skulls a vastly greater difference, or series of differences, occurs, due primarily to man's larger brain and smaller jaws, as shown in Figure 57. Comments on brain development and intelligence are given in chapter 8.

The gorilla's apparently shorter neck is due to its enormous jaws, which extend downward almost onto the clavicles (collarbones), thus giving the impression that the head is set directly on the trunk. The thorax of the gorilla is characterized by huge scapulae (shoulder blades) and a broad and deep (from front to back) ribcage, which flares out at the bottom to match the wide pelvis and to accommodate the capacious abdominal contents. At their medial ends the seven "true" ribs are attached to the four or five segments of the exceedingly broad sternum (breast bone). Lowland gorillas normally have 13 pairs of ribs (12–14), as compared with 12 pairs in man. In the mountain gorilla the average number is 12.6—this being derived from the fact that five out of eight have 13 ribs, and three out of eight, 12 ribs.

The great height from top to bottom of the pelvis in the gorilla, along with the iliac crests being close to the lower ribs, gives to the lower back the straightness and rigidity that prevents any appreciable amount of forward flexion or side-to-side bending of the lumbar spine, especially in full-

---

* Some other writers, in referring to adult males, have used the term "bull ape," or "bull gorilla," or "the bulls of the troop," etc. While the meaning of *bull* in this connection is obvious, the corresponding term for the female of the species should be *cow*, and for infants, *calves*. While these terms apply aptly to horned animals and certain others, they are rather absurd for apes.

Fig. 51. A young Kivu (mountain) gorilla exhibiting the thick, profuse hair-coat present even in infants of this high-ranging (up to 13,000 feet) east African species. (Photo by W. Suschitsky; courtesy Studio Publications.)

Table 11. Computed Living Girth Measurements of Adult and Newborn Gorillas in both Metric and English Units.

	Measurement (girth)	Adult Western Lowland		Adult Mountain		Newborn Western Lowland	
		Males	Females	Males	Females	Males	Females
Millimeters and Kilograms	Neck	645	521	704	552	200	195
	Upper Arm (Straight)	448	362	489	384	97	94
	Forearm	404	326	441	343	104	101
	Wrist	311	274	339	291	89	87
	Chest	1372	1107	1499	1168	292	285
	Belly	1424	1172	1555	1245	305	302
	Hips	1077	883	1174	936	229	225
	Thigh	673	551	734	584	142	140
	Knee	449	368	489	390	118	116
	Calf	404	330	441	351	95	93
	Ankle	320	282	349	300	89	87
	Standing Height	1600	1342	1683	1397	439	427
	Body Weight	156	85	195	100	2.18	2.00
Inches and Pounds	Neck	25.40	20.50	27.70	21.75	7.87	7.68
	Upper Arm (Straight)	17.65	14.25	19.25	15.12	3.82	3.70
	Forearm	15.90	12.82	17.35	13.52	4.10	4.00
	Wrist	12.25	10.80	13.35	11.46	3.50	3.42
	Chest	54.03	43.60	59.00	46.00	11.50	11.20
	Belly	56.08	46.16	61.24	49.00	12.00	11.90
	Hips	42.40	34.75	46.20	36.85	9.00	8.86
	Thigh	26.50	21.70	28.90	23.00	5.60	5.51
	Knee	17.67	14.47	19.25	15.34	4.65	4.57
	Calf	15.90	13.00	17.35	13.80	3.74	3.66
	Ankle	12.60	11.10	13.73	11.80	3.50	3.42
	Standing Height	63.00	52.85	66.25	55.00	17.30	16.80
	Body Weight	344	188	430	220	4.81	4.40

grown individuals. The absence of the "lumbar curve" characteristic of the spine of man also makes difficult or impossible the attainment of a fully erect standing posture in the gorilla—or for that matter in the chimpanzee and orangutan. A contributing factor to the relatively long trunk (sitting height) in the gorilla, which in adults of the lowland species averages 63 percent of the standing height, as compared with 52 to 53 percent in man, is the great height of the lower part of the pelvis (ischium), or the distance below the pubic arch of the two ischial tuberosities or *sitz* bones (German), the dimension of which is nearly double the height of the same bones in man (Figs. 27 and 40).

With reference to the lower limbs, while the crural index (ratio of tibia length to femur length) in the gorilla of 80.4 (averaging the two sexes) is not greatly below that in man (83.4), the relationship of both leg bones (femur plus tibia) to both arm bones (humerus plus radius) is markedly different in the gorilla and other higher anthropoids. An "index" is customarily the ratio of a smaller measurement to a larger one, yielding a percentage of less than 100. In man the intermembral index (humerus plus radius, divided by femur plus tibia) follows this pattern by averaging 70.4 in men and 69.1 in women, even where both whites and blacks are represented.[27] In contrast, this index in the gorilla (western lowland race) averages about 117 in adults and 116.5 in newborns—very little change in the index throughout the period of growth. In man at birth, due to the relatively short legs of the infant, the index is 89 or 90, as compared with about 70 in adults. Thus, one of the basic characteristics of growth that distinguishes anthropoid apes in general, including the gorilla, is the virtual retention of the infantile body build (so far as length, rather than girths, are concerned). In man, contrariwise, there is a markedly greater increase in the length of the legs than there is in the arms, which thereby causes a reduction in the intermembral index from newborn to adult.

One of the indications that the mountain gorilla (*G. beringei*) is closer to man in some of its physical proportions than is the typical western lowland gorilla (*G. gorilla*) is that in the former species the intermembral index averages only about 112.5, as compared with 117 in the western lowland and 116.25 in the eastern lowland races (Table 6).

When consideration is given to the foot of the gorilla, the difference from man is so great that no prolonged comparison is necessary. Briefly, the gorilla's foot, in common with that of other anthropoid apes, is adapted for grasping, the same as is the human hand; whereas in man there is no wide separation of the hallux (big toe) from the others and consequently little or no lateral grasping power in the foot. Here again, as in the matter of having shorter arms than the lowland gorilla, the mountain gorilla is closer to man because its foot is less adapted for grasping and more for walking on the ground. When performing the latter function, the gorilla brings the big toe close to and in line with the four smaller toes; while if the walking is being done along a branch, the big toe is naturally spread apart from the others so that the foot may be used to grasp the limb.

From Table 8 it will be noted that the weight of an average-sized newborn male lowland gorilla is 4.8 pounds, while that of the mother ape (Table

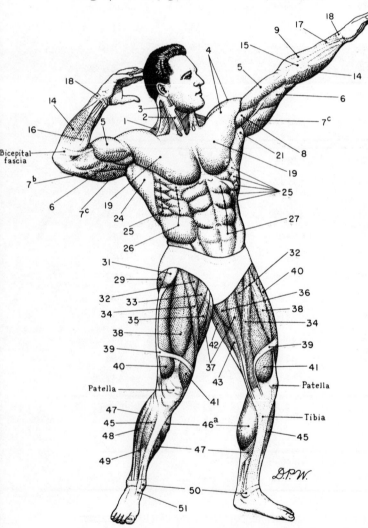

Fig. 52. Drawing by author showing the principal superficial muscles of the human body in front view. Compare with Figure 53. The numbers on the drawing refer to a tabulation of the muscles, which is not included here.

Fig. 53. Muscular anatomy of a male gorilla as depicted by the eminent paleontological artist Charles R. Knight (1945). The voluntary muscles in the anthropoid apes are essentially similar to those in man, the chief differences being in their relative size and adaptation to the largely quadrupedal posture and locomotion in the apes.

9) is 188 pounds. The ratio of the weight of the newborn ape to the mother is therefore 4.8:188, or about 2.5 percent. In the chimpanzee the ratio is 1.85:40.5, or 4.6 percent; in the orangutan, 1.52:37, or 4.1 percent; and in the gibbon (*Hylobates lar*), 0.40:5.3, or 7.55 percent. If these body weights of newborn and mother apes are plotted on a graph, the expected birth weight of a male human infant having a mother weighing, say, 60 kilos or 132 pounds, would be about 2 kg (4.4 pounds), which is the average weight of a newborn female gorilla. Actually, the newborn weight in a well-nourished male human is about 3.5 kg (7.7 pounds), or 75 percent greater than the ratio would be for an anthropoid ape born of a mother of the same body weight as the woman. Thus, the

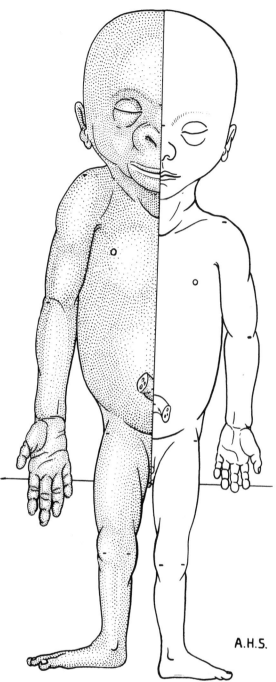

Fig. 54. Gorilla fetus (left), age 98 days, compared with a human fetus of the same sitting height (88.5 mm). Note in the gorilla the larger head (yet smaller brain space); thicker neck; higher shoulders, nipples, and elbow; longer arm, hand, and foot; and shorter leg—compared with man. At birth, the proportions of height and length in each specimen are quite similar to those shown here, although the body build is quite heavier, and the widths and girths are proportionately greater. (Drawing after A. H. Schultz—1927; Plate 3.)

95

relative body size of mother and infant in humans falls completely outside the course of this relationship among the anthropoid apes from gibbon to gorilla.

Just as the skeleton of the gorilla is of the same general pattern as that of man, so are the ape's muscles. If Figure 53 is compared with Figure 52, it will be seen that the same muscles are present and that they differ only in their relative length and bulk, in conformity with the differing body builds in man and gorilla.* It should not be necessary here to list and describe all the muscles in either man or ape, since our object is simply to indicate, in a general way, the similarities and differences of the physique of man and that of his nearest existing primate relative. Additional physical comparisons, with the chimpanzee and the orangutan, are made in chapter 6.

Skeletal Dimensions in Relation to Dental Age

In this book about gorillas, one of the writer's chief objectives has been to correlate the available body measurements of living (i.e., captive) gorillas with the more extensive data that have been published on the measurements of the skeleton in these giant primates and, from this correlation, to arrive at the height, weight, lengths, and girths of all parts of the body in both sexes from birth to maturity, the same as has been done in man.

The basic statistics given in various tables herein—particularly Tables 7 and 12—have enabled the aforementioned objective to be attained. Although the number of measured specimens of living gorillas at any given age has been limited, the graphic correlation of these measurements throughout the range in age covered by the data has shown clearly that while vast differences in size occur during growth in gorillas from newborn to adult, the accompanying changes in the proportions of the body, hands, feet, and limbs are comparatively slight and can be established for the various ages with an acceptable degree of accuracy. The virtual identity of body build in gorillas of both sexes between the ages of one and four years (see Table 7) provides an example of this.

Briefly, once the typical lengths of the long bones of the limbs at various ages have been deduced from the known standing and sitting heights as secured from living (captive) gorillas, and the body weights of these specimens have likewise been recorded, all the other body and limb dimensions, particularly the girths, can thereby be derived. Thus, the detailed measurements presented in Tables 8, 9, and 11, for example, which may at first appear highly speculative, are on the contrary only the result of an application of known ratios of body and limb proportions (Table 10) to reliably recorded measurements of height, weight, span, chest girth, belly girth, and hand and foot length—along with a sufficient number of recordings of neck girth, arm and forearm girth, wrist girth, hip and thigh and calf girth, etc., to enable the latter girths to be derived either directly from chest girth or deduced from the ratio of *general girth* obtained by taking the square root of the weight divided by the height.

Reliable data on the typical body measurements of adult living gorillas have evidently been desired by primatologists and/or physical anthropologists ever since the gorilla was discovered and a limited number of skeletal or embalmed specimens were collected by various natural history museums. Adolph Schultz, writing in 1937, remarks:

> Since it would be exceedingly difficult, if not impossible, to secure statistically adequate series of data on the size and weight of the *entire* bodies of adult male and female great apes, we must utilize instead the large series of adult apes in our collections.[28]

Francis Randall, writing in 1944, adds:

> Unfortunately, growth cannot be studied in terms of absolute time in *Gorilla gorilla* as yet, but must be treated entirely upon a rather rough basis of assessed stages in the growth processes.[29]

The growth stages to which Randall refers are six in number. They are listed in Table 12, together with the lengths of the long bones found in the present study to be commensurate with the living standing heights established from the raw data in Table 7 and systematically applied to the ages listed in Tables 8 and 9. Thus, the information so desired by Schultz, Randall, and other primatologists—namely, the correlation of living body measurements with age in the gorilla—may, through reference to Table 12, be obtained.

Note, in Table 12, that the least variable measurement in both male and female gorillas from birth to the adult stage is the estimated living stature, and that the most variable measurement in both sexes is the width of the pelvis. The adult-birth ratios in the limb bones and the clavicles

---

* Yet there evidently occur individual exceptions in the muscles of the gorilla, just as in man; for instance, an adult male mountain gorilla reported by Philippa Gaffikin was found, upon dissection, to have no Sartorius muscles (see *East African Med. Jour.* 26, no. 8 (August 1949):2.

Table 12. Skeletal Dimensions in Western Lowland Gorillas in Relation to Dental Age.* ( ) estimated

Sex	Dental Age Stage	Dental Age Yrs.	Length of Long Bones, mm (Av. of Right and Left)						Width, mm, of		Estimated Living Stature, mm
			Clavicle	Humerus	Radius	Femur	Tibia	Total, H+R+F+T	Shoulders (bi-acromial)	Pelvis (bi-cristal)	
Males	—	Birth	39.4	112.7	85.2	92.6	76.4	366.9	105.1	73.6	439.4
	I	24.5 mos. (18–30)	79.0	201.8	158.8	168.6	135.3	664.5	210.8	164.9	801.6
	II	4.5 (3.5–6.0)	106.9	285.6	228.3	240.0	194.8	948.7	285.1	250.8	1087.1
	III	6.5 (5–8)	126.6	330.6	265.6	278.3	226.9	1101.4	337.7	297.0	1252.2
	IV	9.0 (6.5–11.5)	145.7	386.0	311.5	328.8	265.4	1291.7	388.6	353.8	1432.6
	V	12.0 (9–15)	162.6	432.2	349.9	371.0	297.5	1450.4	433.6	401.1	1585.0
	VI	12.5 and over	164.3	436.2	352.0	374.6	300.2	1463.0	438.0	405.2	1600.0
	♂ Adult / Birth		4.17	3.87	4.13	4.05	3.93	3.99	4.17	5.51	3.64
Females	—	Birth	37.4	109.4	81.7	89.9	74.2	355.2	98.7	71.5	426.7
	I	24.5 mos. (18–30)	76.4	203.6	158.9	170.9	137.8	671.2	200.7	168.1	801.6
	II	2.75 (2–4)	87.9	231.5	182.7	197.8	158.9	770.9	230.8	199.3	900.5
	III	6.5 (5–8)	110.1	306.3	243.5	263.0	210.4	1023.5	289.0	274.2	1143.0
	IV	9.0 (6.5–11.5)	124.2	340.0	271.9	287.6	232.8	1132.3	325.9	309.3	1262.4
	V	11.33 (9–14)	131.4	362.8	292.0	305.0	248.6	1208.4	344.7	333.7	1338.7
	VI	11.5 and over	131.7	364.0	293.0	306.0	249.5	1212.5	345.5	335.0	1342.4
	♀ Adult / Birth		3.52	3.33	3.59	3.40	3.36	3.41	3.50	4.68	3.15

Based mainly on data from Randall, 1943–44.

range only between 3.87 and 4.17 in males and 3.33 to 3.59 in females, with the positions of the five bones within these respective ranges being almost identical in the two sexes. The range of variation in females averages only 0.855 times that in males, due partly to females maturing about a year earlier, but mainly because of the more rapid growth of males after stage I is passed.

Ages of Tooth Eruption in Man and Apes

We may now turn briefly to Tables 13 and 14 for a consideration of the times (ages) of eruption of the teeth in man and the higher anthropoid apes. The deciduous or temporary ("milk") teeth in both man and apes are 20 in number. They

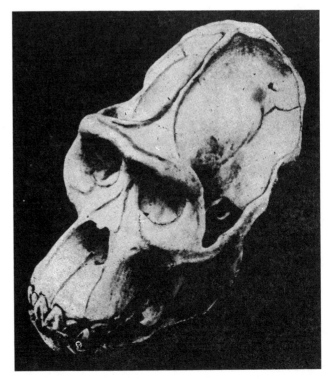

Fig. 55. Skull of an adult male mountain gorilla as viewed in frontal-profile position. (From Philippa Gaffikin—1947; photo by M. Jaffe.)

consist of 2 incisors, 1 canine, and 2 molars, on each side, above and below. What is known as the "dental formula" is accordingly: i$\frac{2}{2}$, c$\frac{1}{1}$, m$\frac{2}{2}$. The milk teeth differ from the permanent teeth by their smaller size, slender necks, and, in the molars, by the wide separation of their roots. By "eruption" of the teeth is meant the age at which the entire top surface of the crown has emerged from the gum.

In Table 13 are listed the mean or average eruption ages, in days, of the deciduous dentition, along with "early" and "late" ages as commonly recognized. However, in exceptional cases, the range in age may be a great deal more extensive than is given in this table.* Generally, however, in human children all the milk teeth have erupted by the end of 2 to 2½ years. Then follows a period of from 3 to 4 years during which no visible change takes place within the mouth, although active preparation for the eruption of the permanent teeth is proceeding beneath the gums. In the apes listed both in Table 13 and Table 14, it will be seen that these developmental changes take place at considerably earlier ages than in man. There is, however, no universal agreement among odontologists as to a specific average age at which the teeth erupt in man, let alone in the higher apes, on which there is far less information. Table 13 is based, for man, on the ages generally listed in medical texts. For the chimpanzee the average ages are those indicated (as percentages of the adult age) by Gavan, and the relative ranges are those given by Nissen and Riesen.[31] In Table 14, ages for the gorilla appear to fall about one-sixth of the way (timewise) between those for the chimpanzee and for man; while in Table 13 the remarkably early ages for the eruption of the deciduous incisors should be noted. For the orangutan there appears to be only a limited amount of data on eruption ages. Schultz gives the figures on one specimen only, a male observed by Brandes;[32] while Harrisson gives average ages of eruption as derived from an unspecified number of specimens.[33] In view of the enormous differences occurring in the eruption ages for both the deciduous and the permanent dentitions, Tables 13 and 14 should be regarded simply as rough guides to dental maturation in man and apes. If, in Table 14, the eruption ages for all 16 teeth are averaged, that for man is 10.60 years, for gorilla 6.98 years, and for chim-

* In rhesus monkeys (Macaca mulatta), Schultz[30] gives a variation in the appearance of the first 6 deciduous teeth (the 2 upper middle and 4 lower incisors) ranging from 18 to 42 days—a plus or minus of 40 percent, as compared with the 42 to 48 percent given in Table 13.

panzee 6.25 years. Thus the gorilla matures dentally 1.52 times as fast as man (reciprocal age equals 0.66), and the chimpanzee 1.70 times as fast (reciprocal age equals 0.59). The orangutan doubtless comes somewhere between the gorilla and the chimpanzee in this respect. Thus, all three apes exhibit dental ages close to, but not identical with, the corresponding ages derived from measurements of the postcranial skeleton.

Table 14 is based for man on data from Hurme; for the gorilla, from Schultz; and for the chimpanzee, from Schultz and Nissen and Riesen.[34] The very conservative ranges given in this table follow Hurme's allowance in boys and girls of one standard deviation *plus* (for "late") and one standard deviation *minus* (for "early") tooth eruptions. In all cases the ages listed are the average of the sexes. According to Hurme's data, in boys the dental age is equivalent to 1.046 the age in girls; and in girls, accordingly, to 0.954 the age in boys.

In the gorilla, the dental age of males equals $3.46 \sqrt{\text{age of man}} - 4.22$ years; and in the chimpanzee, $3.03 \sqrt{\text{age of man}} - 3.53$ years. Accordingly, the gorilla's dental age equals 1.14 times the age in the chimpanzee minus 0.175 years; while the chimpanzee's age equals 0.877 times the age in the gorilla plus 0.15 years. In female chimpanzees, both the dental age and the skeletal age are probably close to that in males. However, among gorillas there is probably a sex difference of about one year in these ages, as is assumed in Table 12.

In the deciduous dentition, the upper teeth erupt at an earlier age than the lower, with the exception of the second incisors in the chimpanzee and orangutan, and the first incisors and second molars in both man and apes. In the permanent dentition this sequence is largely reversed, since at that stage it is the lower teeth in most instances that erupt before the upper.

The Skull in Man and Gorilla

Probably the most distinctive single characteristic in the physical makeup of the gorilla and other anthropoid apes, as compared with man, is the skull. This is especially true of the adult male gorilla, which shall be our chief subject here. The prodigious skeletal development of both the cranium and the lower jaw in the gorilla, in comparison with the relative lack of such development in man, reflects the differing predominant needs in these two primates: in the gorilla, powerful jaws; in man, a large brain. These differences are clearly shown in Figure 57.

Table 13. Time (in days) of Eruption of the Deciduous or "Milk" Teeth in Man, Gorilla, Chimpanzee, and Orang-utan.
(sexes combined)

Eruption Sequence in Man	Teeth	Man		Gorilla		Chimpanzee		Orang-utan (one case only)
		No. of days, av.	Range in days	No. of days, av.	Range in days (est.)	No. of days, av.	Range in days (est.)	
1	1st (Medial) Incisors, lower	227	183–274	44	28–70	92	66–139	134
2	" " " upper	274	243–304	69	44–94	101	72–153	147
3	2nd (Lateral) Incisors, upper	334	260–410	77	49–105	121	94–174	226
4	" " " lower	402	310–494	84	55–113	120	85–177	204
5	1st Molars, upper	476	408–544	159	95–223	136	85–205	168
6	" " lower	477	404–550	167	97–237	161	88–240	161
7	Canines, upper	585	524–646	394	276–512	380	258–502	343
8	" lower	593	522–664	420	307–543	404	277–531	350
9	2nd Molars, lower	795	636–934	270	167–373	256	155–357	282
10	" " upper	840	680–1000	337	206–468	321	194–448	301

Table 14. Time (in years) of Eruption of the Permanent Teeth in Man, Gorilla, and Chimpanzee (sexes combined).

Eruption Sequence in Man	Teeth	Man		Gorilla		Chimpanzee	
		Years, av.	Range, years*	Years, av.	Range, years	Years, av.	Range, years
1	1st Molars, lower	6.07	5.1–7.0	3.50	3.0–4.0	2.92	2.5–3.4
2	" " upper	6.31	5.4–7.2	3.50	3.0–4.0	2.92	2.5–3.4
3	1st (Medial) Incisors, lower	6.40	5.5–7.3	5.75	4.9–6.6	5.16	4.4–5.9
4	" " " upper	7.33	6.4–8.2	6.02	5.3–6.8	5.42	4.7–6.1
5	2nd (Lateral) Incisors, lower	7.52	6.5–8.6	6.12	5.3–7.0	5.50	4.7–6.3
6	" " " upper	8.44	7.2–9.7	6.50	5.5–7.5	5.83	5.0–6.7
7	1st Premolars, upper	10.21	8.6–11.9	7.14	6.0–8.3	6.25	5.2–7.3
8	Canines, lower	10.33	8.6–12.1	7.70	6.4–9.0	7.58	6.3–8.9
9	1st Premolars, lower	10.50	8.7–12.3	7.32	6.1–8.6	5.92	4.9–6.9
10	2nd Premolars, upper	11.03	9.3–12.7	6.95	5.9–8.0	6.58	5.5–7.6
11	" " lower	11.18	9.2–13.2	7.14	5.9–8.4	6.50	5.3–7.7
12	Canines, upper	11.34	9.6–13.1	8.90	7.5–10.3	7.17	6.1–8.3
13	2nd Molars, lower	11.89	10.3–13.5	6.58	5.7–7.5	6.25	5.4–7.1
14	" " upper	12.47	10.9–14.0	6.77	5.9–7.6	6.42	5.6–7.2
15	3rd Molars, lower	18.06	c. 15–21	10.38	8.7–12.1	9.33	7.7–10.9
16	" " upper	20.50	c. 17.5–23.5	11.40	9.7–13.1	10.25	8.7–11.8

* = ± one Standard Deviation (V. O. Hurme).

The pioneer gorilla hunter Paul Du Chaillu (see chapter 3), in his book *Explorations and Adventures in Equatorial Africa* (1868), remarks that one of the lowland gorilla's favorite foods is "a kind of nut with a very hard shell." He goes on to say that this nut is so hard that it requires a strong blow with a heavy hammer to break it, and adds that "here probably is one purpose of that enormous strength of the jaw which seems to be thrown away on a non-carnivorous animal."

Indeed, the jaws of a gorilla, particularly an adult male, are actuated by muscles so powerful that they require a stronger development of the bony ridges on the skull than are present in the skull of a lion or a tiger. And along with this great strength, the canine teeth, or tusks, of the gorilla are fully as thick, and nearly as long, as those in the big cats. It is fortunate, indeed, for the jungle-sharing natives, that the gorilla is a peace-loving vegetarian and not a predatory meat eater!

It should be unnecessary here to present a de-

Fig. 56. Semidiagrammatic scale drawing of the head of an adult male western lowland gorilla, showing the characteristic wrinkles, beetling brows, broad nose, and other facial features by which a gorilla is readily distinguishable from a chimpanzee or an orangutan. Shown also is the outline of the **head beneath the hair. Lay a finger over the "frown lines" of the brow, and it will be seen that the expression is really quite placid.** (Drawing by author; times one-half natural size.)

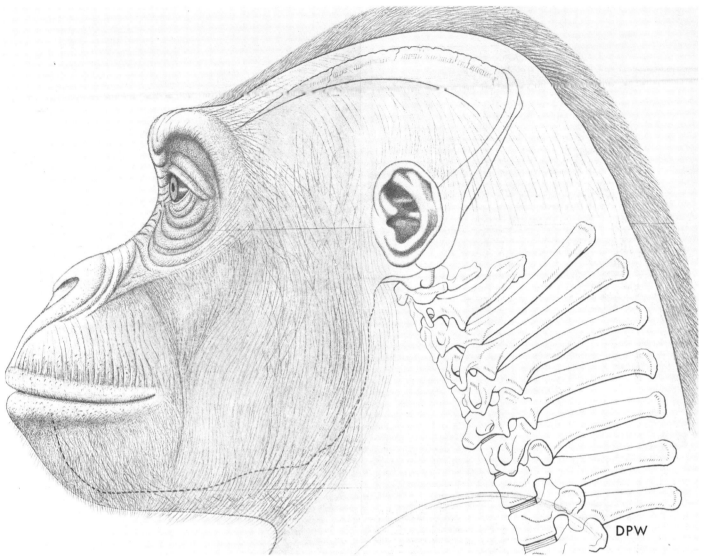

Fig. 57. Semidiagrammatic drawing of the gorilla head in Figure 56, in profile view. Included are the outlines of the midsagittal and occipital crests, along with the seven cervical and the two uppermost thoracic vertebrae. Note the great length of the spinous processes, and the thick, powerful neck. Characteristic also are the heavy, projecting muzzle, bulging supraorbital ridge, sloping facial angle, and relatively small humanlike ear. The dashed line indicates the border of the massive lower jaw. (Drawing by author; times one-half natural size.)

tailed commentary on all the features of the gorilla skull. Suffice it to say that at birth it is very similar to that of a human infant, and that the brutish aspects it takes on later are modifications needed for the mature ape to survive in its surroundings. Table 15 gives some of the dimensions of the western lowland gorilla skull in its development from birth to maturity. Figures 61 and 62 in chapter 6 give profile views of the head and skull in man, gorilla, chimpanzee, and orangutan.

It should perhaps be mentioned that skulls derived from zoo-raised apes may differ significantly in their proportions from skulls typical of wild specimens living under natural conditions. Such a difference has been found also, for example, in the skulls of lions that were raised in zoos. Of the latter animals it is said: "Nondevelopment of the muscles chiefly used in lifting and shaking prey, and the consequent lack of necessity for strong attachments is clearly responsible for the difference in the park (zoo)-reared animals."[35] While in the wild lions the external measurements of the braincase were less than in the zoo-raised animals of equal age, the cranial capacity, due to less thick skull bones, was markedly greater. The teeth also, both canine and molar, were larger in the wild-killed lions. Cranial differences comparable to those observed in lions probably exist also between wild-

Table 15. Some Dimensions of the Skull in Western Lowland Gorillas in Relation to Dental Age. (Compare with Table 12)

( ) estimated

Numbers refer to measurements used by Randall.

Sex	Dental Stage	Age Yrs.	1 Cranial* capacity, cc	3 Maximum length	4 Cranial length	5 Inion-prosthion	8 Maximum breadth	9 Total maximum breadth	13 Bizygomatic breadth	15 Cranial height	19 Cranial index**	Average Skull*** diameter
Males	—	Birth	(318.4)	(96.0)	(96.0)	(117.5)	(76.0)	(76.0)	(72.5)	(65.0)	(79.2)	(79.0)
	I	24.5 mos. (18–30)	453.0	132.7	132.7	161.5	103.0	103.1	98.5	88.9	78.7	108.7
	II	4.5 (3.5–6.0)	486.0	143.1	143.1	195.7	103.3	118.2	120.6	92.9	82.6	118.1
	III	6.5 (5–8)	501.0	152.7	147.0	219.6	103.0	128.6	134.6	95.3	84.2	125.5
	IV	9.0 (6.5–11.5)	516.8	166.3	151.7	248.0	103.8	142.0	153.0	96.2	85.4	134.8
	V	12.0 (9–15)	538.7	187.7	161.2	280.3	101.0	153.5	172.3	99.0	81.8	146.7
	VI	12.5 and over	547.9	195.7	163.4	297.5	101.8	160.0	178.8	99.3	81.8	151.7
Females	—	Birth	(281.0)	(94.8)	(94.8)	(116.0)	(75.0)	(75.0)	(71.6)	(64.2)	(79.1)	(78.0)
	I	24.5 mos. (18–30)	379.0	123.7	123.7	150.7	97.3	97.4	91.9	82.9	78.7	101.3
	II	2.75 (2–4)	412.5	133.4	133.4	177.2	100.4	110.8	111.1	86.7	83.1	110.3
	III	6.5 (5–8)	446.0	147.0	141.7	206.7	99.2	121.8	132.3	93.7	82.9	120.8
	IV	9.0 (6.5–11.5)	444.0	152.5	142.0	218.5	98.3	127.2	137.5	93.6	83.4	124.4
	V	11.33 (9–14)	457.0	155.8	144.0	227.2	99.2	131.6	143.7	95.6	84.5	127.7
	VI	11.5 and over	454.8	156.4	145.5	228.8	98.9	131.1	145.8	94.2	83.8	127.2

* ♂♂ = $[.00115 \left(\dfrac{\text{measurements } 4 \times 8 \times 15}{1000}\right) + 6.283]^3$; ♀♀ = $[.00127 \text{ ditto}, + 5.969]^3$

** = $\dfrac{\text{measurement 9}}{\text{measurement 3}} \times 100$

*** = skull dimensions $\dfrac{3+9+15}{3}$

Fig. 58. "Alvila," of the San Diego Zoo, "chinning" herself at the age of one month. She is holding onto her nurse's fingers. (Photo by Ron Garrison.)

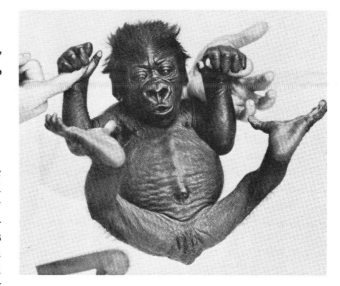

killed and zoo-raised gorillas. In the large male mountain gorilla "Mbongo," for example, I found the breadth of both the skull and the lower jaw to be significantly greater than in wild-killed specimens of the same species. And since Mbongo's cranial capacity (600 cc) was about 80 cc less than would be expected from the size of his external skull, the difference probably resulted from thicker bones.

In Table 15 it will be noted that while most of the dimensions of the skull in gorillas increase markedly from birth to maturity, those dimensions that pertain specifically to the external housing of the brain (i.e., measurements nos. 4, 8, and 15) increase much less. In fact, from Dental Age II onward, the maximum breadth of the cranium (measurement no. 8) actually diminishes (at least, according to Randall's data). In man, there is no diminution of head breadth from birth to maturity, but on the contrary a steady increase in this measurement, as well as in head length and head height. Despite the apparent decrease in cranial or head breadth in the gorilla with age, the average skull diameter increases from birth to maturity about 92 percent in males and nearly 64 percent in females. In man the corresponding increase is significantly less: only about 36 percent in boys and 32 percent in girls. This comparison shows that in the gorilla all parts of the body, including the head or skull, grow at a much more rapid rate than in man. It may be added that the average head diameter (external) in the gorilla is generally from 5 to 6 percent larger than the average skull (dry bone) diameter.

The cranial capacities listed in Table 15, while based on the figures given in Randall's tables of dimensions of the gorilla skull, have been correlated from these formulas:

Readers desiring more extensive information on measurements of gorilla skulls are referred particularly to the three papers by Francis Randall on this subject.[36] Additional data on the skulls of adult gorillas, as well as on chimpanzees, orangutans, siamangs, and gibbons, are given by Mollison (1911); and on gorillas alone by Oppenheim (1912), and Coolidge (1929).[37] The Coolidge data are especially valuable in that they include gorilla skulls of both the lowland (western) and the mountain species.

Table 15 presents—for the first time, I believe—a specific correlation between skull dimensions and age in the gorilla. Thus, the skull dimensions therein listed may be used as a guide to the age and sex determination of gorilla skulls that have not been catalogued as to these particulars. However, as will be noted, there is an extensive range in individual variation at every age. Table 15 should also be compared and used in conjunction with Table 12 on the dimensions of various bones of the postcranial skeleton.

Information on the brain weight in gorillas is given in chapter 7 in connection with the discussion of comparative intelligence in the three higher manlike apes.

Cranial capacity, cc, males $= \left[0.00115 \left(\dfrac{\text{measurements } 4 \times 8 \times 15}{1000}\right) + 6.283\right]^3$

Cranial capacity, cc, females $= \left[0.00127 \left(\dfrac{\text{measurements } 4 \times 8 \times 15}{1000}\right) + 5.969\right]^3$

Behavioral Development in the Gorilla

Fig. 59. "Little John," of the San Diego Zoo, showing how a healthy young lowland gorilla should look at the age of eleven months, weighing twenty pounds. (Photo by F. D. Schmidt, courtesy San Diego Zoo.)

How the great apes compare with man during the early stages of their social and behavioral development is a vast subject involving an extensive literature, and which can only be touched upon lightly here. Information on such development, especially in the gorilla, has accumulated rapidly since this species was first (in 1956) born and successfully raised in captivity, where its actions could be observed around the clock and recorded in detail. Many examples of zoo-born gorillas are listed in chapter 11, while various details and anecdotes pertaining to the early behavior of well-publicized individuals are given in chapter 9.

Here, it may be well to present a brief summary of some of the physical and behavioral distinctions in the newborn or infant gorilla, as compared with those in the chimpanzee and in man. To avoid repetition, the sexes in each case are averaged. The ranges given in the upper tabulation are those commonly regarded as "normal," and by no means do they represent extremes.

The "behavioral" developments just listed, which represent average values taken from the literature, evidently are subject to wide individual variations; for example, the infant female gorilla "Alvila," of the San Diego Zoo, could raise her head while lying in a prone position at the age of only four days. Again, at the age of only eleven days, she was able to sit upright alone, although she helped herself in this by grasping the bars of her crib. At one month of age she could "chin" or pull herself up by hanging onto her keeper's hands (Fig. 58). Some other infant gorillas required another two weeks before they could do this. A detailed review of Alvila's many behavioral developments, from birth to one year of age, compared with those of a female gibbon that was born at the zoo six days earlier than Alvila, are given by Rumbaugh.[38] Coffey and Pook[39] present a very thorough report of the third lowland gorilla born at the Jersey (Channel Islands) Zoological Park, a female that was named "Zaire." In this paper, in addition to accounts of the birth and the hand-rearing of Zaire, details are given on such behavioral observations as "social responses," "play and exploration," and "locomotion." A week-by-week report is made of Zaire's reactions to the zoo people attending her; her means of gaining attention (by whimpering or screaming); her interest in various toys and by degrees in all the different objects within the room; and her range of vision. The behavioral development of an anthropoid infant such as Zaire, therefore, goes through all the stages of a human infant's development, the main differences being (1) that the ape's initial progress proceeds at a more rapid rate than the child's; and (2) that by being "four-handed" the young ape's natural use of its feet in grasping, climbing, etc., is necessarily at variance with that of the young human's.

Following the article by Coffey and Pook, in the same (1974) *Report of the Jersey Wildlife Preser-*

Characteristic, or Developmental Stage	Gorilla	Chimpanzee	Man
Period of gestation, days	265 (236–290)	230 (210–250)	275 (250–310)
Weight at birth, kilograms	2.08 (1.6–2.7)	1.82 (1.4–2.4)	3.44 (2.5–4.5)
Duration of infancy, months (completion of deciduous dentition)	13.8 (10–18)	13.3 (9–17)	27.6 (22–33)
Onset of puberty, years	9.4 (8–11)	8.4 (7–10)	14 (9–19)
Completion of skeletal growth, yrs. (of *length* of long bones)	12 (10–16)	11 (9–14)	18 (15–21)

vation Trust, is a report by two psychologists, Jennifer Hughes and Margaret Redshaw, in which the cognitive, manipulative, and social skills in two infant male gorillas during their first year of postnatal development are analyzed and compared with those that develop at later ages in human infants. An outcome of such studies could be, as these authors express it, "a source of hypotheses concerning primate development and evolution, as well as providing qualitatively different information about human and ape infant development. It could also validate or question current theories of psychological development and the evolution of human behaviour."[40] Such studies, as authors Hughes and Redshaw also point out, should be in addition to others conducted in the wilds rather than in laboratories. However, the opportunity of doing the latter, what with the parent apes being ever-present, obviously imposes many restrictions. Among the most informative studies of gorillas living under natural conditions, including infants up to the age of six months, are those that were made in the 1960s of mountain gorillas by George Schaller.[41]

Additional remarks on the behavior of young gorillas as related to that of their parents in the wilds are given in chapter 8, "Habits and Family Life." For the normal sequence of postural controls and skills in human infants up to two years, a good account is given by Mary M. Shirley.[42]

Notes

1. David P. Willoughby, "The Gorilla—Largest Living Primate," *Scientific Monthly* 70, no. 1 (January 1950):48–57.
2. Thomas Edward Bowdich, *A Mission from Cape Coast Castle to Ashantee* (1819).
3. Richard Owen, "On the Gorilla," *Proceedings of the Zoological Society of London,* 1859, pp. 1–23.
4. Paul Du Chaillu, *Explorations and Adventures in Equatorial Africa,* chap. 21, "On the Bony Structure of the Gorilla and other African Apes" (New York: Harper & Bros., 1868), pp. 427–29.
5. John Edward Gray, "On the Height of the Gorilla," *Annual Magazine of Natural History* 8 (3, 1861):349–50.
6. W. T. Hornaday, *The American Natural History,* 4 vols. (New York, 1914); see vol. 1.
7. C. J. Cornish, ed., *The Living Animals of the World,* 6 vols. (New York, 1917), 5:5.
8. Max Hilzheimer and Ludwig Heck, eds., *Brehm's Tierleben,* 13 vols., 4th ed. (Leipzig, 1922), 13:679.
9. Walter Rothschild, "Notes on Anthropoid Apes," *Proceedings of the Zoological Society of London,* 1923, pp. 176–77.
10. Richard Lydekker, *Wild Life of the World,* 3 vols. (London, 1916), 3:48; see also Hilzheimer and Heck, eds., *Brehm's Tierleben.*
11. Carl E. Akeley, "Gorillas—Real and Mythical," *Natural History* 23 (September–October 1923):441. (See also *The World's Work,* August 1922, p. 397.)
12. Mary H. Bradley, "Among the Gorillas," *Liberty,* February 27, 1926, p. 31.
13. Attilio Gatti, "Among the Pygmies and Gorillas," *Popular Mechanics,* September 1932, pp. 418–23.
14. Attilio Gatti, "Gorilla," *Field & Stream,* October 1932, pp. 18–20, 66–67, 73.
15. H. C. Raven, "Gorilla: The Greatest of All Apes," *Natural History* 31 (May–June 1931):231–41.
16. Harold J. Coolidge, Jr., "Zoological Results of the George Vanderbuilt African Expedition of 1934. Part IV, Notes on four Gorillas from the Sanga (Sangha) River region," *Proceedings of the Academy of Natural Sciences, Philadelphia* 88:479–501; for measurements of limb bones, see p. 482.
17. T. Alexander Barns, *Across the Great Craterland to the Congo* (London, 1923), pp. 128–51.
18. W. L. H. Duckworth, *Morphology and Anthropology* (Cambridge: At the University Press, 1915).
19. Adolph H. Schultz, "The Size of the Orbit and of the Eye in Primates," *American Journal of Physical Anthropology* 26 (March 1940).
20. C. Vogel, "Zur systematischen Untergliederung der Gettung *Gorilla* anhand von Untersuchungen der Mandibel," *Zeitschrift fur Säugetierkunde* 26, no. 2 (1961):65–76.
21. Adolph H. Schultz, "Some Distinguishing Characters of the Mountain Gorilla," *Journal of Mammalogy* 15, no. 1 (February 1934):60–61.
22. George Gaylord Simpson, "Range as a Zoological Character," *American Journal of Science* 239 (November 1941):785–804.
23. Adolph H. Schultz, "The Skeleton of the Trunk and Limbs of Higher Primates," *Human Biology* 11, no. 3 (September 1930):303–438 (specifically, Table 19, pp. 434–35); idem, "Proportions, Variability and Asymmetries of the Long Bones of the Limbs and the Clavicles in Man and Apes," *Human Biology* 9, no. 3 (September 1937):281–328 (specifically, Tables 5–10).
24. Milicent L. Hathaway, "Heights and Weights of Children and Youth in the United States," *Home Economics Research Report No. 2* (U.S.D.A., October 1957).
25. James A. Gavan and Daris R. Swindler, "Growth Rates and Phylogeny in Primates," *American Journal of Physical Anthropology* 24, no. 2 (March 1966):181–90 (Table 9).
26. Richard Lydekker, in *Harmsworth Natural History,* 3 vols. (London, 1910), 1:185.
27. Schultz, "Proportions . . . in Man and Apes," p. 297.
28. Ibid., p. 291.
29. Francis E. Randall, "The Skeletal and Dental Development and Variability of the Gorilla," *Human Biology* 16, no. 1 (February 1941):65.
30. Adolph H. Schultz, "Observations on the Growth, Classification, and Evolutionary Specialization of Gibbons and Siamangs," *Human Biology* 5, no. 2 (May 1933):237.
31. James A. Gavan, "Eruption of Primate Deciduous Dentition: a Comparative Study," *Journal of Dental Research* 46, Supplement to no. 5 (1967):984–88; H. W. Nissen and A. H. Riessen, "The Deciduous Dentition of the Chimpanzee," *Growth* 9 (1945):265–74.
32. Adolph H. Schultz, "Growth and Development of the

Orang-utan," *Contributions to Embryology, No. 182, Carnegie Institute of Washington Publication 545,* 1941, p. 83.
33. Barbara Harrisson, *Orang-utan* (New York: Doubleday & Co., 1963), Appendix 1.
34. V. O. Hurme, "Standards of Variation in the Eruption of the First Six Permanent Teeth," *Child Development* 19, no. 1 (December 1948); Adolph H. Schultz, "Eruption and Decay of the Permanent Teeth in Primates," *American Journal of Physical Anthropology* 19 (1935):489–581; idem, "Morphological Observations on Gorillas," in *H. C. Raven Memorial Volume; The Anatomy of the Gorilla* (New York: Columbia University Press, 1950), pp. 227–53; idem, "Growth and Development of the Chimpanzee," *Contributions to Embryology, No. 170, Carnegie Institute of Washington Publication 518,* 1940, pp. 9–11; Nissen and Riessen, "The Deciduous Dentition," p. 83.
35. N. Hollister, "Some Effects of Environment and Habit on Captive Lions," *Proceedings of the U.S. National Museum* 53, no. 2196:177–93.
36. Francis E. Randall, "The Skeletal and Dental Development and Variability of the Gorilla," *Human Biology* 15, no. 3 (September 1943):236–54; 15, no. 4 (December 1943):307–37; 16, no. 1 (February 1944):23–76.
37. T. Mollison, "Die Körperproportionen der Primaten," *Morphol. Jahrb.* 42 (1911):79–304; S. Oppenheim, "Zur Typologie des Primatencraniums," *Zeitschrift f. Morphol. u. Anthropol.* 14 (1912):1–203; Harold J. Coolidge, Jr., "A Revision of the Genus *Gorilla*," *Memoirs, Museum of Comparative Zoology, Harvard* 50 (1929):241–383.
38. Duane M. Rumbaugh, "The Behavior and Growth of a Lowland Gorilla and Gibbon," *Zoonooz* (San Diego) 39, no. 7 (July 1966):8–17.
39. P. Coffey and J. Pook, "Breeding, Hand-rearing, and Development of the Third Lowland Gorilla at the Jersey Zoological Park," *Eleventh Annual Report, The Jersey Wildlife Preservation Trust* (1974), pp. 45–52.
40. Jennifer Hughes and Margaret Redshaw, "Cognitive, Manipulative, and Social Skills in Gorillas, Part 1, The First Year," *Eleventh Annual Report* (1974), pp. 53–60.
41. George B. Schaller, *The Mountain Gorilla, Ecology and Behavior* (Chicago: University of Chicago Press, 1963), Tables 56 and 57.
42. Mary M. Shirley, chap. 6 in *The First two Years: A Study of Twenty-five Babies,* vol. 1 (Minneapolis, Minn.: University of Minnesota Press, 1959).

6

Two Other Manlike Apes — The Chimpanzee and the Orangutan

While this book is concerned mainly with the gorilla, it seems fitting that a comparison should be made between this giant ape and its nearest living relatives—the chimpanzee and the orangutan. This comparison will be chiefly of the physical or structural similarities and differences existing between the gorilla and the two smaller manlike apes. Only incidental mention will be made of the gibbon. The latter miniature anthropoid (and its somewhat larger form, the siamang) differs so greatly in size, habits, and intelligence from the larger apes just mentioned that it justifies a separate recognition and description.*

* To readers interested in gibbons, the following publications are recommended:
  1. C. R. Carpenter, "A Field Study in Siam of the Behavior and Social Relations of the Gibbon (Hylobates lar)," *Comp. Psych. Mon.*, 16 (1940):5.
  2. R. I. Pocock, "The Gibbons of the Genus *Hylobates*," *Proceedings of the Zoological Society of London*, 1927, pp. 719–41.
  3. A. H. Schultz, "Observations on the Growth, Classification and Evolutionary Specialization of Gibbons and Siamangs," *Human Biology* 5 (1933):212–55, 385–428.
  4. A. H. Schultz, "Age Changes and Variability in Gibbons," *American Journal of Physical Anthropology*, n.s. 2 (1944): 1–129.
  5. R. and A. Yerkes, *The Great Apes* (New Haven, Conn.: Yale University Press, 1929).

Before making physical comparisons, the following particulars relating to the geographical distribution and habitat, life history, and general differences prevailing between (1) the gorilla and the chimpanzee, and (2) the gorilla and the orangutan, are presented for reference.

Chimpanzee (*Pan troglodytes* Blumenbach, 1799)

As several subspecies or varieties of this ape have been identified, the following remarks apply in particular to the typical form native to the forests of central equatorial Africa; that is, the form most commonly exhibited in circuses and zoos. The reason why chimpanzees (or "chimps," for short) are so popular as performing animals is that when in captivity they are generally more tractable, especially as older individuals, than either gorillas or orangutans, the full-grown males of which may prove unpredictable or dangerous because of their size, strength, and changing dispositions. Too, the chimpanzee is a natural-born extrovert or exhibitionist, whose hilarious conduct is in strong contrast to that of the slow-moving

Fig. 60. A "family" of chimpanzees in their native forest, as portrayed by the eminent old-time German illustrator of animals, Friedrich Specht. The mother with her child is on watch for a possible approaching enemy (leopard, snake, or crocodile?), while the father slakes his thirst. Afterward, he will take her place as sentry, while she and her infant likewise have a drink.

orang and the sensitive, introverted gorilla.

Geographically, chimpanzees are far more widely distributed than gorillas, with which they share the tropical rain forests in certain parts of their domain. Whereas the total area presently occupied by gorillas comprises some 270,000 square miles, that in which one or more of the several races of chimpanzees may be found—and which ranges from Sierra Leone on the west coast to Lake Tanganyika in the east—is more than three times as extensive. Consequently, while gorillas—particularly the mountain species—have had to have legal protection in order to survive, chimpanzees are still sufficiently numerous to be in no danger of extinction. Indeed, the nature of the country or terrain inhabited by chimpanzees is so much more varied than that of gorillas—comprising, as it does, not only swamp and mountain forests, but also open grasslands where the trees are widely scattered—that the chimpanzee may be said to be one of the most adaptable of primates. Nevertheless, to this it should be added that in certain areas chimpanzees have been hunted to extinction. Thus, a greater, more enforced, measure of protection for those that remain is urgently needed.

Physically—as will be shown in detail later in this chapter—the typical chimpanzee is only a small, less robustly built model of the gorilla. Adult males average 1311 mm (51.6 in.) in height, and 48 kg (106 lbs.) in weight; adult females, 1244 mm (49.0 in.) and 40.5 kg (89 lbs.), respectively. Thus, among chimpanzees the sexes are more alike in height and weight than are average-sized men and women. In contrast, adult male gorillas average over 9 inches greater in height than females, and are nearly twice as heavy. A comparison of bodily size between adult gorillas and adult chimpanzees is therefore particularly striking if only males are considered.

Chimpanzees are readily distinguishable from gorillas not only because of their much smaller size (and, therefore, greater activity and agility), but also by certain prominent details of their head and face that clearly differ even where two individuals of the same size are being compared (see Fig. 61); for example, in chimps the crown of the head is relatively flat, the skull lacking the often high bony sagittal crest so characteristic of adult male gorillas. Too, in chimps the ears are large and outstanding (see Fig. 84), as compared with the small, close-set ears of the gorilla. In coloration, the coat of the chimpanzee is typically black, although highly variable both in color and density according to age, sex, and environment. In infant chimps the skin of the face (cheeks) is pinkish; but with increasing age the pigmentation becomes dark brown or even blackish. The muzzle, in contrast, is generally lighter in color, although in some (mostly older) individuals it may be as dark as the rest of the face. Young chimpanzees commonly have a small tuft of white hair on the rump, while some old males have a silver grey back, as in the case of adult male gorillas. The eyes of newborn chimpanzees are blue but change to brown within a few weeks. A very odd specimen of chimpanzee was obtained by Louis de Lassaletta in 1956 in Spanish Guinea, in which the hair on the top of the head and on the sides below and in front of the ears was white. The animal, evidently a male of advanced age, was of dark coloration otherwise,

Fig. 61. The facial differences between a young chimpanzee (left) and a young mountain gorilla are clearly brought out in this excellent photo. The gorilla, a male about eighteen months of age, is "Meng"; and his companion, a female chimpanzee named "Jacqueline," is about nine months old. Both lived in the London Zoo in 1938.

Fig. 62. "Sally," a female specimen of the seldom-seen variety known as the "bald" chimpanzee (*Pan calvus*). She was imported to Liverpool in 1883 and lived for some years at the London Zoo. Her odd appearance, along with her exceptional intelligence, made her an anthropoid subject of great interest. Sally had a singular propensity for animal food, and every night she was given a small pigeon, which she killed and ate—skin, feathers, and all! (Photo by London Zoological Society.)

the white hair thus contrasting strongly with the almost black face.[1]

While, as was noted in chapter 3, the gorilla was not introduced to Europeans until 1847, the first chimpanzee to reach the Western world was a specimen presented to the Prince of Orange about 1640. In 1641 this chimpanzee was described by the Dutch physician Nicolaas Tulpius, although at the time it was thought to be an orangutan. Even earlier, in 1598, a description of the chimpanzee as it lived in the wild was brought back from the Congo by a Portuguese sailor named Eduardo Lopez, and the account was published in Frankfurt. As has been mentioned (chapter 3), the English sailor Andrew Battell, while a prisoner of the Portuguese in the Gaboon in 1559, described both the gorilla (which he referred to by the native name *pongo*) and the chimpanzee, the local name of which was *engeco* (probably a corruption of *N'djeko* or *N'schego*). Actually, there is no single "native" name for the chimpanzee but rather a variety of names that have been adopted independently by different tribes. Some of these names are *N'schego* (or *Nshiego*), *M'bouvé, Koola, Baboo, Soko, Ognia,* and *Koola-Kamba* (the latter of which was thought by Du Chaillu to be a distinct species). Again, Du Chaillu's supposed hybrid ape, the *Nshiego-Mbouvé,* was probably a specimen of the variety later called "bald chimpanzee" (see Fig. 63).

Although the first scientific description of the gorilla, which was given by Dr. Thomas Savage in 1847, is fairly well known, it is less known that at an even earlier date Dr. Savage gave a remarkably comprehensive and, for the most part, accurate account of the chimpanzee in its natural forest habitat. Since this interesting account may not be available in its original publication to all readers, it is here quoted at length:

Fig. 63. Profile of the head of "Mafuka," a female chimpanzee that formerly lived in the Zoological Gardens at Dresden. While she was at first regarded as a hybrid gorilla-chimpanzee, Mafuka was later determined to be of the race called by Du Chaillu "Koola-Kamba," which he considered to be a distinct species, both because of its large bodily size and, in the skull, the prominent development of the brow ridges. This excellent engraving was made by the old-time German animal artist G. Mützel.

> The strong development of the canine teeth in the adult would seem to indicate a carnivorous propensity; but in no state save that of domestication do they manifest it. At first they reject flesh, but easily acquire a fondness for it. The canines are early developed, and evidently designed to act the important part of weapons of defence. When in contact with man, almost the first efforts of the animal is—*to bite*.
> They avoid the abodes of men, and build their habitations in trees. Their construction is more that of *nests* than of *huts,* as they have been erroneously termed by some naturalists. They generally build not far above the ground. Branches or twigs are bent or partly broken and crossed, and the whole supported by the body of a limb or a crotch.
> Their dwelling-place is not permanent, but changed in pursuit of food and solitude, according to the force of circumstances.
> When at rest, the sitting posture is that generally assumed. They are sometimes seen standing and walking, but when thus detected they immediately "take to all fours," and flee from the presence of the observer. Such is their organization, that they cannot stand erect, but lean forward. Hence, they are seen when standing, with the hands clasped over the occiput or the lumbar region, which would seem necessary to balance, or ease of posture.
> The toes of the adult are strongly flexed, and turned inwards, and cannot be perfectly straightened. In the attempt the skin gathers into thick folds on the back, showing that the full expansion of the foot, as is necessary in walking, is unnatural. The natural position is upon "*all fours,*" the body anteriorly resting upon the knuckles. These are greatly enlarged, with the skin protuberant and thickened like the sole of the foot. They are expert climbers, as one would suppose, from their organization. In their gambols they swing from limb to limb, to a great distance, and leap with astonishing agility. It is

Fig. 64. "Knuckle-walking" gait of a chimpanzee. Note that on the feet the weight is placed on the outer borders. This excellent drawing is by Raymond Sheppard. Compare with the gorilla in Figure 41.

not unusual to see "the old folks" (in the language of an observer) sitting under a tree regaling themselves with fruit and friendly chat, while "their children" are leaping around them and swinging from branch to branch in boisterous merriment.

As seen here they cannot be called *gregarious*, seldom more than five or ten at most being found together. It has been said on good authority that they occasionally assemble in large numbers in gambols. My informant asserts that he saw once not less than fifty so engaged; hooting, screaming, and drumming with sticks upon old logs, which is done in the latter case with equal facility by the four extremities.

They do not appear ever to act on the offensive, and seldom, if ever really, on the defensive. When about to be captured, they resist by throwing their arms about their opponent, and attempting to draw him into contact with the teeth. *Biting* is their principal act of defence. I have seen one man who had thus been severely wounded in the feet.

They are filthy in their habits. . . . It is a tradition with the natives generally here (Cape Palmas) that they were once members of their own tribe; that for their depraved habits they were expelled from all human society; and, that through an obstinate indulgence of their vile propensities they have degenerated into their present state and organization. They are, however, eaten by them, and, when cooked with the oil and pulp of the palm nut, considered a highly palatable morsel.

They exhibit a remarkable degree of intelligence in their habits, and, on the part of the mother, much affection for their young. The second female described, was upon a tree when first discovered, with her mate and two young ones (a male and female). Her first impulse was to descend with great rapidity and "make off" into the thicket with her mate and female offspring. The young male remaining behind, she soon returned alone to his rescue. She ascended and took him in her arms, at which moment she was shot, the ball passing through the fore-arm of the young one in its course to the heart of the mother. Other instances have been known in which the mother, otherwise timid and fleeing from the presence of man, forsaken by her mate, has fallen a sacrifice to the force of natural affection. In a recent case, the mother when discovered, remained upon a tree with her offspring, watching intently the movements of the hunter. As he took aim, she motioned with her hand, precisely in the manner of a human being, to have him desist and go away. When the wound has not proved instantly fatal, they have been known to stop the flow of blood by pressing with the hand upon the part, and when this did not succeed to apply leaves and grass.

When shot they give a sort of screech not very unlike that of a human being in sudden and acute distress. In their gambols their cry is like the whoop of a native, varied as to volume and strength, which, with the drumming upon logs and other discordant noises and various uncouth movements, make up a scene perfectly unique, defying all description.[2]

The above-quoted account by Dr. Savage, while in most respects as true today as over a hundred years ago, evidently includes statements by an "informant" as well as his own observations as a missionary stationed in typical chimpanzee territory (the Gaboon). Today, as a result of carefully conducted studies continued over long periods of time, a considerably greater amount of information on wild chimpanzees is available than was the case in Dr. Savage's time. Much of the recently acquired knowledge concerning the behavior and habits of chimpanzees in their native haunts is the result of prolonged field work by Henry Nissen, Adriaan Kortlandt, Jane Goodall, and Vernon Reynolds, respectively.[3] Some of the "new" findings of these investigators of chimpanzees, along with verifications of earlier findings by other observers, may here be briefly reviewed.

Dr. Nissen was sent to Africa (Western Guinea) in 1930 by the primatologist Robert Yerkes to observe chimpanzees in the wild and to secure and send some of them to the Yerkes Laboratories of Primate Biology in Orange Park, Florida, where their behavior in captivity also could be studied. As a result of his repeated observations while in Guinea, Nissen saw that the long-held belief that chimpanzees live in humanlike "families" was a fallacy. Rather, the chimps of both sexes, along with their offspring, while traveling in small bands of from two to ten individuals (but sometimes as many as fifty), showed no tendency to pair off "family-wise" and either mingled indiscriminately or split up at any moment. Evidently the only stable association that exists within a troop is that of mother and child. Another observation by Nissen was that while the chimpanzee bands wandered about in search of food, there apparently were no fixed territorial restrictions or intrusions. Groups of the chimpanzees would join or separate at random. One question that remained unsolved by Nissen was why chimpanzees had evolved such high intelligence, when in the wild their life seemed so well regulated and devoid of complexities.

Some thirty years after Nissen's studies—that is, commencing in 1960—the Dutch zoologist Adriaan Kortlandt spent several seasons observing wild chimpanzees both in the Congo and in West Africa. In the eastern part of the Congo (now Zaire) he found a perfect spot for his observations: a plantation given over to the growing of papaws and bananas. To this area the chimps would come daily to feed, "and the Belgian owner did not begrudge them the relatively small amount of fruit they took." First to arrive at the feeding spot were the adult males of the troop. They would poke their heads out very cautiously from the dense foliage to see if they were safe. Once they had decided that no danger was present, "the large males broke out in a wild and deafening display. They chased one another, shrieking and screaming. They stamped the ground with hands and feet and smacked tree trunks with one open hand. Sometimes they pulled down half-grown papaw trees. Occasionally one of them would grab a branch and brandish or throw it while running full tilt through the group. In contrast the females, particularly the mothers, were almost always silent, wary and shy. Indeed, caution was the most conspicuous feature of maternal behavior." Continuing, Kortlandt adds: "Apparently awareness of the dangers of the jungle kept them in check; chimpanzee infants [which in the wild never whined or whimpered and always obeyed their mothers] in zoos behave rather differently in this respect."

Like his predecessor Nissen, Kortlandt found no evidence of "family groups" or "harems." He did observe, however, that mother chimpanzees with their young tended to keep together in "nursery groups." He also distinguished what he called a "sexual group," in which adult males and females without young predominate. These and other enlightening observations were made by Kortlandt—aided by a pair of field glasses—from a blind that, with the aid of pygmy climbers, he installed in the first crotch of a smooth-trunked tree over eighty feet above the ground. According to Kortlandt: "The main problem of primate research today is to explain why the great apes did not become more nearly human than they are."

In the early 1960s, Jane Goodall (later, the Baroness Jane van Lawick-Goodall) commenced a series of studies of chimpanzees in the wild, the region selected being the Gombe Stream Game Reserve on the eastern border of Lake Tanganyika, a protected British area of some sixty square miles. In her long-continued observations the Baroness discovered many new facts concerning wild chimpanzees, perhaps the most important of which is that these alert, suspicious apes can be observed and studied without concealing oneself in a blind, provided that phenomenal patience is exercised and the chimps are allowed plenty of time to become accustomed to the presence of the human invaders of their territory. Applying these initial requisites, Miss Goodall learned (1) that chimpanzees fashion and use tools (an example being their use of properly chosen twigs and blades of grass for removing and consuming termites from their ant-hills), (2) that they capture and eat whatever smaller animals they can overtake (including even half-grown baboons, which they kill mostly by batting them on the ground); (3) that they regulate the behavior of other chimpanzees by various vocalizations (they have a "vocabulary" of some twenty-three different sounds: grunts, hoots, wails, shrieks, barks, cries, sobs, groans, laughter, etc.); (4) that along with their sounds and gestures they have a wide range of facial expressions and grimaces, which may express anything from a friendly greeting to intense antagonism; (5) that they have a long period of care of the young and various forms of play, such as a stylized, carnivallike exhibition, which the author calls a "rain dance."

Regretfully, it appears that the African forest wherein chimpanzees now live may before long be taken over completely by farming, timber-cutting, and other human activities—some inroads of which are already being made. If complete expulsion occurs, the opportunity for making field studies of these interesting near-relatives of man, such as have been made by Ms. Goodall and other dedicated zoologists and conservationists, will be lost forever. The book by Baroness Lawick-Goodall, entitled *My Friends the Wild Chimpanzees*, published by the National Geographic Society, is a superbly illustrated account that summarizes her earlier articles and experiences.

Vernon Reynolds, in his book *The Apes*, presents detailed accounts of the field work of the aforementioned observers of chimpanzees, along with additional information gained from his own and his wife's studies of these apes in the Budongo Forest in Uganda, during 1962.

Comparison of the home life and behavior of the chimpanzee with that of the gorilla may be made by referring to the habits of gorillas as described in chapter 8; while remarks on the comparative intelligence of these two apes and the orangutan are given in chapter 7. A description of the external physical features of the gorilla is given in chapter 5.

Chimpanzees are by nature lively extroverts, although if kept in solitude they can become depressed and listless, whining and moaning. Although they possess lachrymal glands, they cannot shed tears. In one pair of devoted chimps, when the female died the male cried plaintively, tore his hair, violently opposed the removal of his mate's body, and mourned all the next day.

Concerning the senses of chimpanzees, their eyesight surpasses that of man. While they can distinguish three colors, most of what they see is blue green. Their hearing is highly acute (as would be supposed from their large ears), and their taste and smell sensitivity is within the human range; but their perception by touch or temperature or through pain is less keen.

As laboratory subjects, chimps are invaluable because of their susceptibility to all microbic diseases of humans. Their blood is closely comparable with that of man, containing groups O and A. Chimps may contract anything from syphilis and leprosy to tuberculosis and various tropical diseases, along with being the only animals subject to the common cold. The common fatal disease among captive chimpanzees is pneumonia, which causes about one out of five deaths from all causes. Second is enteritis-colitis; and third, diarrhea-dysentery. It has been said that chimpanzees are more properly treated by physicians than by veterinarians. And as everyone knows, they are pioneers in space travel.[4]

As to dangers from animal enemies, particularly leopards, Dr. Livingstone, in his *Last Journals*, states that the chimpanzee "kills the leopard occasionally, by seizing both paws and biting them, so as to disable them; he then goes up a tree, groans over his wounds, and sometimes recovers, while the leopard dies. The lion kills him at once, and sometimes tears his limbs off, but does not eat him."[5]

The sex life of chimpanzees is closely akin to that in humans, the "courtship" involving personal adornment, vocalization, caressing, and kissing. A chimp's kiss is not a quick smack, but a lingering contact. A courting male struts, gesticulates, shouts, and throws out his chest, stamps and pounds the ground. The female, however, makes the decision and is more knowledgeable in the proceedings. Copulation in captive chimps lasts 5 to 30 seconds, but may be longer in the wild. Gestation averages 230 days, and the average weight at birth is 1.82 kg (see p. 55).

The capture of wild chimpanzees is accomplished nowadays largely by the use of tranquilizing guns or by enmeshing the apes in strong jungle nets. It has been done also by using asphyxiating gas, by baiting the chimps with food impregnated with alcohol, and/or by shooting them with arrows dipped in snake venom, afterwards applying antivenom therapy. The chimps thus obtained are sold by dealers at prices from 300 to

Fig. 65. Portrait of an adolescent chimpanzee that once resided in the New York Zoo. Note the characteristic large ears, small nose, less wrinkled brow, and unpeaked head (crown) that distinguish the chimpanzee from the gorilla. Note also the marked difference in appearance between this "typical" chimpanzee and the "bald" variety shown in Figure 62. (Photo by Elwin R. Sanborn; courtesy New York Zoological Society.)

1500 dollars to zoos, research institutions, and trainers.

Concerning the zoological classification of chimpanzees, while Daniel Elliot in 1913 listed no fewer than ten races or subspecies,[6] today only four subspecies are generally recognized. These are:

(1) *Pan troglodytes troglodytes* (Blumenbach, 1799). Common chimpanzee; found in the southern Cameroons, through Congo (Zaire), eastward to the Congo River.
(2) *P. t. verus* (Schwarz, 1934). Western chimpanzee; type locality Sierra Leone.
(3) *P. t. schweinfurthi* (Giglioli, 1872). Eastern chimpanzee; type locality eastern Congo (Zaire).
(4) *P. paniscus* (Schwarz, 1929). Pygmy chimpanzee; type locality "South of the upper Maringa River, 30 km south of Befale, south bank of the Congo.

The first three races or subspecies just listed vary only to a slight extent in general bodily size and are distinguishable (sometimes only by zoologists) mainly by such external features as color and density of hair coat, color of the muzzle and face, amount and direction of the hair on top of the head, and a few other differences resulting in part, at least, from differences in the environment (jungle or mountain, hot or temperate, moist or dry, etc.). The fourth form, however, the pygmy chimpanzee, is so clearly dissimilar from the others that it warrants consideration as at least a distinct species, if not actually a generically different form. *P. paniscus* was first described scientifically by the systematist Ernst Schwarz in 1929. Prior to that time, however, several specimens had been exhibited in various zoos but evidently without having been recognized as other than ordinary chimpanzees. In 1933, Dr. Harold J. Coolidge gave a comprehensive description, including skeletal and bodily measurements, of an adult female specimen.[7]

In 1954, Eduard Tratz and Heinz Heck described and proposed for this species a new genus: *Bonobo*—a name by which the ape is popularly known today.[8]

A male bonobo, which was given the name *Kakowet* (French: Ka' ka weé—"peanut") was obtained by the San Diego Zoo in 1960. Two years later a female bonobo ("Linda") was also procured by the same zoo, making it at that time the only zoo in the United States to possess a pair of these rare anthropoids.* Dr. Coolidge's adult female

* Up until early 1976, Kakowet and Linda had become the parents of five female and one male offspring, the male being the most recent. All are doing well in the San Diego Zoo.

specimen of *P. paniscus* (AMNH No. 86857) stood 1010 mm (39¾ in.) in height and had an arm spread (span) of 1510 mm (59.4 in). However, on the basis of the skull dimensions of this specimen as compared with those of five other adult female skulls, it would appear that the probable typical height of an adult female bonobo is about 1070 mm (42.1 in.), and of a typical adult male, 1125 mm (44.3 in). And if proportioned bodily as in the larger, common chimpanzee, the weight in an adult female bonobo would be about 57 pounds, and in an adult male bonobo, about 67 pounds. However, if more slenderly built—as the bonobo is generally stated to be—these weights would be a few pounds less in each case.

If Coolidge's specimen of *P. paniscus* may be taken as exhibiting typical proportions in the lengths of its limb bones, it is found that in this species the Intermembral Index (= 100 × Humerus + Radius/Femur + Tibia) is only 101.7, while in the female common chimpanzee it is 106.5. In Adolph Schultz's series of sixty female chimpanzee (common species) skeletons, the range in

Fig. 66. "Kakowet," an adult male pygmy chimpanzee, or bonobo, in the San Diego Zoo, 1960. This small-sized species is considered by some zoologists to be generically distinct from the larger, typical chimpanzee so commonly seen. (Photo courtesy San Diego Zoological Society.)

the latter index is 100.1 to 112.0.[9] Thus, while this ratio in Coolidge's specimen is within the range of *Pan troglodytes* sp., it is far from approximating the typical value. A younger, smaller living example of a male pygmy chimpanzee, which was on exhibition in the zoo in Vicennes (France), was measured in 1940 by A. Urbain and P. Rode,[10] who obtained the following figures: standing height 830 mm (32.7 in.); sitting height 520 mm (20.5 in.); span 1190 mm (46.9 in.); hand length 160 mm (6.3 in.); foot length 200 mm (7.9 in.); chest girth 550 mm (21.7 in.); neck girth 250 mm (9.8 in.); weight 11.5 kg (25.4 lbs).

However, while bodily size is a conspicuous feature in separating the bonobo from the larger, common forms of the chimpanzee, even more distinguishing is the general appearance of the smaller ape, especially of its head and face, both of which are very dark. The hair is fine and silky and on the crown of the head does not "part" (i.e., grow in opposite directions), as commonly occurs in *Pan troglodytes*. Rather, in the bonobo, the hair direction is, in general, straight backward, as in the gorilla. Too, the hair is longer and thicker—so much so that in adult specimens it suggests a tam-o-shanter and largely conceals the ears, which are relatively smaller than in the typical chimpanzee. Another distinguishing feature in the bonobo is that the second and third toes are generally joined together for part of their length by a union of the skin. In temperament, the bonobo is said to be the best-natured and least pugnacious of all the apes. Too, in intelligence tests, it appears to rate higher than the common chimpanzee. But some of the foregoing observations may need substantiation via study of a larger number of specimens, which hopefully in due course may come about.

* * *

Orangutan (*Pongo pygmaeus* Hoppius, 1763)

In transferring our attention from the chimpanzee to the orangutan (usually spelled *orang-utan*), we pass geographically from the jungles of equatorial Africa far eastward to the Asiatic islands of Borneo and Sumatra, where the huge red-haired ape, threatened with extinction in the wild, is presently making a last stand. It is interesting to note that if the single letter *g* is added to the word *orang-utan*, which means "man of the woods" (or jungle), the resulting word, *orang-utang*, comes to mean "man who is in debt." As some jokester has remarked, either word would seem to describe a very human creature! Actually, the name used by Dyak (Malayan) natives in referring to the orang in Borneo is *maias* (commonly misspelled *mias*), and to that in Sumatra, *mawas*. Although some zoologists have held that the orangs of these two islands represent different species, the concensus of opinion is that they are not separable even as subspecies, and that certain seeming differences (such as the cheek pads in many adult males, which have given rise to the name *maias-pappan*) are merely external variations that have no effect on the basic skeletal structure of the animal.

While the gorilla was not properly described and classified until 1847, and the chimpanzee until 1799, the orangutan was so identified as early as

Fig. 67. In this scene, which no photographer would likely be able to duplicate, the artist has captured an episode in the daily life of a pair of orangutans. Confronting a large python, the two apes appear to show interest rather than fear. The dense, tangled vegetation of the gloomy swamp-jungle is also depicted, with a clarity that probably no camera could equal. This wood engraving is after a drawing by the great nineteenth-century animal illustrator Friedrich Specht.

1763, although for a long time thereafter, importations of orangs and chimpanzees into Europe continued to cause confusion of the two species. The scientific name of the orangutan, *Pongo pygmaeus*, is in itself a confusing misnomer, since *pongo* is a West African native name for the gorilla, and *pygmaeus* means "pygmy." The generic name, *Pongo*, was not given to the orangutan until Tiedemann, in 1808, used the term *Pongo wurmbi* in reference to the Bornean orang. Prior to that, Linnaeus in 1766 had given this ape the name *Simia satyrus*, an apt designation; while in 1763 Hoppius had been the first to use the species name *pygmaeus*. So again it is confusing to see Hoppius being credited with having given the orangutan its present name, when actually, in making his description of the ape, he contributed only the species name *pygmaeus*, using, as did Linnaeus three years later, the generic name *Simia* (not *Pongo*).

The typical physical proportions of wild-shot adult orangs have been listed by the naturalists Alfred Russel Wallace and W.T. Hornaday, respectively;[11] and the measurements each of these field workers obtained of their specimens are quite similar. Wallace, who measured seventeen freshly killed adult Bornean male orangs, stated that they "only varied from 4 feet 1 inch to 4 feet 2 inches in height, measured fairly to the heel, so as to give the height of the animal if it stood perfectly erect; the extent [span] of the outstretched arms from 7 feet 2 inches to 7 feet 8 inches." This, indeed, is slight variation. Hornaday found the height of ten adult Bornean males to range from 4 feet ¾ inch to 4 feet 6 inches, and the span from 7 feet ¾ inches to 8 feet. He also measured four adult female orangs, the height of which ranged from 3 feet 6 inches to 4 feet, and the span from 6 feet 2½ inches to 7 feet 3¾ inches. Although neither Wallace nor Hornaday weighed their specimens, it is known from other sources, as tabulated by Schultz, that the average weight of adult male orangs is 165 pounds, and of adult females 82 pounds.[12] In the foregoing two series of males, the range in weight, if based on the height in inches cubed, would be from about 145 pounds to 195 pounds. The female orangs obtained by Hornaday should on the same basis range from about 72 pounds to 108 pounds. Some grossly obese male zoo orangs have weighed well over four hundred pounds, but such a poundage cannot be considered a normal, healthy weight in an orang, any more than it would be in a man of average height.

On the basis of a record-sized sitting-height measurement of 889 mm (35.0 in.), the tallest male orang in any museum collection would appear to have been a specimen (cadaver) in the American Museum of Natural History (No. 2281). If in typical proportion to its sitting height, this adult male would have measured about 1450 mm (57.1 in.) in standing height. And if this orang had been proportioned physically the same as a male of average height and weight, his weight would have been 233 pounds. Allowing for a total population of several thousand orangs, it would appear that the maximum normal weight of the tallest male should not be over 250 pounds, and of the tallest female 125 pounds.

In this connection should be mentioned an allegedly "giant" male Sumatran orang reported by W.C.O. Hill.[13] Unfortunately, the external measurements of this specimen were taken on its mounted skin, which evidently was grossly stretched. Fortunately, accurate measurements of the lengths of the limb bones of the same specimen were also recorded, and these bone lengths show that in life this supposedly gigantic orang could not have been much, if any, larger than an average-sized male.

Besides having proportionately much longer arms, another respect in which the orangutan differs significantly from both the chimpanzee and the gorilla is the skull, which is distinctively higher in relation to its fore-and-aft length than in either of the African apes. The orang skull also has less prominent brow ridges, especially in comparison with the immense development of these ridges in the gorilla skull. Another different characteristic is in the bones of the wrist. According to Richard Lydekker:

> In regard to the number of bones in the wrist, we find that the orang-utan possesses the central bone which is lacking in man, the chimpanzee, and the gorilla, and thus has nine, in place of eight, bones in the wrist. In this respect the Bornean ape agrees with the lower members of its order; but in the absence of callosities on the buttocks it shows its kinship with the gorilla and chimpanzee. All these characteristic features clearly indicate that the orang-utan is decidedly lower in the scale than are the other two man-like apes.[14]

The legs in the orangutan are somewhat shorter relative to the standing height than in the chimpanzee, and they are poorly developed in comparison with the long and powerful arms. The legs are somewhat rotated so that the kneecaps face partially sidewise, causing the feet to be set out

Fig. 68. Here is another fine illustration by Friedrich Specht—in this case, of an infant orang. The humanlike expression of the animal, along with the apelike structure of the arms, legs, hands, and feet, are accurately portrayed. The color of the hair was dark reddish brown. This baby orang had its own bed and would drink from its own cup, run and jump, and play all kinds of pranks; and it enjoyed its life as much as would a lively human child.

Fig. 69. Several interesting features of an orangutan are well shown in this photograph of a tame subadult male: 1. the throat pouch; 2. the very broad (as well as long) hands; and 3. the normal position of the feet in standing, which is turned sideways. (Photo from *Animal Life*.)

of line with the long axis of the lower leg. As a result of this adaptation of the feet for use as hands (as in grasping and walking along a tree limb), orangs are ill-adapted for walking or even standing upright on the ground; for, in order to do this, they must rest their weight on the outer sides of their feet rather than on the soles (see Fig. 69).

Although the hands of the orang are usually described as being long and slender, this is by no means always the case. For while they are long, they can also be broad, especially in old and heavy males, where the fingers may be almost as thick as in the gorilla. In the small though nearly adult male in Figure 69, for example, the width of the hand is fully one-tenth of the standing height, which is relatively enormous. As is seen in Figure 70, the fingers and toes of the orang are habitually flexed, in keeping with their constant use as virtual grappling hooks in the ape's passage through the treetops. Indeed, the shortened ligaments in the hands and feet of these primarily tree-living anthropoids make it impossible for the fingers and toes to fully straighten, unless the wrists and ankles are flexed—that explains why it is so difficult to remove an orang from a tree in which it has a grasp on the branches. However, the ape is given credit for having an incredibly strong "grip" (in one account it is said that it took fifty natives to thus dislodge an orang!), when in reality it is the passive resistance of the ape's shortened ligaments that prevents the hand and foot digits from being pulled out straight. And the short, less interfering thumbs of the orang make it easier for the animal to use the remaining four fingers as grasping hooks as it swings from one branch or vine to another. In some orangs, the terminal phalanges of both the thumbs and the big toes are totally lacking, which means that these digits have no nails.

Figure 70 also shows the imposing cheek pads that commonly, though not invariably, develop on the sides of the head in adult male orangs. While the function, if any, of these pads seems to be unknown, the fact that they normally occur only in males suggests that they may have some sexual significance. The pads do not generally appear until the orang is about ten years of age, "but, once started, they grow very quickly, reaching full size in six months."[15] Like the crown pads of adult male gorillas, the cheek pads of the orang are composed mainly of connective tissue. However, the two pads differ in that the crown pad of the gorilla represents essentially a thickening of collagenous (gelatinlike) tissue and is of a dense, fibrous nature; whereas the cheek pads of the orang include "an enormous localized mass of fat with involvement of neighboring facial muscles."[16] Straus considers the cheek pads to be similar to the fatty buttocks of Hottentots, the tails of fat-tailed sheep, and the humps of camels and of certain breeds of oxen (*ibid.*, note 16).

Still another feature that differentiates the orangutan from the chimpanzee and the gorilla is the enormous throat pouch in adult (particularly, old) males, which can hold over 1½ gallons (366 cub. in.) of air and can extend downward to the lower end of the sternum and sidewards well under the armpits. While the pouch also occurs in adult females, it is not nearly so large as in old males. It is even present in undeveloped form at birth, but shortly thereafter grows larger rapidly. The "purpose" of the throat pouch is evidently to give amplitude and resonance to the loud calls, bellows, and squeals with which orangs communicate in the forest. Prior to making the call the throat swells out like a huge balloon. The action is comparable with that which takes place on a smaller scale in the siamang, which inhabits the same forests as the Sumatran orangutan. Mature

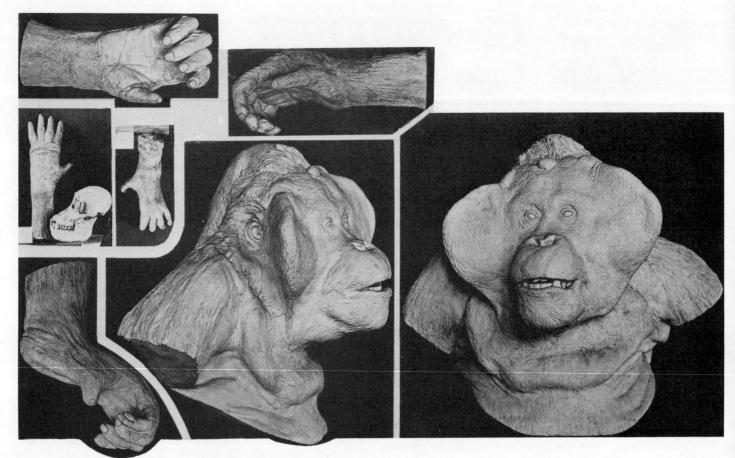

Fig. 70. Cast of the head of an old male orangutan in front and side views, showing the exceptionally wide cheek pads and the deflated throat pouch. At the upper left are shown several casts of orangutan hands and a skull; and at the lower left, a foot. These casts are, or were, in the Natural History Museum, Paris.

male orangs frequently have also a well-leveloped beard on the cheeks and chin, and in some cases a moustache.

Likewise, the coloration of the orangutan is radically different from that of the essentially black African manlike apes. The orang's so-called "red" color may be any hue from reddish brown to orange or even yellow. Often the coarse, straggly hair on the arms, which may reach a length of as much as three feet, is a tangled jumble of both light and dark shades of reddish brown or "carrot" color. The profuseness of the hair, especially in a large old male, is highly conspicuous and striking. Why an animal that inhabits equatorial jungles should develop or require such a heavy coat of hair is a question; but some zoologists feel that it is a protection to the ape from the frequent, heavy rainfall that it must endure in its often roofless arboreal nest. The long hair probably serves also as skin protection against the swarms of mosquitos that infest the orang's swampy jungle surroundings.

To summarize the external physical features by which the orangutan may be distinguished from the gorilla and the chimpanzee, we may note the very flat nose and the exceedingly small ears. The latter enable an observer to distinguish at once between an orang and a chimpanzee (which has very large ears—see Figure 84); and the absence or only slight development of the brow ridges in the orang make it easy to tell one of these apes from either a chimpanzee or a gorilla, in both of which (particularly the gorilla) the brow ridges are massive and overhanging. Then, as has been mentioned, in the orang the hair is very long and coarse and of a reddish brown, ginger, or chestnut hue—a color that is in decided contrast to the jet black of the chimpanzee or the black or ashy tones of the coat of the gorilla. However, the bare skin of the orang appears bluish, slate grey, or chocolate brown.

Further details as to the external appearance of the orangutan are indicated in the adjoining photographs and drawings. As may be seen, there is a vast difference between the facial aspects of an infant, an adult female, and an old male orang,

Fig. 71. Here, for a comparison with Figure 70, is a mature living orangutan possessing fully developed cheek pads. Note also the narrow and bridgeless nose; close-together eyes; moustache and beard; and thick yet tapering fingers characteristic of adult males of this species. (Photo courtesy San Diego Zoo.)

on the ground. Also, if need be, the apes use the fruit as missiles, hurling them down from the treetops with great accuracy and splattering their odiferous contents over the hapless hunters!

A typical orangutan family consists of a mother, usually with a nursing infant, along with several older offspring that have not as yet branched out for themselves. The adult males do not ordinarily join the family group but keep apart in bachelor-

respectively. Owing to the presence of cheek pads (which may be up to 12 or 13 inches in width) in the male, and in the same sex an immense throat pouch and possibly a beard, these individuals are markedly dissimilar from the females and young—indeed, to a degree far surpassing that to be observed between chimpanzees or even between gorillas.

In temperament and actions, the orang is a much more lethargic animal than either the chimpanzee or the gorilla. Most of its time it moves leisurely through the treetops in search of various fruits—especially the durian. This is a fruit peculiar to the Malayan islands, which grows in tall, flowering trees. The fruit is of the size of a large coconut, with a hard, thick rind covered with soft spines. The pulp of the fruit is creamy and, to some persons at least, very tasty. But most Europeans are repelled by it because of its nauseating smell. The orangs break open the tough rind with their powerful fingers after having bitten into it with their long canine teeth. Once a loaded durian tree has been located, the orangs seldom leave until all the fruit has either been eaten or sampled and scattered

Fig. 72. This excellent study of a mother orangutan and her young shows the "dished" face, projecting muzzle, exceedingly small ear, and long hair of an adult female, along with the alert, inquisitive expression of her infant offspring, whose features give little indication of the vast changes they will undergo as the ape grows and develops into an adult. (Photo from *Animal Life*.)

Fig. 73. "Death of a Dyak," a grim tableau by the prominent French sculptor Emmanuel Fremiet (1824–1910). Whether an ape would employ such a means of dispatching a human enemy is doubtful, yet Fremiet's works (see also Fig. 156) are spirited, anatomically correct, and seemingly possible if not probable.

like silence, although it is said that the males do all the nest-making. Such "nests," or resting platforms (see Fig. 79), are usually constructed at the end of each day, in treeforks from thirty to forty feet or more above the ground. However, recent studies of orangs in the wild have disclosed that despite their essentially tree-living nature, occasionally one of them will construct a nest only a few feet above the ground, and that sometimes one will take a nap on the jungle floor during the day.[17] In building a nest, pliant branches are first brought together, then interwoven and covered with smaller twigs and leaves. When these nests are in the treetops, the occupants are completely hidden from any observer on the ground.

Apart from man, about the only creatures that the orangutan needs to fear are crocodiles and pythons. Sometimes, when an orang runs short of fruit in the trees, he descends to the riverbanks to look for young shoots, leaves, and buds. Then a crocodile may try to seize him but the ape usually succeeds in beating the reptile off and, in some cases, even in killing it; It succeeds in doing this also with snakes that attempt to prey on its young, provided the serpents are not too large. It may be supposed that the orang's method of killing a

Fig. 74. At first glance it might appear that this young adult orangutan has placed his hands flat on the ground. Actually, he is folding his hands over a curb, as the structure of his hands makes necessary. (Photo courtesy San Diego Zoo.)

python would be to bite through its neck just back of its head, although no writer has ever reported seeing such an encounter.

Due to the structure of its hands and feet, the orang, as has been mentioned, is highly specialized for living in trees and poorly adapted for walking on the ground, since it has to walk on the sides of its feet. Even in the trees an orang's movements are slow and deliberate, and its weight prevents it from leaping the way a gibbon or a smaller monkey does. But it can swing and walk from limb to limb with wonderful ease and precision, moving along as fast as a man can run on the jungle floor below. Nevertheless, orangs occasionally fall, and a high percentage of the skeletons of these apes that have been collected show broken bones and healed frac-

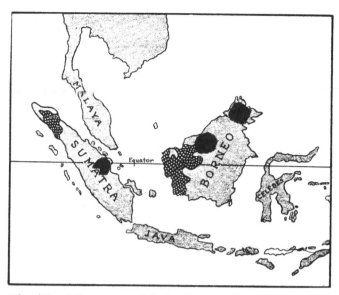

Fig. 76. The geographical range of the orangutan (*Pongo pygmaeus*), which once extended over most of Sumatra and Borneo, and even into the Asiatic mainland, by the year 1940 was reduced to the two dotted areas shown here, and by 1970 to the three blackened areas. At present, only a few thousand of the orangs exist in the wild, and the future of these is precarious.

Fig. 75. Andrew Lang, the famous teller of fairy tales, could also tell factual ones. In his book *The Red Book of Animal Stories* (London, 1899), he tells about some orangutans that came occasionally from their jungle habitat down to the seashore in search of shellfish and particularly of oysters. The latter they would open by putting a stone between the two halves of the shell, to avoid getting their fingers pinched. This drawing of the orangs is by author Lang's talented illustrator, H. J. Ford.

tures. In fighting among themselves, which occurs only infrequently, male orangs sometimes bite off each other's fingers and toes—this evidently being one of their main defensive habits. Like the gorilla, the orang cannot swim, and sinks like a rock if it falls into water above its depth.

The vitality of the orang in its native haunts is remarkable, and specimens have been known to continue climbing and to break off and throw branches at the hunters after they had been fatally wounded. The ape does well in captivity, several zoo specimens having each reached the age of thirty-five years or over. Like the gorilla and the chimpanzee, however, the orang is subject to most of the diseases that cause illness or death in man. Quite recently, a fine adult male Bornean orang died in the Jersey (Channel Island) Zoo from ulcerated colitis, the whole duration of the illness having been only four days.[18]

The natural habitat of the orangutan is being reduced year by year. As Carl Mydans, a present-day field worker with these apes puts it:

orangutans face eventual extinction from the encroachment of Man. With the possible exception of the clouded leopard, Man is their only enemy. Not only has Man from ancient times hunted them for food and, more recently, for his zoos, but today he is inexorably diminishing

Fig. 77. This drawing, by A. de Neuville, depicts an actual encounter between Dyak natives and an adult male orang, as described by the eminent naturalist Alfred Wallace. Although the orang was killed, the attacked native was severely injured and never fully recovered the use of his bitten arm. However, an orang thus encountered in the jungle rarely takes the offensive, unless, as in this case, it was first attacked either by the hunter or his dogs.

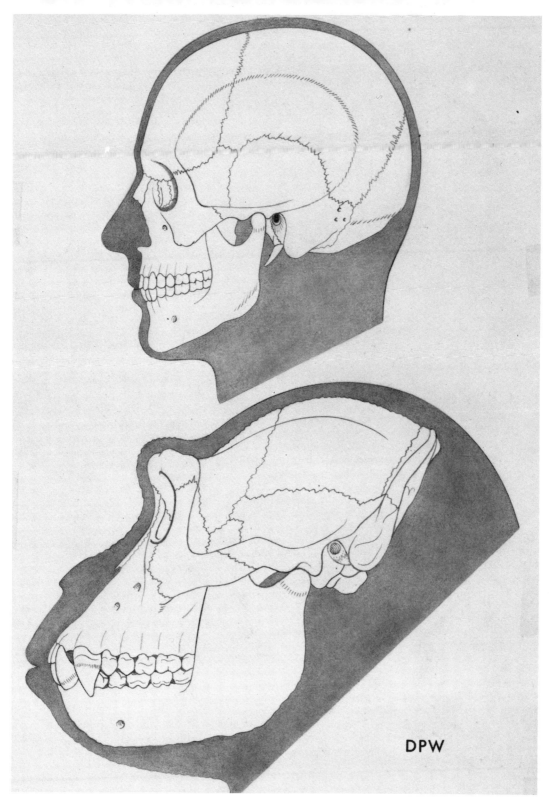

Profile views of the head and skull in man and gorilla. The man is a typical white American; the gorilla, an adult male of the western lowland species. The profound differences between the two skulls result mainly from the reduction of the face and enlargement of the cranium in man, as compared with the immense development of the teeth and jaws and the smaller cranial capacity in the gorilla. The adult female gorilla skull (not shown) has little or no sagittal crest and closely resembles that of an adult male chimpanzee (the following page), except for being about twenty percent larger in its linear dimensions. Both skulls are reduced to 0.40 natural size.

Fig. 78. An orangutan building a sleeping platform. (Drawing by Neave Parker; courtesy *The Illustrated London News*.)

their arboreal world by cutting down their primary jungles for farms and factories and towns, and lacing their forests with roads.[19]

In earlier times the ape inhabited the Asiatic mainland, as is shown by fossil remains of it in northern India and southern China (see Fig. 12). These date from the middle Pleistocene period, some 500,000 years ago. However, when Borneo and Sumatra became separated from the Malayan Peninsula, the apes became confined to the latter two islands. At the present time they are known to occur in the wild state only within the three spots shown on the accompanying map (Fig. 76). While for a long time it was thought that the orangs from Sumatra represented a different species from those in Borneo, it is now, as has been mentioned, generally agreed that the two geographical races are identical anatomically, although they may present minor external differences. On both islands the orangs are now rigidly protected; and by rearing infant specimens in captivity until they are large enough to be released back into the forests, it is hoped that the previously dwindling population of these interesting manlike creatures may gradually be built up. Well-cared-for female orangs may produce offspring every two to three years.

In its natural habitat the orangutan shows a marked preference for low-lying, swampy jungles, although it may range up into the foothills to an elevation of over 4000 feet. Mostly, however, the apes keep to altitudes below 2500 feet. There is evidence that they move about locally in following the seasonal ripening of certain fruits on which they feed. As large clearings are made by natives in the dense forests, the orangs are either destroyed or are forced back into areas in which they can scarcely find enough food. However, European colonists, rather than natives, are said to be the most serious threat to the present orangutan population.

As to the behavioral characteristics of orangutans, a whole book could be written on the subject, just as could be about gorillas or about chimpanzees. Suffice it to say that in all three of these manlike apes the infants are much more like human infants than the adults are like human adults and, likewise, vary individually in their behavior to the same wide extent. Stories about the behavior of various specimens of orangutans in captivity, or under observation in the wild, are given (to name only a few) by the following authors: Wallace, Ulmer, Brindamour, Mydans, Brehm, Yerkes, Conant, Harrisson, and Gage.[20]

* * *

Notes

1. Bernhard Grzimek, "Weisskopfschimpanzen," *Säugetierkundliche Mitt.* 6, no. 2 (April 1, 1958):77.
2. Thomas S. Savage, in *Boston Journal of Natural History* 4 (1844):382.
3. Henry W. Nissen, "A Field Study of the Chimpanzee. . . ," *Comparative Psychology Monthly* 8, no. 36 (1932):1–122; Adriaan Kortlandt, "Chimpanzees in the Wild," *Scientific American* 206, no. 5 (1962): 128–38; Jane Goodall, "My Life Among Wild Chimpanzees," *National Geographic* 124, no. 2 (August 1963):272–308; Jane van Lawick-Goodall, "New Discoveries Among Africa's Chimpanzees," *National Geographic* 128, no. 6 (December 1965):802–31; idem, *My Friends the Wild Chimpanzees* (Washington, D.C.: National Geographic Society Publications, 1967).
4. "Cousin Chimpanzee," *MD*, December 1960, pp. 155–58.
5. David Livingstone, *The Last Journals of David Livingstone in Central Africa* (London, 1874).
6. Daniel G. Elliot, *A Review of the Primates*, 3 vols. (New York, 1913).
7. Harold J. Coolidge, "Pan paniscus. Pygmy Chimpanzee from South of the Congo River," *American Journal of Physical Anthropology* 18 (1933):1–57.
8. Eduard Tratz and Heinz Heck, "Der Afrikanische Anthropoide 'Bonobo,' eine Neue Menschenaffengattung," *Sond.*

aus Säugetier, Mitt. 2, no. 3 (1954):97–101.
9. Adolph H. Schultz, "Proportions, Variability and Asymmetries of the Long Bones of the Limbs and the Clavicles in Man and Apes," *Human Biology* 9, no. 3 (September 1937), Table 8.
10. A. Urbain and P. Rode, "Un chimpanzée pygmée (*Pan satyrus paniscus* Schwarz) au park zoologique du bois de Vicennes," *Mammalia* 4 (1940):12–14.
11. Alfred R. Wallace, *The Malay Archipelago* (London, 1906); William T. Hornaday, *Two Years in the Jungle*, 10th ed. (New York, 1926), pp. 375, 406.
12. Adolph H. Schultz, "Growth and Development of the Orang-utan," *Contributions to Embryology, No. 182, Carnegie Institute of Washington Publication 545*, 1941, pp. 57–110 (see also Table 2, p. 63).
13. W. C. O. Hill, "Observations on a Giant Sumatran Orang," *American Journal of Physical Anthropology* 24 (1939):449–505.
14. Richard Lydekker, in *Harmsworth Natural History*, 3 vols. (London, 1910), 1:191.
15. Frederick A. Ulmer, Jr., "Man of the Woods," *Fauna* 8, no. 4 (December 1946):98–103.
16. William L. Straus, Jr., "The Structure of the Crown-pad of the Gorilla and of the Cheek-pad of the Orang-utan," *Journal of Mammalogy* 23, no. 3 (August 14, 1942):276–81.
17. Biruté G. Brindamour, "Orangutans, Indonesia's 'People of the Forest,'" *National Geographic* 148, no. 4 (October 1975): 444–73.
18. John Cragg and G. B. D. Scott, "Ulcerated Colitis in an Adult Male Bornean Orang-utan," *Eleventh Annual Report, The Jersey Wildlife Preservation Trust* (1974), pp. 100–101.
19. Carl Mydans, "Orangutans Can Return to the Wild with Some Help," *Smithsonian* 4, no. 8 (November 1973):26–33.
20. Alfred E. Brehm, *Brehm's Life of Animals* (English edition) (Chicago, 1896); Ernest A. Bryant, in *Harmsworth Natural History*, vol. 1; Robert M. and Ada W. Yerkes, *The Great Apes* (New Haven, Conn.: Yale University Press, 1929); Roger Conant, in *Fauna* (Philadelphia) 6, no. 1 (March 1944):31; Barbara Harrisson, *Orang-utan* (New York: Doubleday & Co., 1963); Bill Gage, "Painting with Jim," *The Bear Facts* (Topeka, Kan.) 17, no. 2 (March–April 1971).

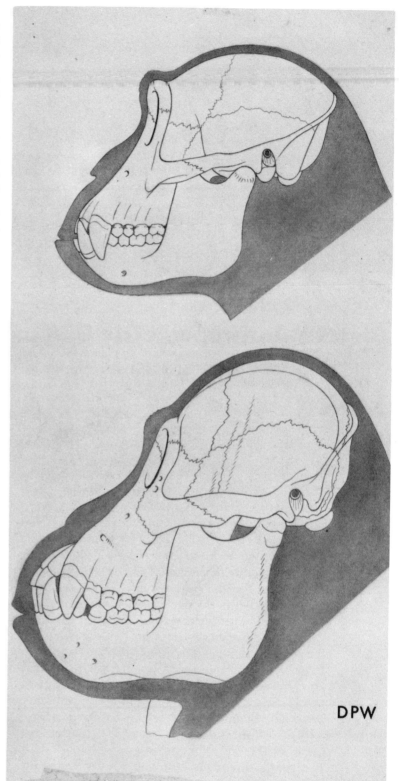

Profile views of the head and skull in an adult male chimpanzee (above) and an adult male orangutan. As is seen, the skull of the chimpanzee is not so heavily developed (as a result of the attached masticatory muscles) as is that of the gorilla (opposite page), and so differs less markedly from the skull of man. In contrast, the teeth and jaws of the adult male orangutan are almost as large as those in the gorilla, although the sagittal and occipital crests are considerably less developed, as are also the brow ridges. Both skulls are reduced to 0.40 natural size, for direct comparison with the skulls of the gorilla and man on the opposite page.

7

An Anthropometric Comparison of the Three Higher Apes and Man

The purpose of this chapter is to compare the physique of the gorilla, as detailed in chapter 5, with that of man and with the gorilla's nearest existing anthropoid relatives—the chimpanzee and the orangutan. In making these comparisons, I shall endeavor to be as "nontechnical" as possible without sacrificing essential accuracy. This, in places, may be difficult!

To start with, the skeletons of man and apes shown in Figure 79 reveal at a glance the chief structural differences prevailing between the simian and the human physique—the latter with its large cranial capacity, erect posture, relatively shallow (vertically) pelvis, and (compared with the apes) short arms and long legs. But this summary, pictorial comparison needs to be substantiated with exact measurements, not only of the long bones of the limbs, but as far as possible also with bodily measurements of living apes. Fortunately, sufficient statistics in the latter respect are available from which to define the average or typical bodily dimensions and proportions both in adult man and the three higher apes, and to a certain extent also in the newborn of these species. These bodily measurements are, it is believed, here established for the first time, not only by measurements taken on zoo or laboratory apes, but from a verification of these living measurements through their derivation from the average dimensions and proportions exhibited in large numbers of precisely measured limb bones or skeletal material.

No attempt is made here to list and discuss all the physical measurements that have been recorded of man and woman at all stages of growth and development, since this alone would require a volume of encyclopedic scope; rather, the measurements of man are here restricted to the relative few that have been recorded also of young captive gorillas and of chimpanzees that likewise were sufficiently tractable to handle. These measurements are indicated in Figure 80, which shows where each dimension is to be taken, and in Tables 16 and 17, which permit a numerical or proportional comparison to be made between the species concerned.

Although details of the measuring techniques to be followed are given in chapter 5, it may clarify matters to review briefly these directions here.

Standing Height, or Stature

While this basic measurement is taken on men, women, and children in the erect standing posi-

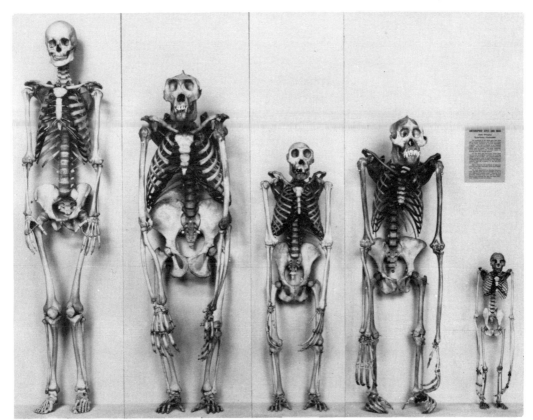

Fig. 79. Skeletons of man, gorilla, chimpanzee, orangutan, and gibbon, respectively. The gorilla and the chimpanzee are both average adult male size. Note the striking differences in the relative lengths of arms, legs, and pelvis in the apes, as compared with man. The gibbon, while more manlike in some respects than the other apes, is more monkeylike in its small size, lesser intelligence, and extreme adaptation for tree-living. (Photo courtesy Field Museum of Natural History, Chicago; times 0.06 natural size.)

tion, on human infants and all apes it is customary (if a reasonably reliable height is to be obtained) to place the subject in a supine (i.e., lying face up) position and to measure in a straight line between upright standards (such as a wooden block or a carpenter's square) placed against the crown of the head and the soles of the feet. While in adult humans such a lying "height" is generally a fraction of an inch greater than one taken in the standing position, in apes the lying dimension may be regarded as essentially identical to the standing height, since in it the back and legs may be straightened to the maximum possible degree (which they cannot be when standing). If for some reason the lying height cannot be secured on a particular gorilla with acceptable accuracy, and the span (distance between the fingertips of the outstretched arms) can be so taken, the height may be estimated with sufficient accuracy by deriving it from the span, using one of the following formulae (which apply to both sexes of the lowland gorilla at all stages of growth):

Height, inches = 0.646 × Span, inches, + 2.8
Height, millimeters = 0.646 × Span, mm, + 71

This method of deriving the probable "standing" height of a wild-shot gorilla, as measured in the field, is particularly applicable in view of the span—in such cases, generally being taken with greater accuracy than is the height, the measurement of which, because of the variable "crest" on the crown of adult male gorillas, along with other difficulties, may be in error by as much as several inches. An adult male lowland gorilla having a span of 8 feet (96 inches), for example, would thus be expected to have a living "standing" height of 0.646 × 96 plus 2.8, or 64.8 inches. For mountain gorillas the corresponding formulae are:

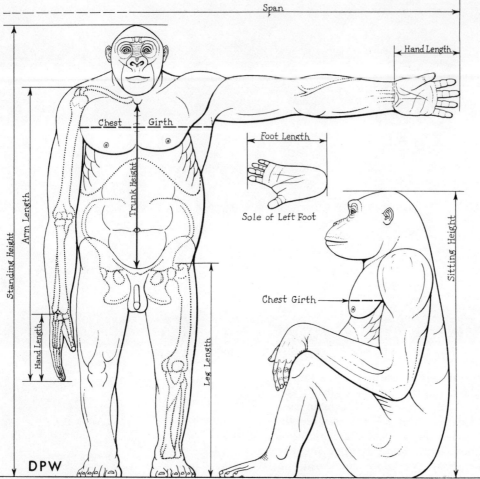

Fig. 80. Outline drawings of a typical adult male lowland gorilla, with hair removed, to show where some of the most important external body and limb measurements are to be taken on anthropoid apes and on humans. In Table 16, these measurements are listed for both sexes at maturity and also at birth. The dotted lines indicate some of the underlying skeletal structure. The erect standing posture, which is not normal for apes, has here been assumed in order to make direct comparisons with man. For details, see text. (Times 0.075 natural size.)

Height, inches = 0.680 × Span, inches, + 3.0
Height, millimeters = 0.680 × Span, mm, + 77

For a given span, therefore, a mountain gorilla should be expected to have a "standing" height about 1.054 times that of a lowland gorilla.

Individual gorillas of both species may show variations in height of as much as ± three inches if estimated by these formulae, but on the average the estimations are reliable.

Sitting Height

Like standing height, sitting height in apes is

Table 16. Comparative external measurements of the Body and Limbs of Man and the three higher Anthropoid Apes, in Newborn and Adults of both sexes.

To be used in connection with Fig. 80. See also Table 17.

(the upper rows of measurements are in millimeters; the lower rows, in inches).

Primate	Standing Height	Sitting Height	Trunk Height	Chest Girth	Span	Arm Length	Leg Length	Hand Length	Foot Length**	Body Weight, Kg. and lbs.
*Man (U. S. White) ♂	1753	917	536	1067	1803	779	915	191	267	75.75
	69.00	36.10	21.10	42.00	71.00	30.66	36.03	7.53	10.51	167
" " " ♀	1626	859	509	864	1628	710	841	175	241	56.23
	64.00	33.83	20.03	34.00	64.10	27.97	33.10	6.87	9.47	124

Adult	Lowland Gorilla	♂	1600 63.00 1342 52.85	1008 39.69 846 33.30	573 22.56 486 19.13	1372 54.02 1102 43.40	2367 93.19 1968 77.50	1040 40.96 874 34.40	752 29.61 621 24.45	241 9.47 207 8.15	291 11.45 248 9.75	156 344 85 188
	"	♀										
	Mountain Gorilla	♂	1683 66.25 1397 55.00	1074 42.28 887 34.90	681 26.80 570 22.45	1559 61.37 1238 48.73	2362 93.00 1938 76.30	1026 40.39 843 33.18	765 30.12 636 25.04	223 8.78 184 7.23	273 10.75 222 8.74	195 430 100 220
	"	♀										
	Chimpanzee	♂	1311 51.60 1244 48.98	813 32.00 780 30.71	485 19.08 470 18.48	847 33.35 795 31.29	1872 73.70 1779 70.06	834 32.83 791 31.15	613 24.13 595 23.42	238 9.37 225 8.86	245 9.65 232 9.15	48.0 106 40.5 89
	"	♀										
	Orangutan	♂	1292 50.87 1131 44.53	792 31.18 715 28.15	483 19.00 435 17.13	1011 39.80 756 29.76	2268 89.27 1985 78.14	1010 39.78 901 35.46	574 22.60 511 20.11	265 10.44 240 9.45	317 12.46 281 11.06	74.9 165 37.2 82
	"	♀										
	*Man (U. S. White)	♂	513 20.20 506 19.92	346 13.62 341 13.43	180 7.09 178 7.01	323 12.72 320 12.60	515 20.28 507 19.96	223 8.78 220 8.66	199 7.82 197 7.75	64.5 2.54 63.3 2.49	80.0 3.15 79.0 3.11	3.50 7.72 3.38 7.45
	" " "	♀										
Newborn	Lowland Gorilla	♂	439 17.30 427 16.80	305 12.00 297 11.70	154 6.10 152 6.00	292 11.50 285 11.20	571 22.50 551 21.70	251 9.90 244 9.60	188 7.40 183 7.20	74.0 2.90 71.0 2.80	97.0 3.80 94.0 3.70	2.18 4.80 2.00 4.41
	"	♀										
	Chimpanzee	♂	424 16.70 416 16.40	289 11.40 284 11.17	150 5.91 145 5.71	279 11.00 277 10.90	533 20.97 523 20.60	237 9.33 233 9.16	169 6.65 166 6.52	73.0 2.87 70.5 2.78	73.7 2.90 71.2 2.80	1.85 4.10 1.79 3.95
	"	♀										
	Orangutan	♂	389 15.32 385 15.14	265 10.43 262 10.31	129 5.08 128 5.02	261 10.27 258 10.17	601 23.66 590 23.23	270 10.63 267 10.51	155 6.10 153 6.04	86.5 3.41 85.6 3.37	98.6 3.88 97.5 3.84	1.50 3.30 1.45 3.20
	"	♀										

* These figures are for well-developed, *athletic* subjects. The infants also are superior to the general average.

** Foot length in man here is *weighted*; in apes, *unweighted* (*i.e.*, taken in the lying position, as in human infants). To convert unweighted foot length to weighted (as if the foot were measured when the ape was standing upright), multiply by 1.018. The structural differences between man and apes shown in this table are better expressed as *proportions* than as absolute dimensions. These *proportions* (relative to standing height) are given in Table 17.

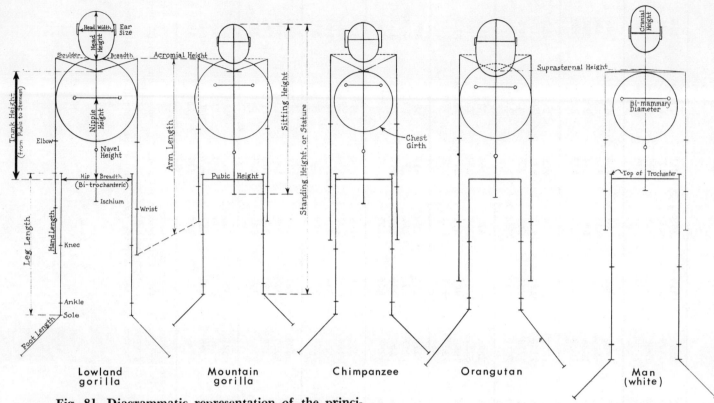

Fig. 81. Diagrammatic representation of the principal body proportions in man and the four higher species of anthropoid apes (all adult males). In these simplified linear diagrams, exact comparisons may be made of the various limb and body length and breadth measurements, as related to the anterior trunk height, which is here reduced to the same absolute length in all species. The trunk height of 573 mm, which actually applies to the adult male lowland gorilla, is here reduced to one-twentieth natural size. For details, see text.

generally taken with the subject in a face-up, lying position (preferably on a measuring board), with the hips flexed, the trunk flat on the board, and the distance then measured between blocks or uprights pressed against the buttocks and the crown of the head, respectively. Of course, if a large or mature ape is sufficiently tame and "cooperative" with its handler, as was "Susie" of the Cincinnati Zoo (see chapter 10), the animal may be taught to sit on some level surface, such as a bench, and the sitting height may then be taken vertically in that position—the same as for an adult human subject. The raised-knee posture shown in Figure 80 is therefore simply illustrative of the points between which the sitting height is to be taken, and not of the proper position for doing this.

Trunk Height (Anterior Trunk Length)

Although this is a valuable basic measurement from which to derive phylogenetically significant ratios of bodily proportion (see Fig. 81 and Table 18), the difficulty of securing the measurement with accuracy is practically confined to the measuring of preserved (cadaveric) or freshly killed specimens rather than living apes, even young and manageable ones. The difficulty consists in properly locating on the external body the skeletal points known respectively as the symphysion (symphysis pubis) and the supersternale (superior border of the sternum), as shown in Figure 81. While the latter or upper point may in most subjects be located by palpation and thus measured, the symphysis pubis is often covered by a layer of skin, fat, and connective tissue so thick that its exact location is practically impossible to ascertain without the use of an X-ray picture of the area. The measurement of trunk height is, therefore, best left to primatologists, like Dr. Adolph Schultz, who through long experience are capable of securing the measurement with the necessary degree of accuracy. However, on the basis of gorilla bodies in which the trunk height has already been taken accurately, these formulae—from which (at all ages) the trunk height may be derived from the readily taken sitting height—have been evolved:

Lowland male gorillas: Trunk Height = 0.594 × Sitting Height − 1.03 in. (26.2 mm)

Lowland female gorillas: Trunk Height = 0.608 × Sitting Height − 1.12 in. (28.5 mm)

Mountain male gorillas: Trunk Height = 0.662 × Sitting Height − 1.19 in. (30.2 mm)

Mountain female gorillas: Trunk Height = 0.681 × Sitting Height − 1.32 in. (33.5 mm)

Chest Girth

With reference to Figure 80, it should be noted that this girth measurement should be taken with both arms hanging (or lying) at the sides of the subject, and not as shown in the illustration, in which one arm is raised in order to indicate how the span is taken. The only other admonition is, when measuring an ape, to pull the tape sufficiently snug to compensate for the effect of hair.

Belly Girth

This measurement is omitted from Tables 16 and 17 because no data is possessed by the writer with respect to the chimpanzee and the orangutan. In gorillas the belly girth is usually from 3 to 6 percent larger than the chest girth, the superiority being greater in females. Actual girths of both the chest and the belly in living gorillas over a wide range of ages are listed in Table 7 of chapter 5.

Span, or Horizontal Arm Stretch

The span, as mentioned previously in connection with standing (or lying) height, is a useful measurement from which to derive the probable or proportionate height in gorillas in which the accurate recording of the height presents difficulties. It may be added that in some accounts of wild-shot gorillas the "arm stretch," or "reach," of the ape was taken with one or both arms raised high overhead, and the measurement made from the soles of the feet to the tip or tips of the upraised fingers. The vertical "reach" taken in this manner averages about 1.7 percent greater than the horizontal reach or span in adult lowland gorillas and about 3.2 percent greater in mountain gorillas. In a few instances, where evidently the hunter was eager to make his gorilla appear as large as possible, the "height" of the ape was taken, in the lying position, from the middle fingertip of one arm as stretched above the head, to the tip of the longest (i.e., second) toe with the foot extended!

Arm Length

This measurement, as explained in chapter 5, is the length from the bony tip of the shoulder (acromial process) to the end of the middle finger, with the arm held straight. Twice the length of this measurement plus the bony (acromial) width of the shoulders times 0.94 gives the horizontal span or arm stretch in most apes. By a reverse procedure, when the span and the shoulder width are known, the arm length may be derived. Figure 81 shows the great difference in arm length compared with trunk length in the mountain gorilla, as compared with the lowland gorilla. It shows also that, among the five primates depicted, the relatively shortest arms are those of man, and the relatively longest arms are those of the orangutan. Even longer, relatively, are the arms of the gibbon (not shown in Fig. 81), the fingertips of which reach clear to the ground when the ape is standing erect. And even longer are the arms of the gibbon's larger relative, the siamang, the fingertips of which, when the ape is standing erect, extend more than an inch below the level of the feet!

Leg Length

This measurement in apes is taken between the same points ("landmarks") as in adult man, with the exception that in apes the subject is measured in a lying position. It is the same measurement that in man is designated trochanter height and is taken with a sliding calipers or anthropometer from the soles of the feet to the uppermost (superior) border of the greater trochanter of the femur or thigh bone (see Fig. 80). Leg length as here adopted is therefore a combination of femur length (from the trochanter), tibia length, and foot height. As is seen in Figure 81, in comparison with trunk height the shortest leg length is in the mountain gorilla and the longest leg length in man, with the relative leg lengths of the other depicted primates ranging between these ratios. While in actuality leg length in the mountain gorilla is somewhat greater than in the lowland gorilla (see Table 16), it is relatively shorter due to the mountain gorilla's significantly longer trunk. For other comparisons, see the discussion of this measurement in chapter 5.

Hand Length

Here, as in arm length, the proportionate length is considerably less in the mountain gorilla than in the lowland gorilla. Indeed, in the mountain species, relative hand length is even shorter than in man. In contrast, in the orangutan, hand length, considered either in relation to trunk height or absolutely, is exceedingly long, the hands in an adult male orang being about an inch longer than

Table 17. Relative external measurements of Man and the three higher Anthropoid Apes as compared with the Standing Height (= 1000). Derived from Table 16. To be used in connection with Fig. 80.

	Primate		Standing Height	Sitting Height	Trunk Height	Chest Girth	Span	Arm Length	Leg Length	Hand Length	Foot Length*	Relative Body Build**
Adult	Man (U. S. White)	♂	1000	523	306	609	1029	444	522	109	152	1000
	,, ,,	♀	,,	529	313	531	1002	437	517	107	148	968
	Lowland Gorilla	♂	,,	630	358	857	1479	650	470	150	182	1650
	,, ,,	♀	,,	630	362	821	1466	651	463	154	184	1589
	Mountain Gorilla	♂	,,	638	405	926	1404	610	455	132	162	1712
	,, ,,	♀	,,	635	408	886	1387	603	455	131	159	1619
	Chimpanzee	♂	,,	632	370	646	1428	636	468	182	187	1234
	,,	♀	,,	627	377	639	1430	636	478	181	186	1225
	Orangutan	♂	,,	613	373	782	1755	782	444	205	245	1576
	,,	♀	,,	632	385	668	1755	796	452	212	248	1356
Newborn	Man (U. S. White)	♂	,,	674	351	630	1004	435	387	126	156	1361
	,, ,,	♀	,,	674	352	633	1002	435	389	125	156	1367
	Lowland Gorilla	♂	,,	694	353	665	1301	572	428	168	220	1357
	,, ,,	♀	,,	696	357	667	1292	571	428	167	220	1357
	Chimpanzee	♂	,,	683	354	659	1256	559	398	172	174	1321
	,,	♀	,,	681	348	665	1256	556	398	170	170	1332
	Orangutan	♂	,,	681	332	670	1544	694	398	223	253	1349
	,,	♀	,,	681	332	672	1534	694	399	223	253	1352

* See note on foot length under Table 16.
** Body build is here derived from the formula: Build $= \sqrt{\dfrac{\text{Weight, lbs.}}{\text{Standing Height, in.}}} \div \text{Height, in.,} \times 44516$. The latter multiplier was chosen in order to yield a Build of 1000 for Man. In the adult male lowland gorilla, for example, the figure 1650 indicates that, at a given standing height, the girths and the horizontal breadth and depth measurements average 1.65 times as large as in man, etc. Again, since body weight is equal to girth squared times height, if an adult male mountain gorilla were the same height (69 in.) as a "typical" man, as here defined, the gorilla would be expected to weigh 1.712 *squared* times 167 (the man's weight), or 489 pounds. In metric units Body Build $= \sqrt{\text{Weight, Kg.}} \div \text{Height, cm.,} \times 2676$.

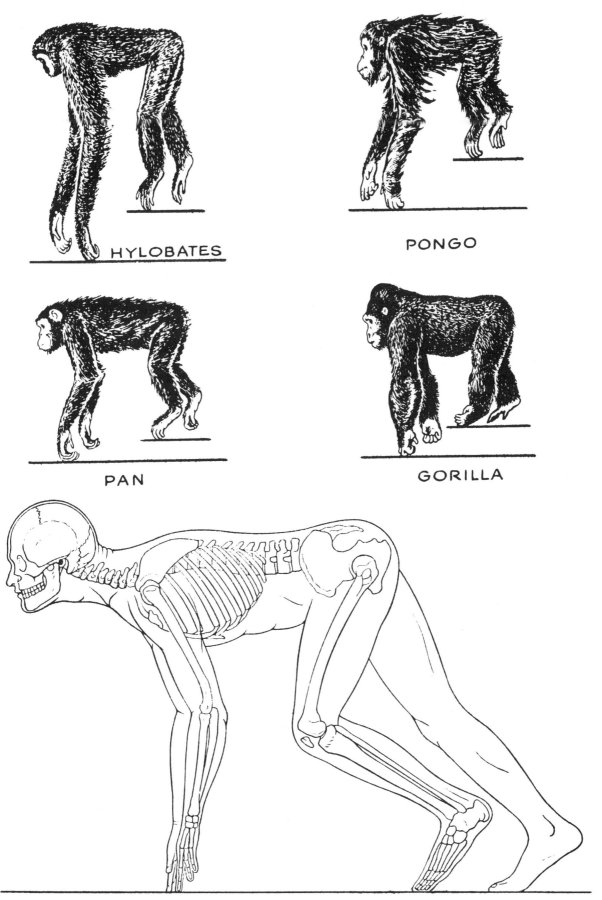

Fig. 82. *Above:* profile views of gibbon, orangutan, chimpanzee, and gorilla, respectively, reduced to the same trunk lengths to show comparative lengths of arms and legs. *Below:* man in a similar posture, but with legs bent, showing markedly shorter arms and longer legs. (Drawings of apes from Erikson, 1963; drawing of man by author.)

in the generally much larger and heavier lowland gorilla (see Table 16). An index, which may be termed "Ratio of Arboreal Adaptation," may be derived by relating hand length to leg length. In man this index averages 20.9; in the mountain gorilla, 29.1; lowland gorilla, 32.0; chimpanzee, 38.8; and orangutan, 46.2. Here, then, is another indication of the largely terrestrial nature of the gorilla—particularly the mountain species—and of the extreme tree-living adaptation of the orangutan.

Foot Length

As Figure 81 indicates, foot length, like hand length, is relatively shortest in the mountain gorilla and in man, and relatively longest in the orangutan. Too, whether a higher primate is mainly terrestrial or mainly arboreal in its living habits is to a certain extent indicated by the relative shortness or longness of its feet.

Body Build

Various formulae for expressing this important index have been proposed over the years by physical anthropologists concerned mainly with its differential application to the various races of man and in the changing of the index during the developmental period from infancy to adulthood. In this connection, this writer in 1942 devised an index that is as adaptable here to apes as it was then to humans. This index relates, in essence, the general width or girth of the body to the height.[1] As is described in the footnote under Table 17, the formula for the index is: body build equals the square root of the bodyweight in pounds divided by the height in inches—this quotient again being divided by the height in inches and then multiplied by the constant 44516. Expressed as an equation: body build $= \sqrt{\dfrac{\text{Weight}}{\text{Height}}} \div \text{Height}$, $\times$ 44516. If the metric units of centimeters and kilograms are used (in place of inches and pounds), the terminal multiplier becomes 2676 (rather than 44516).

As thus defined, the formula yields a body build of 1000 for the optimal white adult male, 968 for the adult female, and for the heaviest built of the anthropoid apes, which is the adult male mountain gorilla, a build of 1712. However, in some presumably overfed and underexercised zoo gorillas, the latter index has been much higher. In the male mountain gorilla "Mbongo," for example, the build was at one time (when Mbongo weighed 660 pounds) no less than 2080; and a number of other over-sized captive gorillas (and a few orangutans and chimpanzees) have likewise had excessively high indices of body build (see chapter 10).

Table 18. Ratios of Various Body and Limb Measurements to Trunk Height (Anterior Trunk Length) in Typical Examples of Adult Male African Anthropoid Apes.

Ratio (to Trunk Height)	Lowland Gorilla	Mountain Gorilla	Chimpanzee	% Deviation from Lowland Gorilla	
				Mountain Gorilla	Chimpanzee
Shoulder Breadth (bi-acromial)	76.44	67.58	62.95	−11.59	−17.65
Hip Breadth (bi-trochanteric)	67.19	58.46	54.59	−13.00	−18.75
Av. Head Diameter	28.75	24.50	26.20	−14.78	−8.87
Arm Length	181.50	150.73	172.94	−16.95	−4.72
Hand Length	42.06	32.76	49.12	−22.11	+16.79
Leg Length	131.24	112.38	126.52	−14.37	−3.60
Foot Length	50.78	40.11	50.57	−21.02	−0.41
Average of 7 deviations, percent				16.26	10.11

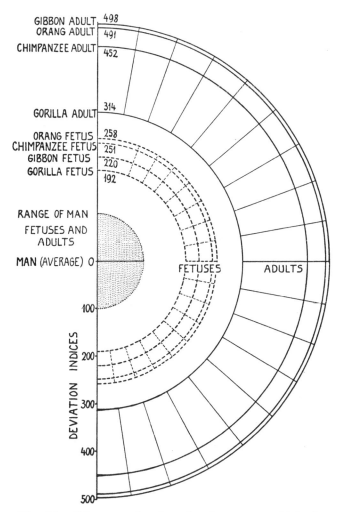

Fig. 83. Diagram showing the degree to which the manlike apes deviate in their physical proportions from man. Note that the adult apes are farther removed from man than the fetal apes, and that during the fetal stage, apes and man are less differentiated than as adults. (After A. H. Schultz, 1927, p. 57.)

An earlier designer of the type of diagram shown in Figure 81 was the German primatologist T. Mollison, who (in 1911) was one of the first to use specifically the anterior trunk length (suprasternale-symphysis pubis) as a basic dimension from which to determine the proportions of the "appendages" (arms and legs) in relation to the body.[2] Although Mollison measured one hundred living men (Germans), he had available (as cadavers) only one immature female gorilla, two immature male and four immature female chimpanzees, and three male and two female orangutans, also immature. Nonetheless, his findings pro-

An index of 2000, for instance, means that the general range of girth in relation to height is just twice that of a typical, well-proportioned male human athlete. The fattest woman, incidentally (who stood 65.5 inches and weighed just over 800 pounds), had an index of body build of approximately 2400!

With reference again to Figure 81, it should be mentioned that although the diagrams shown therein are original with the present writer and encompass a greater range of measurements and proportions than do earlier diagrams by other authors, the latter diagrams have been of great help to the writer in establishing the proportions adopted in Figure 81, which, wherever possible, are based also on measurements taken of living men and anthropoid apes.

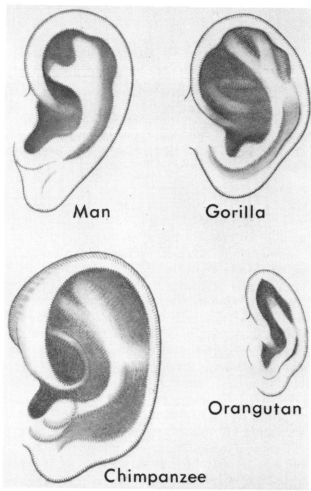

Fig. 84. Semidiagrammatic drawings of the ears of the four higher primates (male), showing relative sizes. While the average heights and widths of the ears in man and apes can be established, it should be noted that in all these species, great individual variation in the form and complexity of the ear occurs. In general, it may be said that the ears of apes appear more "corrugated" and irregular, and with smaller lobes, than those of man. Noticeable also is the large ear of the chimpanzee and the small one of the orangutan. These drawings, by the author, are reproduced life-size.

Fig. 85. A splendid photo of a splendid zoo gorilla, the identity of which, along with the photographer, was not stated.

vided a basis for the more extensive work of later investigators, notably Adolph Schultz, who measured hundreds of primate skeletons and bodies.

Even before Mollison, Thomas Huxley, in his paper "Man's Place in Nature," had made similar comparisons of the relative lengths of arm, leg, hand, and foot in apes and man, although, as in Mollison's case, the number of specimens was small.[3] Too, Huxley's "trunk height," rather than being the length between suprasternale and pubis, was a measurement taken on the skeleton (vertebral column), presumably from the base of the skull to the end of the coccyx. This has made Huxley's figure (diagram) not strictly comparable with those of later primatologists.

By far the most extensive information on the physical dimensions and proportions of the body and limbs (cadaveric and skeletal) of primates has been that published in the numerous studies made by Dr. Adolph H. Schultz. Since practically every one of Schultz's papers contains more or less data of this kind, here I shall refer only to one of his later publications, in which he includes a reference list of twenty-eight additional titles, seventeen of which are his own.[4] (See also the references to Schultz's papers in earlier chapters of this book.)

Concerning chimpanzees, one of the most definitive studies to date, as far as comparative physical measurements and proportions are concerned, is that by Dr. James A. Gavan, who at that time carried on his investigations at the Yerkes Laboratories of Primate Biology, Orange Park, Florida.[5] Gavan presents diagrams comparing the body and limb proportions of man and chimpanzee in both sexes from birth to adult. Unfortunately, because of differences in the measurements employed, a correlation of Gavan's diagrams with those used herein cannot be made without making allowances

Fig. 86. Jim, of the San Diego Zoo, demonstrating how high he can reach. The usual height attained in this upward reach, by gorillas both adult and subadult of both sexes, is just 1½ (1.50) times the erect standing height (as is being done here by Jim). In man, the ratio is only about 1.27 times the standing height. (Photo by F. D. Schmidt; courtesy San Diego Zoo.)

for the differing techniques applied in the two studies.

Figure 82 represents another way of showing diagrammatically the relative differences in arm length and leg length occurring in man and four genera of the higher primates. Whatever means of comparison is adopted, it is found that the arm length-leg length ratio among these primates is highest in the extremely arboreal gibbon and the orangutan, followed by the more terrestrial gorilla and the chimpanzee. The index is pronouncedly lowest in man, who normally does all his moving about on his legs, not his arms. This, no doubt, is one of the reasons for the radically different proportions of man's physique as compared with that of even the closest of his anthropoid cousins. The four upper drawing in Figure 82 are borrowed from G.E. Erikson (1963), who in his paper includes a valuable bibliography on the morphology of primates, listing some fifty titles.[6]

The measurements and proportions which in the diagrams in Figure 81 differentiate the mountain gorilla from the western lowland gorilla are listed numerically in Table 18 to show the same differences. Listed in this table also are the cor-

Fig. 88. The exceedingly long arms and hands of apes of the Malaysian genus *Hylobates* ("tree walkers") are well shown in this study of a white-handed (and -footed) gibbon (*Hylobates lar*). The aerial acrobatic powers of these diminutive "great apes" would put to shame those of the most skilled human gymnast. (Photo courtesy New York Zoological Park.)

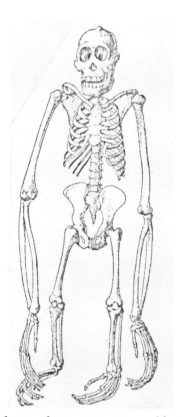

Fig. 87. Skeleton of an orangutan—evidently an immature specimen—showing the exceedingly long arms and hands of this almost completely arboreal ape. Approximately one-twelfth natural size.

responding proportions of a typical adult male chimpanzee. As is shown in the final two columns of the table, the summated differences of the mountain gorilla from the lowland species are approximately 60 percent greater than the differences of the chimpanzee—a generically distinct primate—from the lowland gorilla. While this comparison of course does not imply that the chimpanzee is closer biologically to the lowland gorilla than is the mountain gorilla, it should lay to rest the assertions of those who casually regard both forms of the gorilla as belonging to the same species. No graph is needed to demonstrate from these figures the specific distinctness of the mountain gorilla (*Gorilla beringei*) from the common or western lowland species (*Gorilla gorilla*).

Although it is true that if a sufficiently large number of specimens of the lowland and the mountain gorillas were to be compared, a considerable degree of "overlapping" of body proportions would be encountered, it is the comparison of the average or typical specimens of each series—and not necessarily the range among individuals—that is of

Fig. 89. Although the chimpanzee is not generally considered as having exceedingly long arms, as have the gibbon and the orangutan, the chimp's arms *are* far longer relative to his trunk length than are man's. This is clearly shown in this photo of a mounted chimpanzee in the Field Museum of Natural History, Chicago.

prime importance in the determination of specific or generic status. A parallel situation prevails among many other types of mammals—the great Indian rhinoceros and the lesser one-horned (Javan) rhinoceros, for example—where a small specimen of the greater might be found indistinguishable from a large specimen of the lesser, unless a bone-by-bone comparison of the two were made by an expert in the matter. It is not enough simply to look at two animals that are known, from an osteological standpoint, to be specifically distinct and say that no differences can be "seen" between them.

The point of all this is that—unless skeletal (osteometric) criteria are to be ignored—the mountain gorilla (*Gorilla beringei*) stands as a distinct species (not merely subspecies) of its genus. Through eons of its evolution from a less specialized ancestry, the mountain gorilla has become adapted for a more terrestrial and less arboreal form of existence than the lowland, jungle-living species. In thus adapting to its differing environment the mountain gorilla has evolved the many distinguishing cranial, vertebral, limb-bone, and external bodily features that clearly entitle it to be recognized as a separate species of its kind.

Visitors to the zoo often ask the question, "Which of the great apes is closest to man?" Perhaps the questioner is looking for an answer that will make him least repugnant of his remote ancestry! No unqualified reply can be given to the question, since in some respects one species of ape will come closest, and in other respects another species.* This is indicated, in Figure 81, by a comparison of the bodily proportions of the various apes with those of man.

One of the earliest papers written by Adolph Schultz[7] includes a chart in which is shown the degree to which the proportions of the body in general deviate in apes from man. This chart, by the kind permission of Dr. Schultz, is here reproduced in Figure 83. Considered in this comparison are not only the body proportions of adult anthropoid apes, but also fetuses of the same species. As the chart shows, the ape physically nearest to man, both in the fetal and the adult stages, is the gorilla (western lowland species). Surprisingly, the fetal gibbon is closer to man than is either the fetal orangutan or even the fetal chimpanzee. Among adult examples, the chimpanzee is about 44 percent farther removed from man than is the gorilla, and the orangutan about 56 percent farther. According to the data herein presented on the mountain gorilla (Fig. 81 and Table 18), this species would come on Dr. Schultz's chart somewhere between man and the lowland gorilla.

Quite apart from the physical differences of apes from man are the differences in temperament, behavior, "intelligence," expressiveness, response to captivity, etc., which characterize not only the different genera of anthropoid apes, but which also vary widely among the individuals of a given species. A discussion of "intelligence" in apes is made in chapter 8; while the reactions and behavior of various captive gorillas that have been studied at length are commented upon in chapter 10.

---

* In a comparison of blood-groupings, it would appear that the chimpanzee is somewhat closer to man than is the gorilla. In behavior, also, the chimpanzee may be closer. However, as Dr. Bernhard Grzimek expresses it: "Despite this fact, we are more attracted to the gorilla than to the behavior of the chimpanzee, which seems to reflect the more unfavorable qualities of man as well, in a rather embarrassing manner for us."[8]

To sum up the gist of the present chapter: while almost bone-for-bone man and the manlike apes are similar, the two families—*Hominidae* and *Pongidae,* respectively—are still far from being even close in an evolutionary sense. Moreover, as is shown in the next chapter, the gulf that separates man from apes in the development and potential of the brain is far wider than that which separates the two in their respective physical (i.e., locomotor) structures.

Notes

1. David P. Willoughby, "An Extraordinary Case of Obesity and a Review of Some Lesser Cases," *Human Biology* 14, no. 2 (May 1942):166–77.
2. T. Mollison, "Die Körperproportionen der Primaten," *Morphologie Jahrbuch* 42 (1911):79–304.
3. Thomas Henry Huxley, *Man's Place in Nature* (London, 1863).
4. Adolph H. Schultz, "Age Changes, Variability and Generic Differences in Body Proportions of Recent Hominids," in *Folia Primatologia* (Zurich, 1973), 10:338–50.
5. James A. Gavan, "Growth and Development of the Chimpanzee; a Longitudinal and Comparative Study," *Human Biology* 25, no. 2 (May 1953):93–143.
6. G. E. Erikson, "Brachiation in New World Monkeys and in Anthropoid Apes," *Symposium of the Zoological Society of London,* 1963, pp. 135–63.
7. Adolph H. Schultz, "Studies on the Growth of Gorilla and of Other Higher Primates with Special Reference to a Fetus of Gorilla, Preserved in the Carnegie Museum," *Memoirs, Carnegie Museum* 11 (1927):1–86.
8. Bernhard Grzimek, "The Gorilla," chap. 22 in *Grzimek's Animal Life Encyclopedia,* 13 vols. (New York: Van Nostrand-Reinhold, 1972), 10:543.

8

Intelligence, Brain Weight, Physical Powers, Longevity

Cranial Capacity and Brain Weight

To determine a reliable measure of "intelligence," even of a human and cooperative subject, is a highly subjective and uncertain matter; and where this endeavor is extended to a lower animal, the result may be even more subject to question. For this reason it is possible that a more reliable, objective means of gauging the "intelligence" of a species—if not of an individual animal—is by relating the size of the brain to the size of the body.* Hence, before discussing intelligence tests, it may be of advantage to look into the brain-body relationship. Wherever the weights both of the brain and of the entire body are known, this ratio can be established. Fortunately—since the brain-body relationship in man has long been a point of study among physical anthropologists—there is an abundance of statistics on human brain weight. Fortunately also for our comparison, there is a large amount of data on cranial capacity and brain weight in primates other than man—particularly the anthropoid apes.

In Table 19, the brain-body index has been calculated for a series of primates of both sexes of a given species, ranging in size from man and the gorilla down to the tiny marmoset. That the index here adopted is rationally based and therefore reliable is demonstrated by the fact that it yields closely comparable figures for both sexes, even in species where the females are markedly smaller than the males, as in the gorilla and the orangutan. The formula for this index is:

$$\text{Brain/Body Index} = \frac{\text{Brain Weight, grams}}{\sqrt[3]{\text{Bodyweight, kg.}}}$$

A further step is to convert this index to the figure 100.00 in man (white), which is done by multiplying the index by 0.298. When this is done, a series that may be termed "Relative Brain Potential" is derived. This series is listed in the right-hand column of Table 19. Whatever the actual

---

* Note that the term used here is *size*, not *weight*; for in evaluating various capacities of the body, the "size" involved may be either volume (weight), length, surface area, or a derivative of one or more of these measures.

Table 19. Brain Weight Relative to Body Weight in Man, Apes, and Lower Primates.

Primate	Normal Body Weight, kg.		Cranial Capacity, c.c.		Brain Weight, gms.		Brain/Body Index*	Relative Brain Potential**
	Average	Range in 1000	Average	Range in 1000	Average	Range in 1000		
Man (European)	75.75	52.6–113	1500	1100–2000	1420	1034–1902	335.6	100.0
Woman "	56.25	40–80	1361	1000–1800	1286	939–1748	335.6	100.0
Man (Negro)	75.75	52.6–113	1440	1056–1920	1362	992–1825	321.9	95.9
Woman "	56.25	40–80	1307	958–1725	1234	900–1675	320.4	95.5
Man (Australian Native)	63.5	42–90	1310	960–1747	1237	903–1660	310.0	92.4
Woman " "	51.0	36–72	1215	889–1600	1145	838–1558	308.7	92.0
? Java Ape-man	c. 68?	––	940	––	840	––	c. 206	c. 61
♂ Gorilla (Lowland)	156	105–236	543	380–760	485	328–694	90.1	26.9
♀ " "	85	57–127	461	323–645	406	273–583	92.3	27.5
♂ Chimpanzee	48	32–72	396	277–555	343	229–497	94.4	28.1
♀ "	40.5	27–61	377	264–528	325	216–471	94.6	28.2
♂ Pygmy Chimpanzee	30.5	20–46	295	208–416	254	170–366	94.5	28.2
♀ " "	26	17–39	280	197–394	241	160–347	94.5	28.2
♂ Orangutan	75	50–112	424	297–594	373	248–534	88.5	26.4
♀ "	37	25–56	366	256–512	314	208–455	94.2	28.1
♂ Siamang	11.1	7.4–16.7	126	88–176	110	75–154	49.3	14.7
♀ "	10.2	6.8–15.3	123	86–172	107	73–150	49.3	14.7
♂ Gibbon (H. lar)	5.7	3.8–8.5	104	73–146	91	62–128	50.9	15.2
♀ " " "	5.3	3.5–7.9	101	71–142	88	60–124	50.5	15.0
♂ Capuchin	3.77	2.5–5.7	74	52–104	66	46–95	42.4	12.6
♀ "	3.70	2.4–5.6	73	51–102	65	45–93	42.0	12.5
♂ + ♀ Marmoset (common)	0.614	0.40–0.90	10.65	7.5–14.9	9.5	6.6–13.3	11.2	3.3

* Brain/Body Index = $\dfrac{\text{Brain Weight, gms.}}{\sqrt[3]{\text{Bodyweight, kg.}}}$

** Relative Brain Potential = Brain/Body Index × 0.298

The Standard Range (1000 cases) in Body Weight is based on a Coefficient of Variation of 15.5 (percent), which means that in 1000 cases the minimum weight would be expected to be about two-thirds of the average weight, and the maximum weight about 1½ times the average weight or approximately twice the minimum weight.

Similarly, the Coefficient of Variation for Cranial Capacity is found to be about 11.0 percent. This results in the minimum capacity in 1000 cases being in man approximately seventy-three percent of the average capacity and in the apes and monkeys about seventy percent of the average; while the maximum capacity is twice the minimum.

Brain Weight in man averages 0.964 × Cranial Capacity −26.0; in the higher apes, ditto −38.5. In many instances, including those in small monkeys, the ratio of Brain Weight to Cranial Capacity is about 0.89.

relationship of intellectual power prevailing between man and the lower primates, the ratings listed under "Relative Brain Potential" would appear to give a fair idea of the range of this ratio among the primates here considered.

The cranial capacities and brain weights adopted in Table 19 as representing typical values have been derived and correlated from many sources. Among the latter (in chronological order) are papers by Selenka, Pearl, Oppenheim, Todd, Hrdlička, Connolly, Quiring, Weidenreich, Schultz, Martin, and Jerison.[1]

Most authors who discuss the subject of brain weight in relation to body weight point out that when the respective logarithms of these two quantities (rather than the actual weights) in mammals of different sizes are plotted on a double logarithmic grid, they result in an approximate straight-line correlation. This, while true enough, nevertheless requires differing amounts to be added to the figures resulting from the formula, in order that animals of different sizes may be evaluated. And within a given species, this formula invariably shows the smaller individuals, such as the females, to have relatively larger brains than the larger individuals. The fallacy here is the assumption that brain weight must necessarily be correlated primarily with body weight. Why not with body surface, or with body length? As Table 19 reveals, when brain weight is related, not to body weight per se, but to the cube root of body weight (a linear derivative), men and women, despite their difference in bodily size, are found to have brains of proportionately (by this formula) the same size, along with (as should be expected) brain power of the same magnitude.

That the formula just mentioned yields brain-body ratios that are valid from a geometrical standpoint is shown by all three higher anthropoid apes, irrespective of sex (in which there are great sex differences in bodily size in the gorilla and the orangutan), having essentially the same Relative Brain Potential, the average of which is 27.7. This figure also compares well with the ratio resulting in man, which among Caucasians is assumed as 100.0. Next to man, as other investigators (e.g., Daniel Quiring) also have found, the highest brain-body potential* is in marine mammals (*Cetacea*), such as whales and dolphins, rather than in subhuman primates. When the formula used in Table 19 is applied to the largest mammal, the blue whale (*Balaenoptera musculus*), it is found that a specimen 100 feet in length, weighing 166,000 kg, and having a brain weight of 5560 grams has a Relative Brain Potential of 30.15, which is several percent higher than that in the great apes. The brain-body formula presented here also fits well such diverse animals as dogs, cats, rats, mice, and shrews, among some of the latter of which are the smallest existing mammals.

To get back to anthropoid apes, Figure 90 shows diagrammatically where these higher subhuman primates stand in brain power in comparison with modern man and the Java ape-man above them, and with the smaller and less brainy apes and monkeys below. It is seen that the Java ape-man

* A conventional term for this ratio is "Encephalization Quotient" (E.Q.).

(*Pithecanthropus erectus*) stands in relative mentality about halfway between the highest apes and the Australian aboriginals, the latter of which are generally regarded as one of the most primitive existing races of mankind.[2]

While the comparisons made in Table 19 and other such studies are useful for indicating phylogenetic relationships, there is no positive correlation within a given species between brain size and

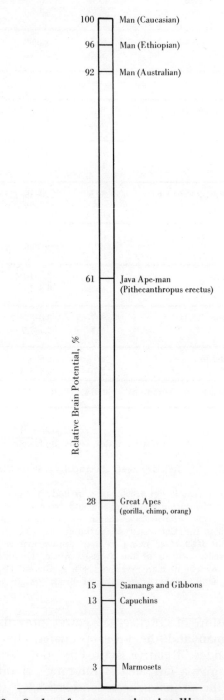

Fig. 90. Scale of comparative intelligence as calibrated from "Relative Brain Potential." (Table 19)

intelligence. For while many men of high intellect have had brains larger than the average, it does not necessarily follow that large brains are superior, since some very large human brains have belonged to diseased or psychopathic individuals. Some larger-than-average brains among famous men include that of Daniel Webster (1805 gms); Cuvier, the French naturalist (1826 gms); and Ivan Turgenev, the Russian novelist (2095 gms). Lord Byron had a very large brain weighing 2237 gms; and Oliver Cromwell a simply gigantic one weighing 2548 gms; however, Cromwell's brain was definitely diseased, and Byron's probably so. Assertedly the largest brain on record for a woman was one of 1742 gms. She was insane and died of tuberculosis. But it would certainly appear that the normal upper range in women's brain weight is over 1742 grams.

In Schultz's series of 400 gorilla skulls the largest has a cranial capacity of 752 cc. As increased statistically to 1000 skulls, as in Table 19, the latter would increase to about 760 cc and the corresponding brain weight to 694 grams. The latter weight may be assumed to represent the approximate upper limit of size in the brain of an anthropoid ape (i.e., adult male gorilla).

In the *growth* of the lowland gorilla brain, as indicated by the data of Randall on cranial capacity,[3] the latter would appear to be about 430 cc at two years of age, and about 240 cc at birth. Thus, as would be presumed from the respective body sizes of newborn humans and gorillas, the gorilla brain at birth is much larger compared with that of a human newborn (c. 240 cc as compared with c. 400 cc), or about 60 percent, than in the adult, where the ratio in the male gorilla as compared with man is only 543:1500, or about 36 percent. Accordingly, the size of the brain in the gorilla increases during the period from birth to maturity (say, 12 years) from about 240 cc to 543 cc, or times 2.26. In comparison, the brain in man, during a growth period approximately fifty percent longer than that in the gorilla, increases from about 400 cc to 1500 cc, or times 3.75. From this it might be theorized that if only the growth period in apes were longer, the development of their brain might continue to a higher level.

Although Table 19 shows a definite relationship between brain size and body size in man and various lower primates, it is, of course, questionable whether actual intelligence can be inferred from what is here termed "Relative Brain Potential." The reason for this is that some animals have relatively large brains possibly because of abilities derived from a high development of certain of their senses: dogs, for example, from their senses of smell and hearing; elephants from the varied uses of their trunk; man from stimuli induced by the use of his hands; etc. That, at least, was the belief advanced by the Dutch paleontologist Eugen Dubois.[4] However, this theory is rejected by the anthropologist Franz Weidenreich, who points out that dogs do not have a better sense of smell than foxes of the same size, even though they have 15 percent more brain than foxes. He goes on to mention that monkeys of the South American genus *Cebus* (capuchins), while not having as prehensile a tail as do spider monkeys (genus *Ateles*), have developed hands that are equal to those of anthropoid apes in utilizing objects as tools. Weidenreich continues:

> That the development of the hand and nothing more should be the underlying cause of the extraordinary enlargement of the human brain, as Dubois suggests, is entirely out of the question. With much more justification can the claim be made that, on the contrary, the enlargement of the brain constitutes the cause underlying the special accomplishment of the hand.[5]

This leads to the question: if the development of the brain and intelligence in man preceded the free use of his hands, why have not apes (the hands of which are only partially freed from their locomotor functions) similarly developed their brains to a higher level? Weidenreich endeavors to answer this question by giving his opinion on the cause of man's cerebral development: "But the enlargement of the [man's] brain is certainly in some way connected with the adoption of the erect posture and the corresponding transformation of the entire skeleton."[6] Specifically, Weidenreich alludes to the extensive structural or mechanical differences between the skulls of man and apes, which made possible the large brain and small face of man and the smaller brain and enlarged muzzle of apes. And at this point we must leave the complex question of brain size versus intelligence and proceed by considering some of the "I.Q." tests that have been given to the chimpanzee, the gorilla, and the orangutan in an endeavor to probe the reasoning powers of these physically near yet mentally remote relatives of man.

Intelligence in the Manlike Apes

It is probable that ever since the first living

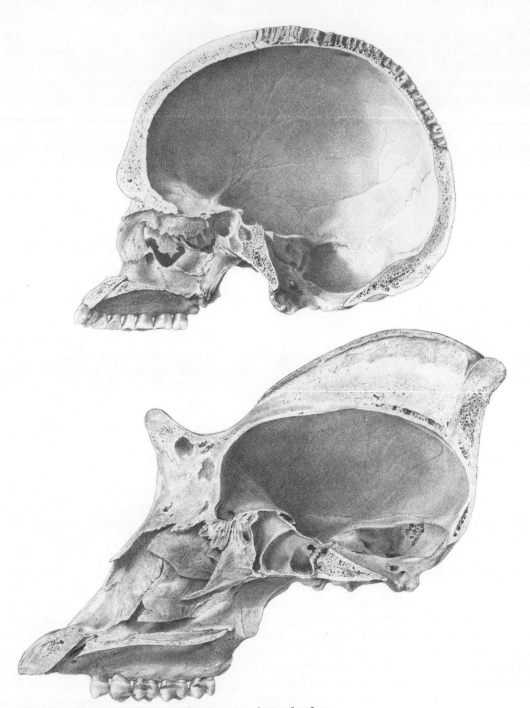

Fig. 91. Cross-sectional (sagittal) cuts through the skulls of an adult male lowland gorilla (below) and a man (Melanesian), showing relative sizes of the cranial cavity of braincase. The gorilla skull is an historic one. It was described in great detail anatomically by Professor Richard Owen in *Transactions of the Zoological Society of London* 4, p. 3 (1851): 75–88. The skull, which belonged to a specimen obtained in the Cameroons, had a cranial capacity of 32.6 cubic inches (534 cc). The capacity of the Melanesian (Papuan) skull was about 1230 cc. Times approximately one-half natural size.

Fig. 92. This illustration, which is reproduced from *Brehm's Life of Animals* (1896 English edition), shows two young trained chimpanzees that had mastered many of the accomplishments of boys and girls of the same age. The chimps went to "school" in the zoo at Stuttgart, Germany, where this documentary illustration was made by the animal artist Friedrich Specht some time in the 1870s.

specimens of chimpanzees and orangutans were brought into zoos and taught to perform, there has been a desire on man's part to learn just how near to him these anthropoid relatives are in intelligence, and whether this intelligence is of a degree that can reasonably be called *reasoning*.

Gradually the desire to gain this information gave rise to the devising, by psychologists, of various tests by which the mental capacity of apes could be observed, compared, and in some tests even measured. While, in due course, tests of a comparable nature were administered to animals lower in the scale of brain development, it was soon realized that the most interesting animals to test were those that most closely resembled man physically; namely, the higher anthropoid apes: gorilla, chimpanzee, and orangutan. However, since chimpanzees can safely be worked with and tested to a more nearly adult stage of physical and mental growth than can the much larger (and potentially dangerous) gorillas, or even male orangutans, "chimps" became the favored subjects for so-called I.Q. testing. But prior to the use of sophisticated laboratory tests for this purpose, some interesting observations were made of the behavior of chimpanzees by zoo keepers who recorded all acts that appeared to indicate actual reasoning. The adjoining illustration (Fig. 92) shows two young chimpanzees that resided in the zoo at Stuttgart, Germany, about a hundred years ago, and which even then delighted audiences with their many humanlike accomplishments.

The female chimpanzee "Sally" (see Fig. 62) of the London Zoo in the 1880s was adjudged highly intelligent, even though that judgment was made through observation of her everyday actions rather than by laboratory testing, which at that early date had not come into use:

> She is an expert rat-catcher, and has caught and killed many rats that had entered her cage during the night. Her intelligence is far above that of the ordinary chimpanzee [Sally was of the rare variety known as bald chimpanzee]. With but little trouble she can be taught to do many things that require the exercise of considerable thought and understanding. She recognizes those who have made her acquaintance, and pays marked attention to men of colour, by uttering a loud cry of *bon, bun, bun*. She is never tired of romping and playing, and is generally in a good temper.[7]

Before discussing some of the laboratory tests that have been given to gorillas and orangutans as well as to chimpanzees, it may be helpful to describe some of the first simple procedures that were used in probing the mental powers of the more available and cooperative anthropoid subjects: "chimps." When one of these apes is to be tested (without having had any previous training), it is put inside a cage and given two bamboo sticks. The diameters of the sticks are such that the end of one can be fitted inside the other like a fishing pole. An attractive bit of food, usually a banana, is placed outside the cage, at such a distance that if the ape wants to reach it he can do so only by putting one of the sticks inside the other and using this double-length pole to draw the food to him.

At first the chimp tries to reach the banana (or other food) with his outstretched hand; then with

147

one of the sticks. Failing in both attempts, the ape will sit down, play aimlessly with the sticks, then apparently start to reason. All of a sudden he will fit the two sticks together, end to end, then rush to the side of the cage and happily draw in the food. Moreover, the ape will *remember* how he got his results. Another test is to place the food high out of reach, by suspending it from the top of the cage, and to leave on the floor several tables or boxes that can be piled one on top of the other. Chimps, orangs, and gorillas all soon learn to place the boxes so that they can climb on top of them and reach the food.

Some amusing (and factual) stories are told that illustrate the exceptional mental powers of that comic extrovert of the ape family—the chimpanzee. Vance Packard has related several of these experiences.[8] In one test a psychologist filled a room with toys, led in a chimpanzee, and sneaked out. How would the chimp react to the various playthings when alone? The psychologist peeped through the keyhole to see. What he saw was one of the chimp's bright eyes looking right back at him! Another chimp, who had been left some boxes on which to stand to reach some fruit, instead pulled the psychologist (Professor Kohler) over directly under the food, then climbed up onto his shoulders and grabbed the morsel! In a third test, a chimp, rather than using a long pole to knock down the fruit, stood the pole on end and ran up it before it had time to fall, thus obtaining the fruit unbruised!

But in emphasizing the mentality of the chimpanzee, it should be borne in mind that the emotional makeup of an animal (or a human) is almost inseparably interwoven with its apparent

Fig. 93. This young chimpanzee is evidently performing satisfactorily in placing blocks of various shapes into their proper sockets. In this procedure, he may be aided by the manlike shape and adaptability of his hands, in which the fingers and the thumb are more readily apposable than in either the gorilla or orangutan, in which the thumbs are relatively shorter. (Photo by Lilo Hess.)

intelligence. Accordingly, where an intelligence test is given to such a subject, what is measured may be the manifestation of intelligence, rather than the intelligence itself. Chimpanzees have an alert, restless, playful type of personality—and this in some respects makes them appear even brighter, in relation to their cousins the gorilla and the orangutan, than they really are. Both the latter apes, in contrast to chimps, are slow, deliberate, and placid in disposition and action, and from this it is often inferred that they must be correspondingly dull in mentality. According to Professor Robert M. Yerkes, one of the foremost students of the psychobiology of apes, the chimpanzee ranks highest in curiosity, attention, imitation, intuition, perception of relations, insight, foresight, and imagination. In these qualities, according to Yerkes, the gorilla and the orang are about evenly matched for second place. But, to repeat, in all so-called "intelligence" tests, the emotional or dispositional makeup of the individual must be borne in mind. Gorillas, for example, except in their childhood, are undemonstrative creatures compared to chimpanzees, and they may possess capacities and potentialities that they decline to exhibit. The following instance is an apt example of this. At the San Diego Zoo, during the 1960s, a series of I.Q. tests was initiated and applied to all three forms of the higher manlike apes: gorilla, chimpanzee, and orangutan. One of the pieces of apparatus used in the testing procedure is described thus:

> The animal will view from its side of the apparatus a large half-inch-thick plate of "Herculite." Along the bottom of this transparent tempered glass it will see three containers, each 6 x 6 inches. The front of each container will be of half-inch-thick plate glass. Through this glass the animal will be able to see the objects which constitute the varied object-quality discrimination problems presented to it during the course of our learning-set experiments. The animal will indicate its choice of objects, on a given trial, by pushing back on the plate glass which forms the front of the chosen container. As it pushes on the glass the container will swing back and will bare a food well in which a reward, such as a grape, will be available. Immediately behind these three containers there will be a piece of steel. It will arc 90 degrees upward so that even if a container were to break for any unexpected reason, it will be impossible for the animal to reach through to seize the experimenter.[9]

After this test had been given in 1963 to some of the apes that were then mostly young or immature, and which revealed, as would be expected, favorable responses from some of the "pupils" and unfavorable reactions from others, it was again employed in 1965.[10] At the latter date, Albert, the zoo's best-known male (lowland) gorilla, and one of the apes tested, was fifteen years of age and presumably fully adult. To make a long story short, Albert was so discomfited by his repeated failures to choose the right container and so receive the desired reward (food) that he struck the front of one of the food bins with his left fist with such force that the glass was shattered and an imprint was left of his thumbnail and the hair on the back of his hand. The shattered pane was formed of two thicknesses of 3/16-inch laminated glass, 5½ inches wide by 5¾ inches high. Fortunately, Albert's hand showed no injury, and apparently he suffered no discomfort as a result of the violent blow he had delivered. However, the experiment left unresolved Albert's degree of reasoning power or "learning ability" as compared with that of the other gorillas, chimps, and orangs that had been tested by the same means with varying degrees of success.

Although by means of such tests, applied countless times over a period of months, the "learning ability" of apes or other animals may be improved, the question is whether such learning (knowledge-gaining) is likewise indicative of a corresponding increase in basic brain power or reasoning ability. In short, if an acquired capacity (on the part of an individual animal), such as learning how to solve a puzzle, has no transmittable or phylogenetic value, and cannot be used even as a gauge of ability in other directions (such as those encountered under the natural living conditions of an animal in its native habitat), of what biological significance is the acquired ability? Because of the individual (rather than specific or generic) nature of all so-called I.Q. tests—which are mostly tests of knowledge rather than of inherent braininess—it would seem to the writer that actually measurable cranial capacity and/or brain weight, as derived from the average of large numbers of individuals (and as listed, for example, in Table 19), are more useful and reliable indices of the course of primate evolution than any recordings of presumed reasoning power made under laboratory conditions. Viewed from this perspective, there is no significant difference between the natural intelligence possessed by the gorilla, the chimpanzee, and the orangutan. Too, among members of the order Primates there can be no question

Fig. 94. A banana-seeking chimp is quick to see the connection between a batch of boxes and some otherwise unreachable objects of his appetite. Gorillas and orangutans also, though perhaps a bit slower and more deliberate in their thinking, similarly solve this box-stacking problem. (Photo by Lilo Hess.)

that these anthropoids are next to man in intelligence.*

One more reference to Table 19: the "Relative Brain Potential" of gorillas, chimpanzees, and orangutans, as derived from the formula underlying this table, is in conformity with the conclusion reached by the primatologist Adolph H. Schultz, who says that "there are no true phylogenetic differences in relative cranial capacity among the great apes, besides those due entirely to age and body weight."[11]

Another anthropologist, Dr. D.M. Rumbaugh, in his study *Chimpanzee Intelligence*[12], compares the great apes thus: "collectively, great apes are superior in learning to gibbons, though gibbons are not necessarily superior to the more advanced monkeys. Also, great apes are probably superior to all monkeys so far as capacity *to innovate* is concerned."

Rumbaugh feels that chimpanzees are not necessarily more intelligent than either gorillas or orangs but have been more publicized because of their availability in larger numbers for testing in behavior, etc. Chimps are more docile, while orangs are "melancholy" and gorillas "reserved, if not frankly negativistic and obstinate" (cf. Albert, above!). These characteristics refer more to personality and temperament than to either intelligence or learning skills. All apes are highly imitative.

(Some "anecdotal" instances of noteworthy intelligence in captive gorillas are related of some of the individual apes discussed in chapter 10.)

The Physical Powers of Anthropoid Apes

To start on this inadequately documented and rather debatable subject, let us skim over the factors of speed and stamina and give our chief attention to strength. Speed among the primates is the prerogative mostly of smaller forms than adult gorillas, who by their very nature are introverted, leisurely moving members of their order. Under stress, however, their tremendous power enables them to take off or charge with astonishing speed. But a small monkey can be incredibly quick in its movements even under ordinary circumstances. With a tiny hand it can, for example, catch a fly every time, even though a person might fail to do so by using a flyswatter. As to stamina or endur-

Fig. 95. "Meng," a young male mountain gorilla formerly (1938–39) in the London Zoo, tracing the outline of his shadow on the wall with outstretched forefinger. This action on Meng's part has been considered an "artistic" advance on the usual aimless scribblings and paintings of other anthropoid apes when provided with art materials. (Drawing from life by Nina Scott Langley.)

ance, it is represented or confined among chimpanzees and gorillas mainly to the distances the apes are obliged to traverse in their search for food. Indeed, it is perplexing why a gorilla should be endowed with the obviously enormous muscular strength that it possesses, when seemingly it needs no such degree of strength to survive in its natural jungle habitat.

Accounts of aggressiveness, or defensive action, by gorillas in the wild, in which the tremendous physical strength of the ape is a conspicuous element, have been made by hunters since the time of Du Chaillu, over a hundred years ago, to the present day. As is shown in one of the illustrations published by Du Chaillu (see Fig. 18, chapter 3), an adult male lowland gorilla is depicted in the act of bending a rifle double (using its hands only), after having killed the native hunter who attempted to fire the weapon. But it would seem more likely that the ape would have bent the gun by using its jaws as well as its hands in the process.

---

* But only until, and if, it is conclusively established that such subhuman primates as the Yeti ("Abominable Snowman") and the Sasquatch ("Bigfoot") do not presently exist (see chapter 2).

Fig. 96. Statuette by William Umlauff of Hamburg, Germany, depicting an actual occurrence in German east Africa (now Tanzania), in which two native hunters were killed by a maddened gorilla. In view of the location described, the gorilla must have been either of the mountain species (which the statue does not resemble) or the eastern variety of the lowland gorilla.

Also, the single-barreled, muzzle-loading guns used in those days (c. 1855) were doubtless more "bendable" than the rifles made some years later by Winchester and other manufacturers. On the other hand, the gorilla may have been using only a fraction of its full strength in performing this impressive feat.

Over a century after Du Chaillu's experiences, the professional hunter Tony Sanchez corroborated the gorilla's rifle-bending ability by remarking that "I have personally seen the barrel of a .303 rifle bent double, just like a hairpin . . . by a maddened gorilla." And he adds: "Just a twist of his [the gorilla's] massive hand can tear the head right off a human being, as more than one unfortunate native has learned."[13] Sanchez also states that although he has shot many a plantation-raiding adult gorilla, he has never measured one that in height came even up to his chin (which would probably be less than 5 feet 7 inches, since Sanchez gives his own height as 6 feet 4 inches). He also states that his heaviest gorilla, "a real monster," weighed 200 kilograms (441 pounds), and that the usual weight is from 150 kg (330 pounds) to 175 kg (385 pounds): "Old males weighing a bit over 200 kilos . . . are extremely uncommon in the wild state." These singularly reliable statements by Sanchez as to the size of wild lowland gorillas give authenticity to his remarks on the apes' physical strength.

Another gorilla hunter, Frank E. Delano, makes the following comments concerning these apes:

Like most animals, the family bull* gorilla prefers to be left alone and, unless bothered, will not attempt to kill. Once confronted by danger, however, he becomes an extremely dangerous animal, whose fantastic strength is terrifying. Gorillas can and have seized a man with their huge arms, lifted him into the air and, grasping one leg in each hand, torn the unlucky individual in half. The great inch-long fangs of a gorilla . . . can crush a gun barrel in an instant.[14]

Delano points out also that when trailing a gorilla in dense jungle it is advisable to use a pistol rather than a rifle, since more than one gorilla,

---

* Possibly it was Edgar Rice Burroughs who, in his *Tarzan* stories, first used the term "bull ape." It should be remembered that wherever "bull" is used in referring to a male animal, "cow" is the corresponding term for the female. And one never hears of a female ape (e.g., gorilla) being called a "cow ape."

when approached at close quarters, has been known to snatch the gun from the hunter's grasp, "sometimes taking a trigger finger," and with it beat the hunter to death. Delano's guide, a French professional hunter named Maurice Patry, remarked that the adult male gorilla shot by Delano —which stood (or rather lay) 5 feet 8 inches and weighed (presumably by estimation) 450 pounds —was the second-largest gorilla that he (Patry) had seen during sixteen years in the Gabon. This practically corresponds with the opinion of Tony Sanchez, given above, as to the maximum size attained by wild (not zoo) adult male gorillas of the western lowland species.

Although the fighting abilities of gorillas in the wild may often constitute awesome demonstrations of physical power, such actions can hardly be taken as measurable tests of the apes' muscular strength. And the latter tests, or examples, as far as captive gorillas are concerned, are few and far between. One reason for this, perhaps, is because the apes are trained to be manageable rather than prompted into exhibiting their strength. Many a zoogoer has seen the ease with which an adult gorilla can handle and bend a large truck tire, but this is far from being an indication of the ape's maximum strength. One of the most impressive feats I have heard of in this connection is where an adult male gorilla lifted up a steel cage door that weighed over 600 pounds, hoisting or flinging it up with such force that it jammed! This incident, which occurred some years ago, was related by one of the keepers at the San Diego Wild Animal Farm, although he may have witnessed the action in some other zoo.

Orangutans, too, have enormous strength in their long, muscular arms, and particularly in their fingers and grip. In some instances even the steel bars of an orang's cage have been loosened from the ape's continual working on them. Apparently the skin of an ape's hands and fingers is practically immune to pain. However, there are limitations to the strength of orangs and the other species of great apes, although in the following account this is not evident! (I shall mercifully refrain from giving the name of the author.) It seems that some years ago, in Borneo, a large adult male orangutan, which had been singled out for capture alive, climbed up a tall tree, the first limb of which was allegedly a hundred feet above the ground. After much trouble, the team of native hunters managed to get a noosed rope around the orang. They then attempted to pull the ape out of the tree. But according to the author, who assertedly witnessed the action, it took prolonged pulling by fifty (!) natives on the other end of the rope before the determinedly resisting ape was finally dislodged from the tree. From this it was deduced that the orang had the strength of fifty men combined! It is small wonder that when accounts such as this, by presumably reputable authors, are published as facts, many a reader accepts them at their face value. In this case, further exaggeration was added by the statement that the arm span (horizontal reach) of the orang, which the storyteller referred to as "the largest . . . ever to be taken alive," was over 11 feet (!). Actually, when the ape was later measured in the Bronx Zoo, the span was found to be 7 feet 10½ inches. This, indeed, is the span of a large male orang, but still far from a record measurement. This orang's weight, if in proportion to its height (about 4 feet 6 inches) and span, would almost certainly have been less than 200 pounds. And its general strength, accordingly, was probably no more than that of two strong men of the same body weight. As to an orang's great strength of grip, it should be remembered that in effect this arboreal species possesses four hands, since its feet are essentially equal to hands in grasping the limbs of trees. And as has been mentioned, since the fingers of an ape do not fully extend when the wrists are straight (as in pulling), the short tendons enable the hands to act as veritable unbreakable hooks. Thus, any ape has singular strength in retaining its grip, as in pulling on a rope; and if such pulling is done against one or more human contestants the ape has another advantage in the nonslip traction: that its handlike feet are able to secure on practically any surface.*

Another point that should be mentioned in this connection is that when two men presumably unite their strength, as in pulling together on a rope, the power of their pull is not necessarily twice as great as that of one man pulling alone. Complete efficiency, or 100-percent coordination, would be needed in order for this to occur, and such efficiency is hardly ever attained, or applied, in actual practice. Many years ago, for highly practical purposes, this relationship was deter-

---

* A possible factor in the muscular strength of anthropoid apes is that their muscles may contain *more fibers* per unit of cross-section than is the case in man. In the gorilla, for example, the body weight in relation to body volume (*i.e.*, the specific gravity of the body) can be shown to be at least *25 percent greater* than in man. That is, if a man had identically the same physical dimensions as a gorilla weighing 350 pounds, the man would weigh not more than 280 pounds. Here is a field of inquiry which apparently has not been investigated so far by comparative anatomists or primatologists.

mined in its application to draft horses. The findings were that the ratio of the pull of two horses combined, as compared with that of a single horse, is not 2.00, but 1.96, which in this case indicates an efficiency of 98 percent. However, when more than two horses are used together, as in hauling a heavy load, this high degree of efficiency drops off at a surprisingly rapid rate. Thus, when a four-horse team is used, the pulling power is that of 3.2 horses, an efficiency of 80 percent; and when an eight-horse hitch is used, only 3.92, or 49 percent. So, if this rate of decrease applies to men as well as to horses, the fifty (?) natives that assertedly combined their strength to pull a single orangutan out of a tree may have had the effective pulling power of only six or seven men. (Indeed, what with all the excitement and confusion, the highly intelligent ape may have had the natives pulling against each other!)

The point of this digression is that statements made by excited eyewitnesses of anything, especially man-animal combats which may be interpreted as tests of physical strength, should be taken with a large grain of salt.

Of the muscular strength of captive chimpanzees, a number of tests (which were practically confined to pulling power) have been made by various investigators. Of these tests, possibly the most reliable were those conducted by the mammalogist G. Finch, some years ago.[15] In his experiments, Finch used four men and eight chimpanzees, the latter consisting of four males and four females. Strength was measured by pulling (presumably using both hands together) on a fixed spring dynamometer. The best pull performed by a chimpanzee ("Bokar," an adult male weighing 107 pounds) was 487 pounds. This pull was duplicated by one of the men tested, who weighed 145 pounds. It was the best pull, in relation to body weight, made by any of the men.

Since in the latter two pulls the weight per inch of height (which may be taken as representing the general cross-section of the body) in both the man and the chimpanzee must have been approximately equal (being in each about 2.1 pounds per inch of height), it would appear that the muscular strength—so far as lifting or pulling power is concerned—of both man and ape was about the same relative to muscular cross-section.

From the foregoing it may be inferred, with due reservation, that the general strength (as in lifting or pulling) of an average-sized adult male chimpanzee is approximately the same as that of an average-sized man in athletic condition.

On the same basis, it would appear that an adult male gorilla of average size, weighing say 350 pounds, should be expected to be from 2½ to 3 times as strong in pulling and lifting as either an average-sized male chimpanzee or an average-sized, athletic male human. This estimate, in turn, would make the gorilla about as strong generally (as in back and leg strength) as a 350-pound champion weightlifter (such, for example, as the Soviet world champion, Vasili Alexeev, who in the 1976 Olympic Games competition raised no less than 562 pounds from the ground to arms' length overhead). Of course, in some respects—such as strength of neck, forearms, hands, and grip—no human athlete of any size could approach the colossal power possessed by a male gorilla, not only because of the ape's having larger muscles, but in some respects also better bony leverage. Equally, of course, the fighting equipment of a gorilla—namely his lion-sized fangs and jaws, along with fingers and hands of gigantic dimensions—would make it suicidal for any man to engage in combat with one of these apes without using a battle axe.

On the other hand, however, it might be possible for a hard-hitting heavyweight boxer to knock out a gorilla with a punch to the jaw. The animal trader Frank Buck once delivered such a punch to an unruly adult male orangutan, a species of ape that has jaws not much smaller than those of a gorilla. And with reference to an adult male and pugnacious chimpanzee, former zoo director William Mann tells this story: "A specimen with a circus on the Pacific coast . . . was so intractable that the circus owner employed a pugilist to handle him. The boxer met the frontal attack of the chimpanzee by a knock-out blow, and afterward the chimpanzee not only respected him, but they became great friends."[16]

Gorillas in their native habitats have few enemies outside of intruding man, and evidently they are able to cope with an occasional leopard, a python, or even a crocodile sufficiently well to maintain a balanced population. A leopard seldom runs the risk of attacking an adult (and awake) male gorilla but sometimes is able to carry off a young ape, or even to kill an adult female. A few years ago a zoo-kept black panther escaped from its cage and killed* a male gorilla, but one cannot assume from this singular occurrence that the outcome would have been the same if the panther had encountered a wild, alert gorilla, rather than a

---

* Actually, the gorilla had to be shot because of the severity of the wounds inflicted on it by the infuriated panther (all of which were delivered within a few seconds!).

Fig. 97. A drawing by the old-time German animal artist Friedrich Specht, showing the manner in which it was then thought that the gorilla attacked the leopard. Present-day natives, however, say that the gorillas kill the leopards by an open-handed blow on the side of the head, swinging their long arms like clubs and with terrific force. In rare fights among themselves, a similar technique is said to be used.

relatively inactive zoo specimen. As to big snakes and crocodiles, one can only speculate on what would occur if either of these reptiles took the offensive, since there is no record, so far as I know, of anyone, even a jungle-living native, having witnessed such an encounter. P.T. Barnum in one of his books tells about a battle between a gorilla and a crocodile, but the story is so obviously fabricated that I have placed it in chapter 12, along with other flights of the imagination.

To summarize: the physical strength of apes, no more than their intellectual capacities, can only be partially apprehended by man, due to the unbridgeable gulf between the communicative and interpretative faculties of the two distinctive primate types.

Longevity in Man, Apes, and Various Lower Mammals

Just as in the matter of brain size and mental capacity, man has long pondered the causes of longevity and has naturally focused his attention on such types of animals as have exhibited the longest spans of life. Various theories have been advanced by anthropologists, biologists, gerontologists, and zoologists as to why some persons and some forms of animal life succeed in living longer than others. If and when a practicable key to this problem—such as the determination of an optimum course of living—is positively attained, man may become able to extend not only his own lifespan, but that also of the numerous lower animals whose destinies he largely controls.*

Table 20, following, gives average statistics on longevity, not only of man and other primates, but of various familiar domesticated animals, the ages of which are reasonably well documented. These average or (for each species) typical ages have been drawn from many sources, those on human longevity being from Thoms, Young, and Bowerman.[17] On lower animals the ages are those quoted by Lankester, Chamberlain, Flower, Comfort, and Jones.[18]

In order that the aging process in apes and monkeys (on which reliable data are meager) may be determined—or at least estimated—Table 20 includes the known ages of various other mammals from which comparisons may be made; for example, if an index is used in which the maximum age is related to the average age at death minus the average age at maturity, the range of this index, among all the mammalian species listed, is only from 2.00 to 2.31, with an average value of 2.11. The highest index is in man. If for him a maximum age of 120 years is assumed, the ratio of maximum age to average minus mature age is 2.31. However, if the oldest human age so far certified—one of 113 years, 124 days—is taken as the maximum, the latter ratio becomes 2.18, which is still the highest among the species here listed. If, in view of these ratios, one of 2.18 is adopted for the higher anthropoid apes—gorilla, chimpanzee, and orangutan —the average age at death of these apes becomes 40 years and the maximum age 60 years. Considering that in man—even among a population of

---

* In some quarters this course is regarded as already indicated, in that experiments made on laboratory rats have scientifically confirmed the long-held concept that "length of life is inversely related to the amount of food consumed." The same thought is expressed in the ancient maxim "A lean horse for a long race."

Table 20. Comparative Ages, in Years, of Man and Various Lower Animals. ( )assumed

Species (all male) (unless noted)	Average Age at Maturity	Average Age at Death	Maximum Age Reported	Maximum Age Estimated
Man, ♂	20	72	115+	120
" ♀	18	77	112	116
Elephant (Indian)	22?	(60)	77+	83
Whale (Blue)	(6)	22–32 (30)	37	50
Horse	5	30	40+, 54 (?)	52
Cow	4	19 (22)	32, 39	39
Sheep	1.5	(11)	20	20
Dog (Av. Size)	1.5	11.5, 13–17 (18)	34	34
Cat	1	12 (15)	27+, 36	28
Rat (Albino)	0.20 (72–74 days)	2.5	3.8	4.6
Mouse (house)	0.18 (66–67 days)	2.2	3.5	4
Gorilla	12.5	26 (40)	46+	60
Chimpanzee	12.5	(40)	51+	60
Orangutan	12.5	(40)	34	60
Gibbon	(5.0)	(22)	32+	35
Baboon (Chacma)	(5.0)	11 (27)	45	45
Rhesus Macaque	(3.2)	15 (17)	29	29
Capuchin	(3.0)	9 (18)	37	37

Relative Ages (years):
(not to be confused with Average Age at Death)

Age, Gorilla, Chimp, or Orang = .475 Man's age +3.0   Age, Man = 2.10 Ape's age   −6.0
"   Gibbon = .30 Man's age −1.0   "   "   = 3.33 Gibbon's age +3.3
"   Horse = .47 "   "   −4.4   "   "   = 2.13 Horse's age +9.3
"   Cow   = .35 "   "   −3.0   "   "   = 2.86 Cow's age   +8.5
"   Dog   = .325 "   "   −5.0   "   "   = 3.08 Dog's age   +15.5
"   Cat   = .27 "   "   −4.4   "   "   = 3.70 Cat's age   +6.3

Example: A cat is known to be ten years and three months (10.25 years) old. What is the corresponding age in a man? Answer: 3.7 × 10.25 + 6.3 = 44.2 years. Again, if a man is seventy years old, what is the corresponding age in a gorilla? Answer: 0.475 × 70 + 3.0 = 36¼ years.

billions—the ratio of maximum-average age rarely exceeds 1.5, it should be safe to assume that the same ratio in apes is likewise not over 1.5, which would make the maximum age somewhere close to 60 years. Too, an average age at death of 40 years for the larger apes would appear reasonable, since the actual recorded ages in this respect for both apes, baboons, and smaller monkeys are far lower than would be the case if infant mortality among these primates were reduced (which it steadily is).

As to the gibbon, the baboon, and the smaller monkeys listed in Table 20, the only positive figures are those that were actually recorded for maximum age, the figures (in parentheses) in the first two columns being simply "educated guesses." The 32-year maximum age for a gibbon and the 34 years for an orangutan are both reported by Jones, as is the 37 years (!) for a capuchin. The latter age in particular makes the elsewhere-quoted average age at death (9 years) in this species seem ridiculously low. The 45-year age given for the baboon (*chacma* species) is by Flower. The longest-lived chimpanzee to date is evidently one that reached 51 years in 1971, and which may still be living. The oldest gorilla is the male, "Massa," of the Philadelphia Zoo, who was still living in 1976 at the age of 46 years, his estimated birth date having been December 30, 1930. Among female captive gorillas the oldest specimen appears to have been "M'Toto," who was owned by Mrs. Maria Hoyt of Havana, Cuba, and who died in 1968 at the age of 37 years.

Other relatively "aged" gorillas, such as "Bamboo," of the Philadelphia Zoo (who in 1961 reached 34½ years), and "Guy," of the London Zoo (who was still living in 1972 at the age of 26), could be quoted at some length; but such ages are so far below those estimated in Table 20—and which in due course will probably be reached—that they fail to convey an idea of the true potential maximum age in the gorilla, chimpanzee, and orangutan.

Approximately a hundred years ago the German naturalist Alfred Edmund Brehm considered forty years to be the probable "average life" of the gorilla and the chimpanzee; and today the validity of this early estimate is being confirmed. With greater knowledge being gained daily about all aspects of ape-keeping—especially feeding, proper exercise, comfortable surroundings and companionship, and the eradication of various common diseases—an increase in the average longevity of apes and monkeys comparable with that recorded for man during the last hundred years should surely be attained.

Notes

(Cranial Capacity, Brain Weight, Intelligence)

1. Emil Selenka, *Menschenaffen (Anthropomorphae). Studien über Entwickelung und Schädelbau* (Weisbaden, 1898-1903); Raymond Pearl, "Biometrical Studies on Man. I. Variation and Correlation in Brain Weight," *Biometrika* 4 (1906): 13-104; Stefanie Oppenheim, "Zur Typologie des Primatencraniums," *Zeitschrift fur Morphologie und Anthropologie* 14 (1911):1-203; T. Wingate Todd, "Cranial Capacity and Linear Dimensions, in White and Negro," *American Journal of Physical Anthropology* 6 (1923):97-194; T. Wingate Todd and Wilhelmine Kuenzel, "The Estimation of Cranial Capacity—a Comparison of the Direct Water and Seed Methods," ibid. 8 (1925):25-259; Ales Hrdlicka, "Weight of the Brain and of the Internal Organs in American Monkeys, with Data on Brain Weight in other Apes," ibid. 8 (1925): 201-11; C. J. Connolly, "Brain Indices of Anthropoid Apes," ibid. 17 (1932):57-69; Daniel P. Quiring, "The Scale of Being, According to the Power Formula," *Growth* 5, no. 3 (1941):301-27; Franz Weidenreich, "The Brain and its Role in the Phylogenetic Transformation of the Human Skull," *Transcripts of the American Philosophical Society*, n.s. 31, pt. 5 (August 1941):321-442; Adolph H. Schultz, "The Relative Size of the Cranial Capacity in Primates," *American Journal of Physical Anthropology* 28, no. 3 (September 1941): 273-87; idem, "The Recent Hominoid Primates," in *Perspectives on Human Evolution* (1), ed. S. L. Washburn and P. C. Jay (Berkeley, Calif.: Society for Study of Human Evolution, 1948); Rudolf Martin, *Lehrbuch der Anthropologie*, 3d ed. (Jena: Fischer, 1958); Harry J. Jerison, "Paleoneurology and the Evolution of Mind," *Scientific American* 234, no. 1 (January 1976):90-101.
2. T. D. Campbell and C. J. Hackett, "Adelaide University Field Anthropology: Central Australia," *Transcripts and Proceedings, Royal Society of South Australia* 51 (1927): 65-75 (see their Bibliography for earlier studies of Australian natives).
3. Francis E. Randall, "The Skeletal and Dental Development and Variability of the Gorilla," *Human Biology* 15, no. 4 (1943):325-36; and ibid. 16, no. 1 (1944):68-70.
4. Eugen Dubois, "Über die Abhängigkeit des Hirngewichtes von der Körpergrösse. I. Bei den Säugetieren," *Arch. f. Anthrop.* 25 (1898):1-128; idem, "II. Beim Menschen," ibid. 25 (1898):423-41.
5. Weidenreich, "The Brain and its Role," p. 406.
6. Ibid., p. 435.
7. A. D. Bartlett, in J. Fortune Nott, *Wild Animals Photographed and Described* (London, 1886), p. 548.
8. Vance Packard, "What Do You Mean—DUMB Animals!", *American Magazine*, January 1949, pp. 81-82.
9. Duane M. Rumbaugh, "Comparative Primate Learning," *Zoonooz* (San Diego) 37 (July 1963):4-6.
10. Duane M. Rumbaugh, "Problems in Testing an Adult Male Gorilla," *Zoonooz* (San Diego) 38 (February 1965):10-15.
11. Schultz, "The Relative Size," p. 282.
12. Duane M. Rumbaugh, "Chimpanzee Intelligence," in Rumbaugh, *The Chimpanzee* (Baltimore, Md.: University Park Press, 1971), 4:19-45.

SEE ALSO:
Frank A. Beach, "Payday for Primates," *Natural History* 56, no. 10 (December 1947):448-51.
Wolfgang Kohler, *The Mentality of Apes* (New York: The Humanities Press, 1956).
Carl J. Warden, "Animal Intelligence," *Scientific American* 186, no. 6 (June 1951):64-68.
James Wiley, *Beasts, Brains, and Behavior* (New York: Scholastic Magazines, Inc., 1963).

(Physical Powers)

13. Tony Sanchez, "Gunning Gorilla in Africa," *Safaris Unlimited*, January 1961.
14. Frank E. Delano, "Gabon Gorilla Hunt," *Sports Afield*, August 1963, pp. 42, 70.
15. G. Finch, in *Journal of Mammalogy* 24 (1943):224-28.
16. William M. Mann, *Wild Animals In and Out of the Zoo*, Smithsonian Scientific Series, vol. 6 (Washington, D.C., 1943), p. 33.

(Longevity)

17. W. J. Thoms, *Human Longevity–Its Facts and Its Fictions*

(London, 1873); T. E. Young, *On Centenarians* (London, 1899); Walter G. Bowerman, "Centenarians," *Transcripts of the Actuarial Society of America* 40, pt. 2, no. 102 (September 1939):360–78.

18. E. Ray Lankester, *On Comparative Longevity in Man and the Lower Animals* (London, 1870); A. F. Chamberlain, *The Child and Childhood in Folk-Thought*, 1900. Table reproduced by G. Stanley Hall in *Adolescence* . . . (New York, 1904), p. 473; Stanley S. Flower, "Contributions to Our Knowledge of the Duration of Life in Vertebrate Animals. V. Mammals," *Proceedings of the Zoological Society of London* (1931);145–234; Alex Comfort, "The Life Span of Animals," *Scientific American* 205, no. 2 (August 1961):108–19; idem, *Ageing: The Biology of Senescence* (New York, 1964); Marvin L. Jones, "Longevity of Primates in Captivity," *International Zoo Year Book*, no. 8 (1968), pp. 183–92.

See also a preliminary study in this field by the author: David P. Willoughby, "Animal Ages," *Natural History* 78, no. 10 (December 1969):57–59.

9

Habits and Family Life of the Gorilla

Social Relations

Despite the fact that European and American naturalists and hunters have, for over a century, sought to learn by direct observation the habits and behavior of the gorilla in its native haunts, various obstacles have hindered the acquisition of exact knowledge on how these giant anthropoids live day by day in their dense jungle stronghold. One difficulty has been that the habitat of the more common, or western lowland, species of gorilla is conducive to "jungle fever" (malaria) and other maladies that may incapacitate the white hunter or explorer before he even gets near his objective. Again, while exact knowledge may be gained by the study of the body of a dead gorilla, it is quite another matter to approach and possibly photograph a living (and naturally either wary or threatening) individual or band of these apes.

Only during comparatively recent years has success in the latter objective been attained; and this has been mainly in connection with the mountain (eastern), rather than lowland, gorilla. Thanks to the dedication and perseverance of several independent investigators at various times, a considerable gain in knowledge of the living habits of mountain gorillas in the wild has been made. Some of these findings have been found applicable also to the western, lowland-living gorillas. From the observations reported by a number of on-the-spot field-workers—mainly George Schaller, Paul Zahl, Bernhard Grzimek, and Dian Fossey, respectively—the following items of information on gorillas in the wild have been obtained.

As has long been suspected, the gorilla is essentially a social or "family" animal, living and traveling in troops of from five to as many as thirty individuals, the average number being about thirteen. Each troop is composed of one, two, or three older and dominant males, which are commonly referred to as "silverbacks" because of the light grey coloration on the lower half (at least) of their backs. Of inferior rank to these troop leaders are two or three black-backed (and therefore younger) and less dominant males, several adult or subadult females, and a few juveniles and infants. It should be added, however, that some observers have found the usual size of a troop to be only five to nine individuals, led by a single old male.

Fig. 98. In this nineteenth-century illustration from *Brehm's Tierleben*, artist Friedrich Specht has shown "A family of gorillas," as living by themselves in the forest. Today it is known that gorillas are more sociable than this, and that their family bands generally include several mothers and young, even if only one "daddy" or "boss."

According to Schaller,[1] if there are several adult males in the troop, one is still the leader or director. The adult males are the most excitable members of the troop, and the reactions of them and all the others depend on the "boss" male. If he is in a nervous mood and moves off, all the other gorillas follow him, even if they had all been sleeping. When excited, the males roar, beat their chests,* and dash about. Females, in contrast,

---

* The chest-beating is not performed with the fists clenched, but with the hands cupped and the fingers partially extended. The sound produced is like the clip-clopping of a horse's hoofs on a hard surface, only much more rapid and more quickly completed. In adult males the sound may be accentuated by virtue of a "sound box" in the larynx or air sacs in the chest. As indicated, gorillas beat their chests not as a threat, but rather to release emotional tension. In the case of an approaching and (to the gorilla) hostile human, the chest-beating may be used to "bluff" the intruder away.

rarely get excited and only occasionally beat their chests. Females, young males, and juveniles are much more curious than the adult males and will approach closer to quiet-sitting human beings in the wild. Too, while the females rank above juveniles, which in turn rank above youngsters that no longer live with their mothers, among the females themselves there seems to be no clear-cut order of rank. Harmony, whenever possible, would appear to be the keynote of wild-living gorillas.

Bernhard Grzimek, who like Schaller spent a long period in closely observing wild (mountain) gorillas, remarks that these apes—particularly the mountain species—are exceedingly difficult to film. They are shy animals that generally hide themselves behind bushes or other dense vegetation; in addition, their black coloration makes them difficult to spot. The few scenes of free-living gorillas that have been filmed in action were actually of specimens that had been caught with the help of hundreds of native beaters. The netted gorillas were then transported to fenced areas, released, and filmed. Grzimek and his assistants worked in a location between the Congo (now Zaire) and Ruanda, in the Virunga Mountains northeast of Lake Kivu (see Fig. 22, chapter 4). He adds that the gorillas in this region are constantly endangered by the clearing of forests and the expansion of farming. Also, the Watusi herdsmen, who have large numbers of cattle that are of little economic value, strive to herd these grazing animals into areas occupied by gorillas.[2]

Food of Gorillas in the Wild*

Generally, as soon as the sun comes up, a troop of gorillas will start feeding upon whatever plants are nearby. Although when in captivity all the species of manlike apes can be taught to eat meat, in the wild state gorillas are almost total vegetarians. The only departure from strictness in this respect is their occasional indulgence in bird's eggs. Meat in any form, even if of a freshly killed animal, is passed by. However, the early gorilla keeper Julius Falkenstein[3] emphatically asserted that the gorilla both in the wild and in captivity takes animal foods, especially insects and birds, as well as bird's eggs. Possibly this conflict with Schaller's recent findings (which indicated a complete adherance to a vegetable diet) is that Falkenstein's observations were confined to the western lowland gorilla, while Schaller's were of the east-

---

* See also chapter 4, some portions of which are here repeated.

Fig. 99. An artist's conception of a band of gorillas (only the leader of which is shown here) "raiding" a native settlement, from which the inhabitants are fleeing in terror. Perhaps more realistically, the male natives would be shouting invectives and hurling spears at the advancing intruders.

ern mountain species. The German investigator Eduard Reichenow suggested that the readiness with which zoo apes take to both eggs and ground meat may result from the disappearance from the intestines of these captive animals, after several weeks' confinement, of certain infusoria, which aid in the digestion of vegetable matter—this changed condition creating a need for animal food.[4] However, in possible contradiction to this is the fact that zoo gorillas, no less than those in the wild state, commonly develop enormous paunches, which would indicate that a voluminous intake of relatively unnutritious vegetable matter is still the normal diet of gorillas.

Lowland gorillas feed upon leaves, bark, berries, nuts (including peanuts), rootstalks of ginger, buds of the parasol tree, plantains (a variety of banana), pineapple leaves, wild sugar cane, and, according to R.L. Garner,[5] a plant called *batuna*, which grows throughout the forest. The wild mangrove, which forms a staple article of food for the chimpanzee, is rarely if ever touched by the gorilla. However, when in their endless quest for food gorillas break into plantations and consume bananas, sugar cane, and whatever other cultivated products are available, the native farmers understandably drive the apes off by whatever means they can employ—including firearms. Thus, in some areas, the natives are torn between obeying the law, which prohibits the killing of gorillas, and protecting their hard-won crops from the depredations of these animals.

Of the mountain gorilla, the food, apart from bamboo shoots, wild bananas, certain berries, and on occasion, honey, consists almost entirely of herbage—docks, sorrels, hemlocks, lobelia, wild celery, leaves, buds, and roots. It does not, however, as a rule grub for roots; neither does it habitually eat fruit.[6] Schaller says that the gorillas of the Virunga Mountains eat about a hundred different kinds of plants. It is noteworthy that of many foods the gorilla prefers the stalk to the fruit, twisting and breaking open the stalk so as to get at the succulent interior of the plant. They are very selective eaters and carefully pick off dead parts or tough bark. When chewing, they keep their mouths open, smack their lips, and grumble contentedly. Each individual ape sits in its own spot and carefully discards the inedible portions of its food into a neat pile. Such eating may continue for several hours, or until midmorning. The need for this long-continued feeding is the unnutritious nature of the apes' herbivorous diet, which in turn accounts for their capacious stomachs and barrel-

Fig. 100. A close-up photo of a Virunga Mountain gorilla taken by Adrien Deschryver, Conservateur of the Kahuzi-Biega National Park, west of Lake Kivu. It shows an adult, "silver-backed" male, with the long, narrow face characteristic of many individuals of the mountain species. (Photo courtesy Drs. Deschryver and Grzimek—*Okapia Films*, Frankfurt a.M.)

Fig. 101. A photograph taken on the northern slope of Mt. Vishoke, looking northeastward toward Mts. Sabinio, Mgahinga, and Muhavura (all extinct volcanoes). Note the profuse vegetation, including alchemilla, lobelia, and senecio—all typical foods of the mountain gorillas. (Photo courtesy *African Wild Life*.)

like bellies. Also, the need to tear apart and often masticate tough or woody substances (particularly, in the mountain gorilla, stalks of bamboo) doubtless has influenced the development of the gorillas' massive teeth and powerful jaws. Although the wear on a gorilla's teeth must be considerable, no author so far appears to have determined the rate of attrition occurring in an adult gorilla's teeth with age.

After eating until satiated, most of the gorillas in a troop will begin their noonday siesta, although a few may continue nibbling. The midday naps take place often within a few hundred feet of where the apes nested the previous night. Settling themselves comfortably in the thick vegetation, they will doze or merely relax. Some pass the time by grooming either themselves or one of their companions. The mother apes groom their young ones. Meanwhile, the juveniles may either be climbing trees or crawling all over the bodies of the older gorillas. If it rains the apes often huddle together; and if the day is sunny they will lie on their backs and enjoy it. After two or three hours of rest they will again start moving about and feeding. The second period of eating continues until 5:00 or 6:00 P.M., when the apes taper off their meanderings and start building their nests for sleep.

As to the use of water, lowland gorillas may drink whenever their wanderings bring them to small, shallow streams, which in the forest are numerous. However, not being natural swimmers, these apes avoid getting into any water "over their depth," as in a large river. Evidently, when stream or lake water is not conveniently available, they get enough liquid from the more succulent of the plants they consume. Perhaps in downpours of rain, which are of frequent occurrence, they may let the water on leaves flow or drip into their mouths, as do orangutans in their high leafy bowers.

Nesting and Climbing

Most observers of gorillas in their natural haunts state that the "nests," or sleeping platforms, of these apes are occupied only once. But at least one observer, R. Akroyd, who was a member of the British Museum Expedition to the Virunga Volcano area in 1933-4, found evidence to the contrary. Of this he writes:

> The gorillas' nest is built on the ground with intertwining of small branches, twigs, etc., the big male having his nest in the centre and the family grouped around him. They generally choose a sheltered place, under the roots of an old tree or under a bank. They are said [by the natives?] to pass one or two nights in one place, and then move on, leaving their nests fouled. They often return later to the same spot and build another nest on top of the old one. The giant Lobelias, Senecios, and giant heather grow over twenty-five feet high.[7]

Akroyd mentions also that the gorillas of the Kayonza Forest, an area of approximately 150 square miles, having elevations up to 7900 feet, "live under different conditions from those on the volcanoes, building their nests in the trees about 10 to 20 feet from the ground. They are probably a different race, perhaps allied to those in the Ituri Forest."

The foregoing notes by Akroyd are particularly significant, since the sole purpose of the expedition of which he was a member was to collect vegetation, including that used in the building of nests, for the background of a gorilla diorama in the British Museum (Natural History). It is thus evident that in certain areas of the mountain gorillas' habitat nests are built at considerable heights above the ground as well as directly on it. Possibly it is a matter of the kind of trees that grow in a particular area, and whether or not they are adapted (as by suitably forked branches) for the support of a nest or sleeping platform.

Another point brought out by Akroyd is his suggestion that the mountain gorillas of the Kayonsa Forest (located in southwest Uganda east of Lake Edward and north of the Virunga Volcanoes) may constitute a different race (i.e., subspecies) from those living on the slopes of the volcanoes. However, since the Kayonsa Forest gorillas are also of the mountain species, it is evident that mountain gorillas in certain areas build nests well up in the trees, the same as is said to be done by certain lowland gorillas. Thus, the matter of nest-building—as well as most other matters connected with gorillas—would appear subject to varied interpretation, depending upon the observer.

In the actual construction of a "nest," the gorilla, standing or sitting, bends in branches, or standing plants, from all directions until a round bowllike nest, from three to five feet across, is formed.* A nest may take anywhere from less than

---

* With reference to plants used in the construction of lowland gorilla beds, Jones and Sabater Pi (1971) list (p. 63, by botanical name and frequency of presence) six different herbaceous materials and four different woody materials that they identified in their examinations of 410 such gorilla beds, or nests, in the forests of Rio Mundi, West Africa. In the same book are listed also (p. 72) some of the many plants utilized for food by the gorillas of this region.[8]

Fig. 102. Two styles of "beds" built by large and small primates, respectively: the sleeping platform of the gorilla, and the nest of the mouse lemur. (Drawing by Neave Parker; courtesy *The Illustrated London News*.)

a minute to fifteen minutes to construct. Generally, by 6:00 P.M. the gorilla sleeps, lying either on his belly, with arms and legs tucked under, or on his side. While some observers declare that gorillas snore, Schaller says that he never heard one do this.

With reference to the claim that gorillas are capable of tying knots, Sanderson says:

> These natives showed the author sleeping platforms constructed by gorillas a few feet off the ground in gnarled dwarf trees made by bending down branches and anchoring them with true knots with a double over-and-under twist—both "grannies" and "reef knots."[9]

Zahl, however, says that of several gorilla pallets examined by him, none showed any knots. He adds that: "before nightfall a gorilla merely gathers leafy branches by armfuls, arranges them in an orderly, comfortable heap, then plunks himself down in the center." Of the nests high up in the trees, he says: "The elevated bamboo beds required more ingenuity. A number of bamboo stalks were bent together and their tops crudely interlocked. Such elevated roosts were never the largest of any group; they could not possibly have supported an adult male. Presumably they were for the protection of females and young."[10]

As to the climbing ability of the gorilla, while it can at any age do so, its ability in this respect is much more limited than that of smaller monkeys and the chimpanzee and orangutan. Most of the gorillas that are seen fairly high up in trees are youngsters rather than adults, since the latter, especially the old and heavy males, in so climbing must proceed cautiously, testing limbs as they go, to make sure that the limbs will bear their weight. Even so, gorillas of all ages, as well as chimpanzees, orangutans, and even gibbons (the latter of which are exceedingly dextrous) occasionally fall from considerable heights to the ground, as is indicated by the proportion of broken limb bones found in the skeletons of these apes. Although some writers have stated that adult gorillas, particularly males, never climb trees, Coolidge[11] mentions a very large [possibly 500-pound] western lowland male, which, even after being speared, climbed about fifty feet up in one of the tallest trees in the forest (see Fig. 34). Adolph Schultz's careful drawings of the feet of gorillas show that the lowland species has greater grasping power than the mountain form. This doubtless helps the lowland gorilla to be the better climber of the two. Indeed, the feet of the mountain gorilla show it to be primarily a ground-walker.[12]

To the foregoing comments about gorillas in general may be added a few of those by Capt. C.R.S. Pitman,[13] who had experiences in particular with the gorillas of the Kayonsa "Impenetrable Forest" of southwest Uganda, of which area he was for some years game warden. Pitman agrees with Akroyd (see above) in stating that gorillas of the Kayonsa area habitually climb trees and build their nests well above the ground, those he examined having been placed anywhere from 6 feet to 20 to 25 feet high. The nests were located so that each was clearly visible from all the others. Pitman adds:

> As in the case of the volcanoes' representatives [i.e., "typical" mountain gorillas] these beds are singularly filthy, and the edges often festooned with excrement. It is probable that the same platforms are used on several consecutive nights. The highest bed seen was nearly 50 feet above the ground. Its foundation consisted of sturdy, upright tree-tops as much as 2½ inches in diameter, which had been snapped like matchsticks. The communal bed was as filthy as usual. . . . One of the droppings was so immense that my gun-bearer naively remarked that it looked more like an elephant's.

Possibly one of the reasons for the Kayonsa gorillas being assertedly "a different race" from those of the Virunga Volcanoes is the marked difference between the two habitats in the matters of elevation, climate, and vegetation (available food). On the volcanoes, the gorillas are not often seen below the 10,000-foot level, whereas in the Kayonsa habitat the elevation is from 6000 to 7900 feet only, and the air is warmer and drier. Pitman remarks:

> There is in the Kayonsa a complete absence of bamboo, wild celery, dock, and similar juicy-stemmed plants such as abound in the humid, high altitudes, forcing the gorilla to confine its diet to a mixture of leaves, berries, ferns, the tender fronds of tree-ferns, parts of the wild banana stems and leaves, and fibrous bark peeled off a variety of shrubs in the undergrowth.

A singular finding reported by Pitman with regard to gorilla skulls is that those from habitats

Fig. 103. A subadult mountain gorilla carrying leaves for a meal. The remarkable feature here is not so much the carrying, but the fact that the ape is doing so while walking bipedally and upright. (AP photo)

in which bamboo is eaten (e.g., the Virunga Volcanoes) are readily distinguishable from skulls coming from areas where bamboo is absent (e.g., Kayonsa). The reason for this is that when tough bamboo is eaten by young gorillas, fragments of the woody fibers become wedged between the teeth, eventually forcing them apart in their spacing. This, of course, is not a valid difference so far as zoological classification of the skulls is concerned; but (if consistently present) it does afford a supplementary means for visually distinguishing Virunga gorilla skulls from others.

Handedness, Locomotion, and Migration

Civilized man, who is mainly right-handed, has arms that on the average are slightly longer (about 0.16 of an inch) on the right side than on the left. In women (U.S. whites) the average difference likewise is 0.19 inch.* Thus, if a longer arm (or bones of the arm) presupposes a greater use of that side of the body, conversely it should be possible to estimate whether a person is naturally right-handed or left-handed, by ascertaining which of his (or her) arms is the longer. And if this identification procedure applies to humans, why not also to the manlike apes?

In application to adult male lowland gorillas, the data presented by Randall[14] show the bones (humerus plus radius) of the left arm to average slightly (4 mm, or about .16 inch) longer than those of the right side, thereby indicating possible left-handedness. This status applies also to female gorillas during the growing stages, but ends, in adults, with the bone lengths being exactly the same length in both limbs. Hence, in view of there being little or no difference in the lengths of the arm bones of the two sides in gorillas, these quadrumanous animals, like most quadrupeds, may perhaps best be regarded as ambidexterous. It may be significant that in civilized man (U.S. white), who is brought up essentially in a right-handed, average-sized environment, is generally right-handed, as far as being longer-armed on the right side is concerned, than are representatives of other human races or nationalities, such as African, Mongolian, Mexican, and Puerto Rican. This, at least, is what is indicated in the respective lengths of the arm bones (humerus, radius, and ulna) on the right and left sides of the body in soldiers that were killed in the Korean War.[15]

Locomotion in the gorilla is that of a typical quadrupedal animal, namely, a walk, a trot, or a gallop. The walk is performed in the usual, transverse quadrupedal manner (i.e., "diagonal walk"), the left forefoot and right hindfoot advancing together, followed by the right forefoot and left hindfoot, etc. (see Fig. 41). The same sequence occurs in the trot. In galloping, there are two styles, known respectively as the lateral gallop—which is natural to dogs, deer, antelopes, and certain other mammals—and the transverse gallop—which is used by horses, cattle, cats, and some other forms. So far as I know, no motion pictures have been taken that show a gorilla or a chimpanzee galloping; so whether in so moving a transverse or a lateral gallop is used remains to be determined. Bipedal movement in the gorilla is confined to walking, and that is rarely for distances of more than a few yards. Some writers have claimed that a charging male gorilla can move at what would seem an incredible speed; but this may be a parallel to statements by emotionally upset hunters that the gorilla they shot was at least eight feet tall. Certainly it would seem unlikely that a 400-pound gorilla could charge with a speed exceeding that of a trained human sprinter over a distance of, say, 50 yards. This distance the sprinter would cover at a rate not exceeding 20 mph. The usual leisurely walking speed in gorillas is only about $\frac{1}{3}$ mile an hour. When chased, a troop will move away at about 5 mph.

As to the distances covered by gorilla troops during their wanderings, the extent is more or less governed by the availability of food, which generally is present in abundance and which grows at a rapid rate. The usual areas roamed by most individual troops are from 10 to 15 square miles in extent, although this coverage may include a certain amount of overlapping of the home ranges of other troops. Generally, however, each troop observes the bounds of its own range or "territory," although this is not defended, and two different troops sometimes even merge and thereafter wander together. Such merging may account for the larger numbers (up to 30) of individuals sometimes seen living together.

Behavior, Temperament, and Expression (in the wild)

The behavior of gorillas is as variable as that of humans. If left undisturbed, and in a placid, contented frame of mind, these apes are by nature

---

* These average figures are taken from an extensive correlation (unpublished) made by the writer of the external dimensions and proportions of all parts of the physique in men and women, based on statistics of white American adults issued in U.S. government studies and other publications on anthropometry.

nonaggressive, even among themselves. But in the face of a threatening enemy (mainly man), a "boss" male gorilla, or leader of a family band, can become a fearsome, pugnacious, and formidable antagonist. Too, just as among humans, an outcast, hounded, or psychopathic gorilla (usually an old male) can become a vengeful, aggressive "killer." This range of temperament in the gorilla may explain the differing reactions of observers of wild individuals of this species, which were encountered under totally different conditions. Thus, to the young hunter Paul du Chaillu—the first white man to face and shoot an aroused male gorilla in its native jungle—the agitated animal (which, after all, sought only to scare the intruder away) became a beast with "fiercely-glaring" eyes, and "a hellish expression of face, which seemed to me some nightmare vision." (See chapter 3.)

In utter contrast to this disturbing picture is that presented by Dian Fossey, a lone woman scientist studying mountain gorillas in their natural haunts, who after months of patient observation and restraint had the unique experience of a subadult male gorilla coming close and gently touching her outstretched hand. Miss Fossey found that much better "relations" with the gorillas developed if she imitated (so as to return) their signs and actions rather than just sat still and stared at the animals, which then might become suspicious. This was true not only of physical actions but of sounds, or vocalizations, which she learned to imitate and to which the apes would respond. She says, "After more than 2,000 hours of direct observation, I can account for less than five minutes of what might be called 'aggressive' behavior."

Dian Fossey has identified some seventeen sounds (hoots, barks, grunts, roars, screams, "wraaghs", etc.) made by mountain (Mikeno) gorillas to express various feelings. A young male she named "Augustus" is the only gorilla she has seen clap his hands (not his chest). She remarks also that she has watched mountain gorillas search in tree trunks for worms or beetles, which is another observation disproving the assertion generally made that gorillas in the wild are total vegetarians.[16]

It is certain that apes, including gorillas, are able to communicate with each other by means of their various vocalizations. Of course, when a "boss" gorilla issues commands, or orders, he usually accompanies them by actions as well, and the subordinate members of his troop are influenced by these actions as well as by his verbal utterances. Yet, apes can readily learn to associate sounds, or human words, without appropriate accompanying actions. An amusing example of this was where a certain visiting zoologist "asked" a trained female gorilla to bring him a mug. The ape brought him about everything else in the room, until the keeper came in and explained that the gorilla had been taught the word *cup,* not *mug!*

Mating, Reproduction, and Infant Care

Sexual activity in the gorilla, in contrast to that in most other apes and monkeys, plays a minor part. The leader of a troop may tolerate another adult male courting one of his several females only a few yards away from him, and copulation may take place between them. Schaller on two separate occasions was able to observe copulation in wild gorillas, and in both cases the male partner was not the leader of the troop. In one case the position taken by the female was to kneel on the ground, supported by her elbows. The male stood behind her and grasped her hips. In the second case the male took a sitting position, holding the female on his lap. While mating gorillas can be very noisy during the sexual act, the other members of the troop pay no attention to them.

In gorilla females the menstrual cycle averages 30 to 31 days (as compared with 28 days in humans), and in the wild state these apes bear a young one every 3½ to 4½ years. Almost fifty percent of the offspring die as infants or juveniles. Within the usual troop there are twice as many females as males. If one takes into account the "bachelors," or the gorilla males living in pairs, the sex ratio within the entire population is probably three females to two males. Also, it is probable that the gorilla males have a higher death rate. Among gorillas in captivity, where the mother ape does not immediately take over the job of nursing her newborn, the baby ape is raised in a primate nursery by trained zoo personnel. In such instances the mother gorilla can again become pregnant within a much shorter time than if she were in the wild and had to "bring up" her offspring over a period of many months. This situation, fortunately, favors the raising of gorillas in captivity, which results not only in a larger population of these comparatively rare and costly animals, but also in a reduction of infant mortality among them.

There is also, however, a possible drawback to the nursery-raising of newborn gorillas that have been rejected by their mothers. Such rejection is said to occur in about 90 percent of all gorillas

Fig. 104. Appearance of San Diego Zoo's female gorilla "Vila" two weeks before the birth, on 3 June 1965, of her first offspring: "Alvila," a female which weighed 4 pounds, 11½ ounces. It was the eleventh gorilla born in captivity anywhere in the world. (Photo courtesy San Diego Zoo.)

born in captivity. The result of nursery-raising is that the mother gorilla is deprived of the experience of caring for her offspring, and the infant likewise of learning from its mother. Such a result, if repeated in successive generations, might conceivably render captive gorilla mothers (1) less reproductive, and (2) less capable of caring for their offspring in case they do reproduce. Too, the young gorillas would thus be deprived of the learning they would receive from association with older gorillas. For this reason, in modern zoos every effort is made to help gorilla mothers—especially primiparous mothers—to learn the requirements of caring for their newborn, and generally "unfamiliar," offspring.

A pregnant state in a female gorilla is often difficult to discern, since even when nonpregnant the paunches of these herbivorous animals are of barrellike proportions. As is mentioned in chapter 5, the gestation period in the gorilla in captivity averages 265 days (8¾ months), ranging usually from 236 days to 290 days (c. 7¾ to 9½ months). Although this period is practically as long as in man, a newborn male gorilla weighs on the average only 2.18 kg (4.8 lbs.), as compared with 3.50 kg (7.7 lbs.) in white American newborns. It is evident, therefore, that a human being *in utero* grows about sixty percent more rapidly than does a gorilla, notwithstanding that postnatally this ratio is reversed, the infant gorilla growing much more rapidly than the human. Surely these completely different sequences in the growth of the gorilla and of man must constitute one of the most significant differential criteria in the ontogeny of these two primates.

Bernhard Grzimek, who as director for many years of the zoo in Frankfurt witnessed a number of births of gorillas, remarks:

> During pregnancy, some gorilla females have swollen ankles at times. During birth, which lasts only a few minutes, the female lies down. Subsequently the female severs the umbilical cord and holds the infant tightly to her chest. In contrast to young of many lower monkeys, the gorilla baby cannot hold onto the mother on its own. Youngsters are born throughout the year; there is no specific breeding season. . . . Gorilla young develop approximately twice as fast as human infants. Within six months they have become active, cheerful youngsters. In the Kabara [mountain gorilla] region the infants started to feed on plant material around 2½ months of age, and it seemed obvious that this was the main source of food at around 6 to 7

Fig. 105. A subadult male lowland gorilla temporarily tranquilized by a shot of *sernylan*. The shot was administered with a dart from a "Capchur" pistol by Deets Pickett, D.V.M., who is holding up the ape's lower jaw. The ape, which was obtained in the central Cameroons, upon regaining consciousness, was released into the jungle. (Photo courtesy Dr. Pickett, a supplier to zoos and circuses of gorillas and chimpanzees.)

months. Nevertheless, some still nursed occasionally from the mother at 1½ years of age. The young are able to crawl by 3 months. By 4½ months they walk very well on all fours, and by 6 to 7 months they are able to climb.

Gorilla mothers which have young of their own are not unfriendly towards others. It can happen that a strange young will come, sit next to the young on the mother's lap, and press itself against her chest without being pushed away. Even the strong males tolerate youngsters playing around them, or permit them to climb up on themselves.

It was observed in zoos that gorilla females which bore young for the first time did not know what to do with it, just like women inexperienced with infants do not know how to handle them. The females are afraid of the infant and are also upset by the birth process. In the end they may show hostility towards the newborn infants.[17]

Dr. Grzimek, in his *Animal Life Encyclopedia* (p. 538), gives details of the birth of gorilla twins —the first known instance either in captivity or in the wild.

Some further details of sexual behavior in gorillas in captivity—which behavior is probably closely similar to that in the wild—are given in chapter 10. Dr. Geoffrey H. Bourne, Director of the Yerkes' Primate Center in Atlanta, Georgia, in his book *The Gentle Giants*, devotes an entire chapter to the intricacies of sexual conduct in gorillas.[18]

Although the time may not be far off when there will be no need to capture and send to zoos gorillas from their native haunts, that time has not yet arrived, and many a zoo is willing to pay anywhere from three thousand to five thousand dollars for a healthy young specimen—preferably a male who may become a fruitful sire. Today, too, there is less difficulty in procuring gorillas, especially large and powerful specimens, than there was before the introduction and use of "capture-guns," which, from a distance, can inoculate an animal with a quick-acting sedative or "tranquilizer." This essentially humane way of temporarily immobilizing a large, powerful, and probably panic-stricken animal is likewise far more efficient than earlier methods in which one or more adult gorillas had to be killed in order to obtain a single young or infant specimen, which in most cases did not live long enough to reach a zoo.

Here is a description of the latter method of capture, as related in the October 1930 issue of *Popular Mechanics*, p. 646:

> A mother with a baby is singled out of a herd [troop!] and followed from a distance for days. Finally the mother is wounded by a poison arrow shot by a native—care being taken that she is not instantly killed, but merely wounded. The herd is then followed, and the mother becoming weaker from her wound, falls farther behind until finally she drops from exhaustion and the baby is then taken from her. Now the real troubles of the animal collector begin, because, like as not, the baby will refuse to eat and will die. No full-grown gorilla has ever been raised, or captured alive.

Today's use of the tranquilizing technique in the capture of gorillas is certainly a vast improvement on the foregoing dismal account of how such capture was effected in 1930. In the latter instance the mother ape must surely have been killed; and since her body was permeated with poison from the native's arrow, it could not be used for food by the tribesmen who so relished gorilla meat! What a waste of time and effort!

A means of capture used later by some gorilla hunters was to dig pits about fifteen feet deep and overlay them with a layer of typical vegetation from the area. A troop of gorillas was then driven by "beaters" (several hundred natives making loud noises) in the direction of the pits. Such a drive could take several days in order to direct the apes unerringly toward the pits. Paths were cut leading into the pits, which some of the gorillas, in their state of alarm and confusion, would follow and so fall into one of the pits. The specimens desired were then forced into strong cages constructed from saplings and tough vines. A long bamboo pole was lowered from the leaf-covered cage into the pit, and as one gorilla after another would climb this pole and enter the cage, natives would slide the bottom (floor) saplings into place and fasten them securely. A final step was to separate young gorillas from their mothers and place them in other cages.

Nowadays, although "beaters" may still be employed to maneuver one or more gorillas into a position where a tranquilizer gun can be used on them, the formerly arduous, and often dangerous, task of getting a gorilla into a cage without injury to either the ape, the hunter, or the native helpers is greatly facilitated.

Notes

1. George B. Schaller, *The Mountain Gorilla: Ecology and Behavior* (Chicago: University of Chicago Press, 1963); idem, *The Year of the Gorilla* (Chicago: University of Chicago Press, 1964); idem, *The Behavior of the Mountain Gorilla*, in I. De Vore, ed., *Primate Behavior* (New York: Holt, Rinehart and Winston, 1965), 1:324–67.
2. Bernhard Grzimek, "The Gorilla," chap. 22 in *Grzimek's Animal Life Encyclopedia*, 13 vols. (New York: Van Nostrand-Reinhold, 1972), 10:525–48.
3. Julius Falkenstein, in *Loango-Expedition* (Leipzig, 1879), p. 51.
4. Edward Reichenow, *Biologische Beobachtungen an Gorilla . . .* (Berlin, 1920); English translation in Robert M. and Ada W. Yerkes, *The Great Apes* (New Haven, Conn.: Yale University Press, 1929), p. 418.
5. R. L. Garner, *Gorillas and Chimpanzees* (London, 1896), pp. 229–30.
6. T. Alexander Barns, "Hunting the Morose Gorilla," *Asia* (New York), February 1928, pp. 116–23, 154–56.
7. R. Akroyd, "The British Museum (Natural History) Expedition to the Birunga Volcanoes, 1933–4," *Proceedings of the Linnean Society of London*, 1934–35, pt. 1, pp. 17–21.
8. Clyde Jones and Jorge Sabater Pi, *Comparative Ecology of "Gorilla gorilla" and "Pan troglodytes" in Rio Mundi, West Africa* (Basel: S. Karger AG, 1971). See also Harold C. Bingham, "Gorillas in a Native Habitat," *Carnegie Institute of Washington Publication No. 426*, August 1932.
9. Ivan T. Sanderson, *Living Mammals of the World* (New York: Doubleday & Co., 1958), p. 102.
10. Paul A. Zahl, "Face to Face with Gorillas in Central Africa," *National Geographic* 17, no. 1 (January 1960):126.
11. Harold J. Coolidge, Jr., "Zoological Results of the George Vanderbuilt African Expedition of 1934. Part IV, Notes on Four Gorillas from the Sangha River Region," *Proceedings of the Academy of Natural Sciences of Philadelphia* 88: 479–501.
12. Adolph H. Schultz, "Some Distinguishing Characters of the Mountain Gorilla," *Journal of Mammalogy* 15, no. 1 (1934): 51–61 (see p. 58).
13. C. R. S. Pitman, "The Gorillas of the Kayonsa Region, Western Kigesi, Southwest Uganda," *Smithsonian Institution Report for 1936* (Washington, D.C.), pp. 253–75 (reprinted from *Proceedings of the Zoological Society of London*, 1935, pt. 3, pp. 477–94).
14. Francis E. Randall, "The Skeletal and Dental Development and Variability of the Gorilla," *Human Biology* 16, no. 1

(February 1944): esp. 49, 50, 55, 72, 74.
15. Mildred Trotter and Goldine C. Gleser, "A Re-evaluation of Estimation of Stature Based on Measurements of Stature taken during Life and of Long Bones after Death," *American Journal of Physical Anthropology* 16, no. 1 (March 1958):79–124.
16. Dian Fossey, "Making Friends with Mountain Gorillas," *National Geographic* 137, no. 1 (January 1970):48–67; idem, "More Years with Mountain Gorillas," ibid. 140, no. 4 (October 1971):574–85.
17. Grzimek, "The Gorilla," p. 537.
18. Geoffrey H. Bourne and M. Cohen, *The Gentle Giants* (New York: G. P. Putnam's Sons, 1975).

10

Some Distinguished Captive Gorillas

NOTE: This chapter is included here not only to record a few noteworthy examples of captive gorillas, but also to throw light on the habits and behavior of zoo gorillas in general by furnishing biographical sketches of particular individuals on which detailed observations were made. Many more examples could be cited; but the diversity exhibited by those chosen here should give a good idea of the never-a-dull-moment lives of gorillas in captivity, not only in the United States, but in a few instances in Europe as well.

Before presenting brief biographies of gorillas that lived in captivity for appreciable lengths of time, it may be appropriate to mention some of the earlier specimens whose tenures in captivity were of lesser, and in most cases regrettably short, duration. This was due primarily to insufficient knowledge on the part of earlier keepers as to how to raise these highly emotional, affection-requiring primates, although in some cases the change in climate from that in their homeland may also have been a contributing cause of dejection and disease.

The first living gorilla to be seen in Europe was a young female (probably not over two years of age), which was exhibited in the winter of 1855–56 in Liverpool and a few adjacent towns by a Mrs. Wombwell, who was the proprietor of a traveling menagerie. The gorilla, which attracted much attention during its brief appearance before the public, unfortunately died after only a few months in captivity, in March 1856. Its skin was preserved in a collection owned by a Mr. Waterton, of Walton Hall; and its skeleton was sent to the Natural History Museum in Leeds. This young gorilla, which was at first thought to be a chimpanzee (possibly because of its small size), was only later recognized for the then almost unknown species of ape it turned out to be.

It was twenty years later before another living young gorilla was brought to Europe. It arrived at Liverpool, in mid-1876, with members of a German expedition that had carried on explorations in Central Congo (now Zaire) since 1873 and were now headed for home. The young gorilla, a male about two years of age, was first known as the "Falkenstein gorilla," from the name of the expedition's physician, who had obtained the animal from a native hunter in the Congo. Upon arriving with the young ape in Germany, Dr. Falkenstein sold it to the Berlin Aquarium for the then munificent sum of 20,000 marks (about $5000), the sum being used to augment funds that had been used by the expedition. At the aquarium, the young gorilla was named "Pongo" (see Fig. 20). He made an instant hit with all who saw him; but like the Wombwell gorilla, he survived captivity only a short time (until 13 November 1877), after having been cared for during nine months in Africa and fifteen months in Berlin. (An extended account of this gorilla, from observations made while it was "living" at the Berlin Aquarium, is given in chapter 3.)

In 1883 another young gorilla was brought to the Berlin Aquarium by the German naturalist Dr. Eduard Pechuel-Loesche. Again, however, the ape did not live long under the conditions customary in zoos of that period. He was given every care then known, by Director Hermes of the Aquarium, but died after fourteen months in captivity "of the same disease [which was not stated] as the first gorilla." Evidently in those days it was not realized how highly susceptible the manlike apes—gorillas in particular—are to various communicable human diseases, and how quickly they decline if not given constant companionship and affection.

Between the years 1887 and 1908, seven young gorillas—two males and five females—all presumably of the western lowland species, were exhibited in the Garden of the Zoological Society of London. None survived for more than a few months. These gorillas, plus seven others that were exhibited either in London or on the Continent at various times during the years 1897 to 1925, are listed in chapter 3.

The first living gorilla to be brought to the United States (Boston) was a young (probably less than a year old) unnamed male that survived for only five days—from 2 May to 7 May 1897.

It was fourteen years later before the second living gorilla to be brought to this country arrived (on 23 September 1911) at the New York Zoo. It was a female, between two and three years of age, and was named Madame Ningo. In the zoo it lived only twelve days.

Some of the many well-known gorillas to have lived, or even thrived, in captivity for appreciable lengths of time, mainly in this country, either privately owned or in a circus or a zoo, may now be briefly reviewed. While the listing of these individual animals is roughly in chronological order, there may be some overlapping of dates.

* * *

1. *Dinah*

Although this young female gorilla was an exception to the others listed below, in that she did not live an "appreciable" length of time in captivity, she nevertheless was notable by far outliving the gorillas previously imported to the United States. Too, various items of information about her species were gained at the Bronx Zoo, where she was a star attraction from 21 August 1914 to 31 July 1915—a period of 11 months, 10 days. Upon arrival at the zoo, Dinah was estimated to be about 3 years of age. Her weight, as of 1 September was 40½ pounds (which was quite light for her estimated age). Her standing height was 42 inches and her span, or horizontal stretch of arms, 50½ inches. According to Dr. Hornaday, then director of the zoo, at 8:00 A.M. the young gorilla was given a raw egg beaten up in milk. At 10:30 A.M. her first full meal of the day consisted mostly of fruits: apples, oranges, bananas, pears, or grapes, along with some crackers and water. At 1:30 P.M. came a dinner brought hot from a nearby restaurant: a meat dish (roast beef, broiled chicken, or lamb), with gravy, mashed potatoes, and bread. She would not eat ordinary vegetables! She scorned boiled potatoes, squash, beans, and sprouts. Thus it would appear that this young ape preferred a humanlike diet! Evidently this "civilized food" had an adverse effect on her health, for her death was attributed to "starvation and malnutrition, complicated with rickets and locomotor ataxia."[1]

2. *John Daniel*

Even as late as the 1920s, a living, captive gorilla was a rarity and a novelty, and in the few places where one could be seen there was certain to be a group of onlookers. That was the situation in March 1921, when a young male gorilla, John Daniel, was placed on exhibition in the Ringling Brothers' Circus in Madison Square Garden, New York. John Daniel was then nearly 4½ years of age. Previously, he had been owned by Miss Alyse Cunningham, of London, who had purchased him when, in April 1920, he was 3½ years old and had been in England six months. As the natural history writer Henry Sheak put it:

> ... under (Miss Cunningham's) tuition he made extraordinary progress. After about two months it was possible to give him the freedom of the house. He had his place at the table, opened doors by turning the knob, and unbolted windows, raised them, lowered them again, and locked them, turned on the lights when entering a dark room, sponged himself when bathing, and adapted himself in many other ways to his urban environment. He became deeply attached

Fig. 106. Dinah, shown here at the age of about 3½ years, was the third gorilla to be brought successfully to the United States (New York Zoo). One or two others died enroute from Europe. Dinah arrived on 21 August 1914 but lived in captivity for only eleven months. (Photo courtesy New York Zoological Park.)

to Miss Cunningham, and when later it was found necessary to sell him and he was sent to New York, he became ill from homesickness and died before Miss Cunningham, who was summoned by cable, had time to reach him.[2]

That this young gorilla was apparently capable of original, constructive "thinking" was shown by the following incident, which occurred while he was still in the custody of Miss Cunningham in her West End London home on Sloane Street. One day, Miss Cunningham was dressed to go out with a number of her relatives and sat down for a moment. John Daniel came to her and wanted to sit in her lap. "Don't let him!," called out one of the

Fig. 107. John Daniel was a much publicized young gorilla in his day (1916–21), probably because he was one of the very few to survive in captivity for any appreciable amount of time. He finally succumbed, however, from loneliness for his former mistress, Alyse Cunningham, just as she was rushing from London to visit him in the Ringling Brothers Circus in New York. (Photo courtesy New York Zoological Society.)

Fig. 108. Miss Alyse Cunningham with John Daniel II. This young gorilla, which Miss Cunningham acquired in 1924, lived with her in London until 1927, when he died at the age of about 7½ years. He never had the publicity nor gained the fame that attended his predecessor, John Daniel I.

other women present, "he'll spoil your gown." When Miss Cunningham said "No," John lay on the floor and, for a brief spell, cried like a spoiled child. Then he stopped, looked about, rose, and found a newspaper. He carried it to Miss Cunningham, spread it on her lap, climbed up, and contentedly sat down. "Even those who saw it said they would not have believed it, had they not seen it themselves," said Miss Cunningham. "Both my nephews, Major Penny and E. C. Penny, and Mrs. Penny were in the room and can testify to the actual occurrence."

While, for a brief month, John Daniel was with the Ringling Brothers' Circus, his diet certainly was not one to be recommended for present-day juvenile zoo gorillas. For one thing, it was considered that in order to prevent melancholia John Daniel should be given at least three good drinks of whiskey a day and sometimes a glass of sherry or port. At noon he was given a half-pound of cooked beefsteak, and he also liked to chew on a piece of smoked fish. He was fond of ice cream; and whenever his keeper felt that John should have some cod-liver oil, the only way the ape could be induced to take the oil was by mixing it in a bowl with some ice cream. Before going to bed, John would be given some more milk (at breakfast he was given a quart) and some fruit. He would then roll himself in a blanket and get into his crib like a small child. At 4½ years, however, John Daniel was far heavier than a boy of that age and weighed 184 pounds at a height of only 4 feet 4 inches. These proportions gave him a build almost as thickset as that of a full-grown male gorilla. Indeed, his combination of height and weight would ordinarily be expected in a male gorilla about 7 years of age rather than only 4½ years (see Table 8). So John Daniel's early demise could not have been due to the common gorilla ailment, malnutrition. Rather, he became so depressed and dispirited by the absence of Miss Cunningham that he just pined away.

3. Congo

This female gorilla is notable for having been the first example of the mountain species (*Gorilla beringei*) brought into the United States. She had been captured in the Virunga Volcanoes area by the gorilla collector Ben Burbridge, and arrived with him in New York City either at the end of 1925 or in January 1926. Later she was taken to the Yerkes Primates Laboratory in Orange Park, Florida, for study. Dr. Yerkes took some body measurements of her on 15 January 1926, at which time she was estimated to be 6¼ years of age. Congo must have been sadly malnourished, since she weighed only 65 pounds. She made remarkable gains, however, and a year later weighed 128 pounds. On 23 April 1928, Congo died, at the closely estimated age of 8½ years. Her weight prior to death had been about 165 pounds. Her preserved body was measured by Dr. Adolph Schultz, who among other measurements found her anterior trunk height (from pubis to suprasternale) to be 523 mm, or 20.6 inches. This was 40.2 percent of her standing height, as compared with 36.3 percent in a typical lowland female gorilla of the same age. Her trunk height thus exhibited one of the features, or body proportions, that differentiate the typical (Kivu) mountain gorilla from the shorter-trunked lowland species, whether the latter is either of the western or the eastern race.

Dr. Yerkes tested Congo's "intelligence" with the conventional stick-and-food technique so quickly understood by chimpanzees, but the gorilla was unable to grasp the idea of using the stick to reach the food. However, in a test that involved memory —as to which of several cans of differing colors contained food—Congo did better than a chimpanzee. Yerkes' conclusion was that the gorilla is less adapted by nature for manipulating an imple-

Finally, during recent years, there were at least two male gorillas named Congo—one in the zoo in Honolulu, and the other in Bristol, England.

4. *Bobby*

At the zoo in Berlin, during the years 1926 to 1935, Bobby, a male gorilla from the Cameroons in west Africa, was the chief attraction and a prized possession, since at that time he may have been the only living gorilla in Europe. Bobby had been captured when he was about a year old; and after being in captivity for several months in Africa, he was brought to Berlin. Despite being raised in the zoo on a diet that was essentially one for humans rather than apes, and being given such tidbits as beefsteak, sausage, weiners, cheese, and beer (!), Bobby did fairly well in captivity, reaching in

Fig. 109. "Congo," at the age of seven years, giving a chest-beating demonstration, presumably at the home of her captor, Ben Burbridge, in Jacksonville, Florida. Congo was the first gorilla of the mountain species to be brought into the United States. This photo appeared in connection with an article about Congo in the August 1926 issue of *Nature* magazine, by Juanita Cassil Burbridge.

ment, such as a stick, than either the chimpanzee or the orangutan—both of which are expert "throwers." Hence, a gorilla's intelligence cannot be evaluated by expecting it to stack boxes or use poles to obtain food, the way chimps and orangs are able to do.

It may be added that in 1936 a lowland female gorilla, which was named Congo II, or Miss Congo, was acquired by the Chicago (Brookfield) Zoological Park, and resided there for 13¼ years.*

* Evidently there was a third female gorilla named Congo, which appeared between the times of Burbridge's mountain Congo and the lowland Congo II. This intermediate Congo came to Germany in 1927. Subsequently she was exhibited in Hamburg and Paris. Her trip to the United States in 1929 was the most unusual ever accorded a nonhuman passenger. Her transportation was booked, for one thousand dollars, aboard the dirigible *Graf Zeppelin,* and she was one of the passengers on the big airship's maiden voyage. (I have just learned that this gorilla, upon her arrival in the U.S., was renamed "Susie" [see No. 8, following.])

Fig. 110. Not all zoo gorillas, especially dominant males, are friendly, loving creatures. "Bobby," a featured attraction at the Berlin Zoo during 1926 to 1935, was a cranky, intolerant specimen of his species. He received a great deal of publicity, partly because he, for awhile, was the only living gorilla in Europe. In this photo, though not fully grown, Bobby weighed over five hundred pounds.

August 1935 the age of 10¼ years, at which time he died from appendicitis (which was hardly to be wondered at!). Bobby's food consumption was enormous, and in less than seven years his weight ballooned from 60 pounds to 595. The latter was close to a record poundage for a lowland gorilla, although a great deal of Bobby's bulk was excess fat. He also developed a rather nasty temper, especially where photographers were concerned: whenever one appeared he would either throw a tantrum or hide his head under a piece of sacking that he carried around for the purpose. Eventually even his keeper was subject to Bobby's sudden, impulsive attacks, but put a stop to them by using an electric, shock-producing club, which Bobby came to respect after having received a single jolt from it. After his death, Bobby's skeleton showed that he would probably have been completely grown, or mature, had he lived to the age of eleven years. As it was, he died at an age that in a human being would, on an age-equivalent basis, have been only 15½ years (see Table 20).

5. *Bamboo*

One way of becoming celebrated is to set a record for longevity—and that is what Bamboo of the Philadelphia Zoo managed to do—although his achievement was certainly aided by excellent care, including a diet that was ample nutritionally without being excessive in quantity. And today his former co-captive, Massa (see No. 9, following), is profiting by the same level of care that made Bamboo the first gorilla in captivity to surpass 30 years (or, for that matter, even the 10¼ years attained by the Berlin Zoo's Bobby, as just related).

When, on 5 August 1927, at the age of approximately 11 months, Bamboo was brought to the Philadelphia Zoo, he is said to have weighed only 11 pounds and was carried comfortably in a suitcase. If actually only 11 pounds at that age, he was truly emaciated (see Fig. 42).* With Bamboo, in another such traveling compartment, was a young chimpanzee who became his playmate for some months and served to keep him alert and physically fit. Since Bamboo was the first gorilla to survive in the United States for an appreciable length of time, it was possible to weigh and measure him on at least two occasions before he grew too large to handle. However, it is evident that for his age Bamboo was a rather small gorilla, since at 38 months of age he weighed only 51 pounds and at death only 281 pounds.[3] Although there is in most gorillas prior to death a marked dropping off in body weight, it is nevertheless difficult to believe that at the age of 13½ years Bamboo weighed as much as 435 pounds, as was asserted.

While it was remarked earlier, in connection with Congo (No. 3, above), that gorillas in general are not naturally given to throwing things, Bamboo was a contradiction of this generality. After years of practice, he developed a great proficiency in hurling objects with an underhand pitch. At this point I quote Zoo Director Roger Conant:

> Usually he confines himself to throwing handfuls of gravel, but occasionally he tosses a banana or a tomato with disastrous results. A wire fence around his cage helps to protect visitors but news photographers, who must go inside the barrier and for whom he appears to entertain a special dislike, are out of luck unless they are adept at dodging. On his last birthday he scored a record by splashing five of them with a single, well-aimed watermelon.[4]

Fig. 111. A close-up shot of "Bamboo" as a young adult of fifteen years. This famous resident of the Philadelphia Zoo went on to establish, in 1961, a then-record for gorilla longevity in captivity by reaching 34½ years—an age that was equivalent to sixty-six or sixty-seven years in a human (see Table 20). (Photo courtesy Philadelphia Zoological Society.)

---

* On or about 12 November 1927, at which date Bamboo was estimated to be fourteen months old, Dr. Adolph Schultz measured the young gorilla and found its weight to be 7.9 kg, or 17.4 pounds. At 23¼ months the weight was 29.5 pounds; and at 38 months, 51 pounds. On this basis of growth, Bamboo on arrival at the zoo should have weighed about 12¼ pounds, which would still have been very light for his age.

After Bamboo had been in captivity alone for over three years and was then just over nine years of age, a young male named Massa, aged five years, was brought in as a possible companion. Although Massa was much smaller than Bamboo, he tried to dominate him. Bamboo endured this for awhile but finally retaliated by giving Massa a cuff on the head. After that, the two males were put in separate living quarters, which each then occupied for over 25 years. Since Bamboo was born presumably early in September 1926, his death on 21 January 1961 gave him a record age of 34 years, $4\frac{1}{2}$ months. Of this time, 33 years and $5\frac{1}{2}$ months were spent in captivity. (A good general account of Bamboo was given in the quarterly magazine *Fauna,* of the Philadelphia Zoological Society, Vol. 1, No. 1, March 1939, pp. 7–9.)

6. *N'Gi*

This brief name, which is a Swahili native term meaning "wild man of the forest," seems rather inappropriate for the meek little gorilla that was given the name when he arrived from west Africa at the National Zoological Park (Washington, D.C.) on 5 December 1928. He was estimated, on the basis of his size, and particularly of his dentition, to have then been about 2 years and 3 months of age. Apparently, like many other imported infant gorillas of those days, N'Gi was undersized for his age, since on the basis of his weight, which was only 30 pounds, an average-sized male lowland gorilla would have been only 21 or 22 months of age, rather than N'Gi's 27 months. But there is a wide range in the size of individual gorillas at all stages of growth, particularly in newly imported infants or juveniles in which a regular pattern of daily living has not yet been established.

N'Gi had been captured on 17 January 1928 by J.L. Buck of Camden, New Jersey, and West Africa. Mr. Buck at the time commuted between his two far-apart homes and, during nine such trips, captured four young gorillas, two of which—N'Gi, and Bushman (No. 7, following)—he succeeded in bringing to the United States. The infant gorilla N'Gi was caught in the manner usually employed in those days, namely, by having a native shoot the mother ape with a poisoned arrow. Mr. Buck took charge of N'Gi for nearly a year and then decided to sell him to Madame Abreu, of Havana, Cuba, who had a noteworthy collection of captive anthropoid apes. On the way, Buck and his young charge stopped between trains in Washington. Taking N'Gi to the National Zoological Park, Buck showed the cute little gorilla to the director,

Dr. William M. Mann, who was captivated by the young ape's antics and "personality" and bought him on the spot.

N'Gi similarly delighted the thousands of visitors who came to the zoo during the years 1929 to 1932. However, during the early months of 1932 the little ape contracted bronchial pneumonia and on 10 March passed away, despite the efforts of a team of medical specialists who fought to save his life. N'Gi at the time was only $5\frac{1}{2}$ years old. Thus, the current gorilla population in the United States was reduced from eight to seven individuals. (A detailed account of N'Gi's life at the National Zoo is given by Dr. Mann in his book *Wild Animals In and Out of the Zoo.*)[5]

7. *Bushman*

Probably the best known and most widely publicized zoo gorilla in the United States—at least during the period 1930–1950—was "Bushman," of the Lincoln Park Zoo in Chicago. It was estimated that each year about 3,500,000 people came to see him—nearly 10,000 people every day! Everything the big gorilla did, or had happen to him, was an item for the newspapers. And when he died, on the last day of December 1950, it was practically a day of mourning for the entire population of Chicago; for the residents of this metropolis had come to regard Bushman as a civic celebrity on a par with any human fellowman. In 1946, when Bushman was 18, he was voted the most outstanding and most valuable (100,000 dollars) single animal of any zoo in the world by a contemporary meeting in St. Louis of the American Association of Zoological Parks and Aquariums.

Bushman had been acquired by the zoo in 1930 when he was an estimated two years of age. He was a western lowland gorilla and had been captured with a net by the animal collector J.L. Buck, "above Moyene, in the French Congo," when he was about a year old. For the following year, missionaries in Africa employed a native woman to nurse the infant ape, since Buck considered him too young and frail to withstand the long ocean voyage. Thus, Bushman was about two years old when Buck returned and brought him to America. And for the modest sum of 3,600 dollars, he disposed of the healthy young ape to the Lincoln Park Zoo. Bushman's weight at that time—38 pounds—was exactly average for a two-year-old lowland male gorilla, which made it evident that he was not underweight from either poor feeding or depression, as so many young captured gorillas had been before him. When, in 1946, Bushman

Fig. 112. These delightful studies of the young gorilla N'Gi were drawn in 1929 by artist Benson B. Moore at the National Zoological Park in Washington, D.C., where the baby ape was the most popular of all the attractions. For some details of N'Gi's history, see text.

was 18 years of age and at the peak of his popularity, the following information about him was issued in a bulletin from the Chicago Bureau of *Wide World Photos*:

> Gentle, playful and affectionate [upon his arrival at the Zoo as a two-year-old], he was turned over to his keeper, Edgar Robinson, who since has been his constant nurse, friend, and trainer. Robinson gave the gorilla a warm shower each morning. A vegetable, fruit and milk diet was served three times a day, with a rest after meals. He exercised on the zoo lawn, played with a football, and wrestled with his keeper.
>
> Robinson continued to take the gorilla outdoors until he weighed 150 pounds, but he ended personal contact when Bushman tipped the scales at 180 pounds. The keeper had been going into the steel-barred cage to pet and feed Bushman. One day the gorilla protested as Robinson prepared to leave. After some persuasion and trickery, the keeper got outside, and has stayed out since.
>
> Once in a while Bushman will get into a huff and pout because Robinson speaks first to a chimpanzee neighbor. Bushman retaliates by going on a hunger strike for three or four days. The sight of a capturing net upsets him. His tantrums are accepted as typical gorilla behavior.
>
> Marlin Perkins, Director of the Zoo, says that Bushman eats and drinks about 20 pounds of food a day—whole wheat bread, a large variety of vegetables and fruits, and two gallons of milk.
>
> Bushman readily obeys the commands of his keeper; a stranger becomes a friend or enemy at the first meeting. He plays with an old automobile tire casing, and chews six or seven multi-vitamin tablets a day.

Notwithstanding all this attention and care, and the probability that for some years of his adult life he was one of the healthiest gorillas ever seen in captivity, Bushman from the age of about 16 years on developed what was called a "heart ailment." While heart failure was probably the immediate cause of his death, an autopsy revealed that Bushman had died, at the age of 22 "after a remittent illness lasting about 6 years of what is believed to be a deficiency, chiefly of Vitamin B Complex."[6] His weight at death was about 430 pounds—120 pounds under his "normal" weight. The only "major" anatomical difference from man, according to the autopsy report, was the presence of laryngeal air pouches.* Bushman's heart, which

* These pouches are evidently the means by which gorillas, particularly adult males, attain additional volume and resonance in their chest-beating sounds.

Fig. 113. Bushman, of the Lincoln Park Zoo, Chicago, as he appeared in 1946, at the age of eighteen years. At that time he was voted the most valuable single animal of any zoo in the world; value $100,000. (Wide World Photo; courtesy Lincoln Park Zoo.)

weighed 805 grams, was "probably hypertrophied," while his liver and sex organs were "probably atrophied."

While during his prime Bushman had been sought by other zoo directors as a sire, Marlin Perkins, who was then director of the Lincoln Park Zoo, refused to let Bushman be "borrowed" for this purpose, since he considered the big male gorilla too valuable a zoo animal to be subjected to the possible wrath of an antagonistic female of his species.

While Bushman was a very large example of a lowland male gorilla, he was, in my opinion, considerably short of the 6 feet 2 inches standing height generally attributed to him. 5 feet 8 inches would be a more reasonable estimate. However, his actual height during life may never be known, since the present whereabouts of his skeleton—the one reliable means of determining this measure-

Fig. 114. Susie, a trained gorilla formerly of the Cincinnati Zoo, using (?) a toothbrush. Susie, who was remarkably tame, was the only full-grown gorilla that permitted her keeper to enter her cage periodically and measure her. Thus, there is a valuable record of her physical dimensions (see series in Table 7).

ment—appears to be unknown. I can only hope that his bones may someday be located!*

8. *Susie*

This charming lady gorilla of the Cincinnati Zoo—where she arrived on 11 June 1931, at the estimated age of about 5 years—was unique among her species, in that as she approached and reached maturity, she was the only gorilla of such age and size to allow her trainer to go into her cage and periodically take her measurements.** Her trainer,

---

* Although Bushman's skeleton may be hidden away in some forgotten location, his mounted skin is on display in a glass case in Chicago's Field Museum of Natural History. To protect the specimen, a liquid poison (carbon tetrachloride) is periodically poured into a container in the top of the exhibit case, where it evaporates and fills the interior space with poison gas. By this means the irreplaceable remains of Bushman are preserved from attack by insects that otherwise would quickly invade and ruin all skins or other organic specimens housed in the museum.

** While this may have been true at the time, it would seem certain that later on, in the 1960s, "Achilla," the small yet adult female gorilla of the Basel, Switzerland, Zoo, would have permitted her keeper, Carl Stemmler, to have measured her, since by his knowledge and patience he had gained her complete confidence and friendship (see Nos. 25 and 26, following).

William Dressman, secured measurements of Susie each year, on or about 1 January, during the years 1936, 1937, 1939, 1941, 1942, 1943, 1944, and 1945. Her approximate age during these years was from $9\frac{1}{2}$ to $18\frac{1}{2}$ years, respectively. Her measurements at the ages of $9\frac{1}{2}$ to $15\frac{1}{2}$ years in height, weight, sitting height, span, chest girth, belly girth, and length of hand and foot are listed in Table 7. Other measurements of Susie that were taken but which are not listed in Table 7 included neck girth, upper-arm girth, wrist girth, calf girth, and standing reach upwards.

In view of the fact that so many gorillas are said to be full-grown at only 10 or 11 years, it is interesting to note that Susie grew in height until the age of seventeen years. At that time she reached a standing (and/or lying) height of 62 inches, which is the tallest I have heard of in a female lowland gorilla, being equivalent to 74 inches in an adult male. Unfortunately, she grew in weight for an even longer period—in fact, up to the time of her death at approximately 22 years of age, when she scaled an obese 458 pounds. This, again, was possibly a record for a female lowland gorilla (or perhaps even one of the mountain species). A poundage of 438 has been attributed to the female gorilla M'Toto (see No. 13, following), but probably either by mistake or through confusion of these two female captive gorillas.

While I have no information on Susie other than the statistics of her bodily dimensions, these have been exceedingly useful to me in formulating averages or standards of growth, along with intercorrelations between height, girth, and weight in lowland gorillas of both sexes. This was made possible by the fact that Susie's maximum weight exceeded that of an average-sized, jungle-living *male* lowland gorilla by at least a hundred pounds.

The data on Susie's body measurements at various ages were kindly furnished to me by Mr. N.S. Hastings, of the Zoological Society of Cincinnati, in personal communications of 2 July and 11 July 1942, respectively. (Since writing the foregoing, I have learned that Susie, before coming to America, had been given the name "Congo," and that she was the gorilla that arrived in New York on the *Graf Zeppelin* in 1929. (See the footnote referring to Congo, No. 3 in this list.)

9. *Massa*

Like the native name N'Gina, or N'Gi, "Massa" would appear to be a local (Gaboon) term meaning, simply, "gorilla," although another interpretation says that it is a native corruption of our word *master*. The little gorilla so-named arrived in New York in September 1931, having made the trip from Africa, along with six young chimpanzees, abroad a freighter piloted by a Captain Phillips. On arrival, Massa was estimated to be about nine months old. He was purchased, along with the six young chimpanzees, by Mrs. Gertrude Lintz, of Brooklyn, who specialized in raising baby exotic animals, especially primates, When Mrs. Lintz received the infant gorilla, it was unconscious and was suffering from pneumonia contracted on the long sea voyage. In all probability, if Mrs. Lintz (whose husband was a physician) had not been experienced in the matter, and had not for the first five days neglected meals and sleep to care for the sick little ape, it would have gone the way of most of its predecessors. But the crisis of its illness was successfully passed, it came out of its coma, and as a result of constant attention gradually became stronger and healthier. However, there were subsequent intermittent periods of illness, and it was not until several years had passed that Massa was on the road to steady good health. The name Massa was given to the little gorilla by Mrs. Lintz, who sincerely believed it to be a female, and always referred to it as such.

When Massa was about five years of age, he discovered that he could dominate Mrs. Lintz physically, and she was forced to dispose of him. Reluctantly, although for the generous sum of $6000, she sold him to the Philadelphia Zoo, at which quarters he arrived on 30 December 1935. That date has since been considered Massa's "birthday," and each year, on 30 December he is given a "party."

As has been related in connection with Bamboo (No. 5 in this review), Massa and Bamboo did not get along amiably after the larger gorilla had been "introduced" to Massa, on the supposition that Massa was a female. As a result, local newspaper headlines of that date were worded: "Zoo's GORILLA WEDDING IS OFF—PROSPECTIVE BRIDE OF BIG BAMBOO TURNS OUT TO BE NO LADY."

At first, and indeed until Massa was 18 years old, he was labeled a "mountain" gorilla. How, having been captured in the Gaboon, this could have been is a mystery, unless the natives who had caught the little ape told Captain Phillips that it had been "in the mountains"—where some elevations worthy of being so-called occur even in the Gaboon.

But today, Massa is the record-holder for gorilla longevity—at least in captivity—and as such has become famous among gorilla lovers (*gina-philes?*) throughout the world. Having attained the age of 46 on 30 December 1976, possibly he may go on to reach the predicted maximum of 60 years (see Table 20). If he does, it will be largely because of the excellent care (including a rational dietary) that he receives as the most distinguished, valuable, and venerable resident of the Philadelphia Zoological Garden.

(For an excellent, detailed account of Massa's life at the zoo, see the article by Frederick A. Ulmer, Jr. in the Garden's quarterly publication, *America's First Zoo*.)[7] Another lowland male gorilla named Massa was acquired by the San Diego Zoo on 28 July 1968 (see Table 25).

10. *Mbongo* and 11. *Ngagi*

These two famous mountain gorillas, which were long-time residents of the San Diego Zoo, are here discussed *together,* in view of the fact that

Fig. 115. Massa, of the Philadelphia Zoo, at the age of ten years, weighing 350 pounds. He has since gained distinction by becoming the oldest gorilla (forty-six years, as this is being written) ever to be raised in captivity. Photo by Mark Mooney, Jr.; courtesy Zooligical Society of Philadelphia.)

they were able to maintain that status with admirable compatibility for a period of over ten years. Mbongo was named after the Alumbongo Mountains of what was then called the Belgian Congo, since that is where he and Ngagi (another native name for "gorilla") were captured with nets by Martin Johnson in December 1930. Some consideration must be given this specific spot of origin, since it has a bearing on whether Mbongo and Ngagi were, as has been generally accepted, true mountain gorillas (*Gorilla beringei*), or whether, on the contrary, they were eastern lowland gorillas (*G. gorilla graueri*), as has been contended by Colin P. Groves.[8] Yet in most differential characters—thick black hair, larger size, long trunk, short neck, distinctive cranial proportions, etc.—Mbongo and Ngagi were almost typically *G. beringei*, or "mountain" gorillas. Kenheim Stott, Jr., of the San Diego Zoo, who was associated with the care of Mbongo and Ngagi throughout their lives in the zoo (1931–1944), regarded the two gorillas as possible hybrids; but if so, "far closer to *beringei* than to *graueri*."[9] An analysis of the measurements and proportions of the limb bones of Mbongo and Ngagi (and of the pelvis of Mbongo), which fortunately were recorded (see Table 4),* shows that Stott's conclusion (which was based on external differences) is confirmed osteometrically, since five out of six long-bone and pelvic indices in these two zoo specimens correspond with *G. beringei,* and only one index (radius-humerus) with *G. g. graueri.* (See also the comments made in this connection in Chapters 4 and 5.)

The significance of this is that true (i.e., Virunga) mountain gorillas may well range northwestward into the Mt. Tshiaberimu region (see Fig. 24), possibly interbreeding with the so-called lowland gorillas of that area, and, from there, range southward all the way to the Mt. Kahuzi district. If this is so, their numbers will be far greater than has heretofore been estimated.

To get back to Mbongo and Ngagi: once these two gorillas had arrived and been housed in their spacious living quarters (a "cage," 18 feet high, 20 feet wide, and 40 feet long) at the San Diego Zoo, they became subjects of great interest to scientists in various fields, as well as to the zoo personnel and the public in general. It was estimated that Mbongo, who weighed 125 pounds upon ar-

---

* Dr. Adolph Schultz secured the measurements of Ngagi's limb bones (Amer. Mus. Nat. Hist. No. 115609) in 1944, while I had secured those of Mbongo (San Diego Nat. Hist. Mus.) in 1942.

Fig. 116. This field photograph, although not very clear, nevertheless is an historic one, since it shows the first male *mountain* gorillas brought to the United States that survived in captivity for a significant length of time. "Ngagi," in the foreground, and "Mbongo," in the rear, are depicted here after they had been caught in nets by the Martin Johnson African Expedition of 1930. They were delivered to the San Diego Zoo, where for over ten years they were a major attraction. Johnson had named the two young apes, one of which he changed from "Congo" to "Mbongo."

rival, was approximately 5 years of age; and that Ngagi, who weighed 147 pounds, was about 5½ years. Thus, the two juvenile gorillas had already passed the particularly vulnerable infantile stage in which so many previous specimens had died within months, weeks, or even days after being brought into contact with humans. Quite probably the close companionship of Mbongo and Ngagi on their long ocean voyage had helped materially to keep either or both of them from lapsing into a state of melancholia. Certainly after they had arrived at the zoo, the constant attention and care given to them—especially by the then-director of the zoo, Mrs. Belle J. Benchley—virtually prevented any feelings of loneliness or dejection from developing. Indeed, Mrs. Benchley became essentially a "mother" to the two gorillas during their lifelong residence at the zoo.

While some of the visitors to the San Diego Garden had difficulty in distinguishing between Mbongo and Ngagi, to anyone familiar with the two gorillas, there were marked differences in their appearance and to a certain degree in their behavior. Ngagi, obviously older than Mbongo, was both facially and bodily more developed, defined, and athletic-appearing than Mbongo; while Mbongo could be likened, rather, to a plump, nonathletic schoolboy. However, in their playful tussles, Mbongo was quite able to hold his own.

When the two gorillas had first been captured, the native hunters had told Martin Johnson that they were a "pair," a male and a female. Because

Fig. 117. Mbongo, who, with Ngagi, was a long-time distinguished resident of the San Diego Zoo. From a five-year-old youngster weighing 125 pounds, Mbongo grew into a genial giant of 600. This photo shows Mbongo at about the age of ten years, weighing about half the poundage he ultimately attained. (Photo courtesy San Diego Zoo.)

of this, the Belgian government permitted the Johnson Expedition to retain the two specimens rather than just one, as would otherwise have been the ruling. But Johnson had his doubts as to the gorillas' sex. In any case, it was not until the gorillas had been in captivity for nearly three years that both were definitely determined to be males. This determination was made by Dr. C.R. Carpenter of Columbia University, who came to the San Diego Zoo in June 1934 to study the two gorillas, which at that time were estimated to be 7 years and 8 months, and 8 years and 2 months of age, respectively. Mbongo weighed about 185 pounds, and Ngagi 205. Dr. Carpenter's observations were carried on over a period of several months. He remarked that the external genitalia in both gorillas was so small, and so obscured by the thick black hair surrounding these parts, that only when the apes assumed certain positions could the penis be detected. And both gorillas, even at their supposedly pubescent age, showed no signs of sexual activity. This may indicate that mountain gorillas, are later in reaching both puberty and maturity than are the smaller lowland gorillas; however, there has not been a sufficient number of observations on mountain gorillas in this respect to establish a difference between the two species.

In Dr. Carpenter's published account of Mbongo and Ngagi,[10] he gives the results of his observations on such points as (1) sexual activity, postures, and locomotion; (2) eating and drinking; (3) types of gorilla play; (4) grooming and care of wounds; (5) vocalizations and gestures; (6) dominance and cooperation; (7) temperament; (8) ability to learn; and (9) nest building. The details in all these activities, as described by Dr. Carpenter, are too extensive to be repeated here. They constitute a valuable record of the habits and behavior of two male gorillas in captivity, as compared with that of a band, or troop, of gorillas of both sexes in the wild, as described by Schaller, Fossey, Zahl, and other field observers (see chapter 9).

Another detailed account of the life together of Mbongo and Ngagi is that related by Mrs. Benchley in her book *My Friends, the Apes*.[11] She has also written about them in several magazine articles (e.g., in *Touring Topics* for October 1932, and in *Nature Magazine* for May 1933, in each of which she refers to Ngagi as a female!).[12] In the booklet about Mbongo and Ngagi that Mrs. Benchley prepared for the San Diego Zoological

Fig. 118. Ngagi, a mountain gorilla who grew to gigantic proportions during his residence at the San Diego Zoo—from 5 October 1931 to his death on 12 January 1944. With his younger companion, Mbongo, Ngagi had been captured in the Belgian Congo in December 1930 by Martin Johnson. (Photo courtesy San Diego Zoo.)

Fig. 119. Ngagi, at about fourteen years of age, in the spacious, sand-floored cage at the San Diego Zoo that was occupied by him and Mbongo jointly for over ten years. Note the characteristic mountain-gorilla features of Ngagi: the long, thick, black hair; pronounced hairy head crest; close-set eyes and very short neck. These external characteristics, along with significant differences in the size and proportions of the skeleton, distinguish this rarer species from both the western and eastern forms of the lowland gorilla. (Photo courtesy San Diego Zoo.)

Society in 1940,[13] there is a list of the amounts of food given on eight successive days to the two gorillas, who were then 13½ to 14 years of age, yet still not fully mature. Mbongo's daily food consumption averaged 31 pounds (!) of fruit and vegetables, plus two quarts of milk, along with one egg on five of the eight days. Ngagi's rations averaged 33½ pounds (!) of fruit and vegetables, two quarts of milk, and one or two eggs. While most of these items were served only once a day, apples, pears, and carrots were given twice a day, and bananas either twice or three times a day (the total ranging from 4 to 11 pounds of bananas a day). Both apes had "voracious" appetites and seemed to relish the drinking of water almost as much as the eating of solid food. "They eat practically everything but the skin of their oranges and grapefruit" (Benchley). They also had the rather repellant habit of regurgitating their food, bringing it up either in their cupped hands or onto a clean floor or shelf, and eating it over again with great relish. Possibly the habit was to facilitate more thorough mastication of coarse vegetable matter, as in the cud-chewing of various cloven-hoofed herbivorous animals. Note that no meat was given, and that the total weight of food and liquid does not include that of the water drunk.

Although the foregoing dietary for Mbongo and Ngagi was considered at the time to be productive of a normal or "typical" rate of growth in gorillas (male), it may be seen—from today's perspective in ape-feeding—to have been approximately double the quantity needed to have produced a mature bodyweight of about 450 pounds in these two mountain gorillas, neither of which measured more than 5 feet 8 inches in height. (The subject of normal feeding in gorillas is discussed in chapter 11, following.)

Knowing of my interest in particular of the growth and physical development of gorillas, Mrs. Benchley kindly prepared for me the following tabulation of the weights of Mbongo and Ngagi at successive stages of their growth. The estimated ages have been added by me. This, I believe, is the first time this full record of weights has been published.

The fluctuations in weight of Mbongo and Ngagi at various periods were due mainly to their having colds or other disturbances that affected their appetites. When Mbongo, in 1938, dropped from 401 pounds to 379, it was because of an abscessed foot that lessened his amount of walking. However, in early 1940, although again "lame," he gained 75 pounds in only two months! During the same period, for some unrecorded cause, Ngagi reduced from 501 pounds to 468, upon which he was given extra amounts of Vitamin B. At the time of his death, on 15 March 1942, Mbongo weighed 618 pounds; but his preserved body (prior to being sent to Ward's taxidermy establishment in Rochester, New York) dehydrated to a weight of 582 pounds. Mbongo had reached a top weight of about 660 pounds early in 1942. Ngagi reached a maximum of 683 pounds in 1943; and at his death, on 12 January 1944, weighed 636 pounds. Thus, Mbongo attained an age of about 15 years and 5 months; and Ngagi, 17 years and 9 months. The cause of Mbongo's death was given as "acute coccidiomycosis," which was a malignant fungus infection of the lungs, liver, spleen, and digestive tract, which could have come either from contaminated food, water, soil, or bedding.[14] Ngagi, in contrast, died from a simple blood clot: pulmonary arterial thrombosis.

Table 21. Body Weights of Mbongo and Ngagi at Successive Ages.

Date	Mbongo		Ngagi	
	Weight, pounds	Age: yrs. & mos.	Weight, pounds	Age: yrs. & mos.
Nov. 1, 1931	122	5 – 1	147	5 – 7
Apr. 1, 1932	138	5 – 6	175	6 – 0
Jan. 30, 1934	168½	7 – 4	190½	7 – 10
May 5, "	184½	7 – 7	—	8 – 1
Feb. 1, 1935	172	8 – 4	205	8 – 10
Apr. 3, "	183	8 – 6	200	9 – 0
July 22, "	216½	8 – 9	248	9 – 3
Oct. 5, "	243	9 – 0	273	9 – 6
Jan. 6, 1936	269	9 – 3	312	9 – 9
Apr. 6, "	293	9 – 6	332	10 – 0
July 6, "	285	9 – 9	305	10 – 3
Oct. 21, "	290	10 – 1	355	10 – 7
Jan. 6, 1937	307	10 – 3	355	10 – 9
Apr. 4, "	344	10 – 6	398	11 – 0
July 7, "	364	10 – 9	405	11 – 3
Jan. 6, 1938	401	11 – 3	410	11 – 9
Mar. 6, "	381	11 – 5	421	11 – 11
Aug. 3, "	379	11 – 10	435	12 – 4
Feb. 1, 1939	431	12 – 4	457	12 – 10
Sept. 30, "	468	12 – 11	484	13 – 5
Feb. 23, 1940	517	13 – 4	501	13 – 10
Apr. 27, "	592	13 – 7	468	14 – 1
May 7, "	602	13 – 7	525	14 – 1
Oct. 6, "	612	14 – 0	539	14 – 6
Feb. 6, 1941	618	14 – 4	578	14 – 10
May 9, 1942	—	——	629	16 – 1
Feb. 1943	—	——	639	16 – 10

Fig. 120. Probably no individual wild animal in history was more maligned than "Gargantua," whose last public appearance was with Ringling Brothers and Barnum and Bailey's Circus in 1949. Cover his acid-disfigured upper lip, and you will see that Gargantua's expression is one of fear, or uncertainty, rather than "murderous intent," as the news media generally put it. This widely published photograph is by Eisenstadt.

12. *Gargantua*

Far more widely publicized than Mbongo and Ngagi—or perhaps even Bushman—was "Gargantua," as a result of the tremendous ballyhoo he received from Ringling Brothers and Barnum & Bailey's Circus, where he was one of their star attractions for nearly twelve years. The story of Gargantua's emergence from obscurity to almost unprecedented fame (for a subhuman creature) is, in brief, as follows.

Late in the year 1932, Gargantua, as an unnamed infant, was being brought from the Cameroons aboard a freighter piloted by Captain Phillips, who, a year earlier, had brought "Massa" (see No. 9, above) to America in the same manner. For some reason, one of the sailors employed on the ship became angered at the little ape and threw some nitric acid into its face, burning the flesh and leaving a horrible scar. Captain Phillips, upon hearing of the incident, immediately discharged the sailor, who may have had too much to drink. But the damage had been done, and the baby gorilla was disfigured for life with a curled upper lip that gave him a perpetual snarling look (which probably had no connection with how the terrified young ape really felt).

One of the passengers on the freighter was Mrs. Gertrude Lintz, who had had marked success in raising pigeons, rabbits, and St. Bernard dogs at her home in Brooklyn. Sorry at the baby gorilla's plight, she asked Captain Phillips if he would sell him to her, upon which Phillips said yes. Shortly after arriving home, Mrs. Lintz gave her new pet the name "Buddy." But the unfortunate animal was evidently destined to have more troubles, for a houseboy, whom Mrs. Lintz had discharged for laziness, retaliated by feeding Buddy a bottle of Lysol, which severely damaged the ape's intestines. However, Mrs. Lintz's devotion and care gradually brought the gorilla into a state of good health and physical condition. But, inevitably, he grew too large and powerful for her to manage or feel at ease with. Accordingly, in December 1937, when Buddy was about $6\frac{1}{2}$ years old, Mrs. Lintz sold him to Ringling Brothers and Barnum & Bailey's Circus for what may have then been the top price for a gorilla: $10,000.

The circus people were elated to obtain the half-grown gorilla, whose unpretentious name they promptly changed to "Gargantua" (after the mighty giant in Rabelais' *Gargantua* and *Pantagruel*). They likewise gave the ape a tremendous publicity buildup, assertedly spending $50,000 a year in the process. This, however, was money well spent, for Gargantua proved such a drawing-card that he virtually lifted the circus out of what had been a hectic period. The publicity was climaxed in 1942 with the announcement that a "wedding" was planned between Gargantua and M'toto, a female gorilla of nearly the same age as Gargantua, which the circus had obtained "on loan" a year earlier from an owner in Havana, Cuba (see No. 13, following).

In order for the two gorillas to become "acquainted," however, it was thought best to put them in communication, but in separate cages. This proved a wise decision! Although Gargantua was interested and friendly, M'toto wanted nothing to do with him. When Gargantua generously tossed one of his sticks of celery into M'toto's cage, she responded by flinging it back smack into his face! That put an end to any plans the circus people may have had of bringing these two separately raised gorillas together.

Gargantua and M'toto were exhibited (but in separate cages!) by the circus from 1942 until 25 November 1949—the last day of the show's season, in Miami—when Gargantua passed away. While the immediate cause of his death was stated to be double pneumonia, he had long suffered from diseased kidneys, tuberculosis, and cancer of the lip. His body was shipped in dry ice by plane to Johns Hopkins Medical School in Baltimore, where the primatologist Dr. Adolph Schultz took his customary measurements of the cadaver. Gargantua in life evidently had a standing height of about 66 inches. The weight of his embalmed body, which was dehydrated to a marked degree, was only 312 pounds. At his height of 66 inches, and if he was of typical lowland gorilla build, his weight in life would have been about an even 400 pounds. That poundage, of course, would hardly have been on a par with his billboard publicity, which attributed to him a weight of 525 pounds. Even outdoing the latter claim concerning Gargantua's size was a list of several of his girth measurements, which listed his chest at 70 inches. At his height of 66 inches (and sitting height, if in proportion thereto, of 41.6 inches), Gargantua's weight would have been $41.6 \times 70^2 \times .00297$ (see Fig. 43), or approximately 605 pounds (!). On the other hand, if accuracy rather than sensationalism had been the objective, Gargantua's chest measurement, at a probable weight of approximately 400 pounds, would have been $\sqrt{\dfrac{400/.00297}{41.6}}$, or not more than 57 inches—which still should have been very impressive.

During his long residence in the circus, Gargantua was given a well-balanced diet. This consisted daily of a dozen assorted vegetables, a dozen assorted kinds of fruit, a loaf of bread, three quarts of milk, and a half-dozen eggs, supplemented with two tablespoons of liver extract and a cupful of chocolate syrup. No meat was given. His feeding hours were 8:00 A.M., noon, 5:00 P.M., and 10:00 P.M. The foregoing quantities of food were spread over these four daily meals.

Had Gargantua not been facially disfigured, he

Fig. 121. "Gargantua," as he was billed on circus posters. It was estimated that during his twelve years before the public, he was seen by more than thirty million persons.

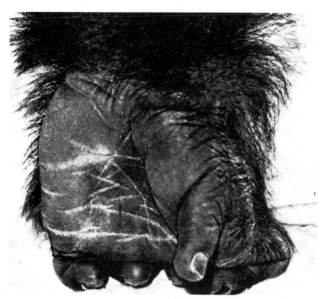

Fig. 122. The hand of Gargantua in knuckle-walking stance. Note the short thumb; broad, hairless palm; and extreme flexion of the fingers.

would have been a splendid specimen of his species. But that disfigurement need not have resulted in his being labeled a malicious, man-killing monster.

13. *M'toto*, or *Toto* (a name which, in Swahili, means "youngster").

This much-publicized female lowland gorilla logically follows Gargantua in these accounts, since she was a contemporary of Gargantua and, as related above under No. 12, was hopefully (but without success!) intended as his "bride." M'toto's first owner was Mrs. Maria Hoyt, of Havana, Cuba. Mrs. Hoyt and her husband, Kenneth, had first brought the young gorilla, which was then (November 1932) about eleven months old, to their home in New York; but they were advised by a medical specialist (who had saved M'toto when she had pneumonia) that if they wanted to keep the young and still delicate ape alive they had better take it to a warmer climate. As the Hoyts had the means, and had become attached to their new pet, they bought an estate in Havana, where the climate was fine and there were no cold winters. M'toto had the run of both the house and the large, fenced-in yard, although neighbors having small children were apprehensive about such an exotic and formidable "pet" being allowed to roam freely about. But the Hoyts had engaged the services of an expert ape keeper, José Tomás, a Cuban, who had previously worked for Mme. Rosalia Abreu, of Havana, until she died in 1930, whereupon her extensive collection of monkeys and chimpanzees was disbanded and sold to various zoos and primate centers. So, Tomas, being unemployed, was happy to work for the Hoyts as the twenty-four-hour keeper of M'toto. Giving M'toto the best of care and attention, Tomas soon gained the confidence, affection, and respect of the ape. Whenever M'toto started to get out of hand, all Tomas had to do was point at her with his battery-charged "correction" rod, one jolt from which she evidently never forgot! He also carried a small snake, curled up in a bag around his waist, the appearance of which would throw M'toto into a panic of fear!

Detailed accounts of the behavior of M'toto, principally while she was living in Havana under the supervision of Mrs. Hoyt (whose husband had since died), were given at the time in several magazines. Two of the best of these accounts are those by Ivan Dmitri, and B.H. McCourt, respectively.[15] Also, there are rather extensive comments on M'toto in the books by Geoffrey H. Bourne and Vernon Reynolds, respectively.[16]

Some of the particular habits of M'toto were that she relished meat and chicken, even when raw; liked to draw "pictures" on the stone floor with pieces of white or colored chalk; liked to slide down banisters; and allowed only the Hoyts, Tomas, and one or two of the Hoyt's servants to come near her. Anyone else she would bite and scratch. Also, she had learned to lock or unlock any door in the Hoyt's home, if necessary searching until she found the right key. Once, she locked her trainer, Tomas, in a bedroom for an hour, finally putting her ear to the door and responding to his urgent command to open it!

As related in connection with Gargantua (No. 12, preceding), Mrs. Hoyt was finally obliged to dispose of M'toto. She turned the gorilla over to the Ringling Brothers and Barnum & Bailey's Circus, in Sarasota, Florida, with the proviso that she could spend as much time with the animal as she wished. That was in February 1941, at which time M'toto was 9 years old, almost adult, and of such size and strength that she could no longer be trusted in a private home, especially if visitors were present. But as related, when M'toto was brought into the circus and was "introduced" to Gargantua, she wanted no part of him; so the "gorilla wedding" that had been planned and given wide and costly publicity by the circus people was never consummated. Incidentally, for a long time—even up to the time she was turned over to the circus—there was doubt as to M'toto's sex. Although the gorilla's relatively small size and

Fig. 123. A photo montage of the gorilla "couple" who never got together! Note the humanlike breasts on M'toto, and the more-pleasant-than-usual expression on Gargantua! This widely circulated publicity shot was released in 1942.

gentleness argued for femininity, newswriters of that period often referred to her as "he"; (even the director of the National Zoo, Dr. William Mann, made such reference in his article "Man's Closest Counterparts," in the August 1940 *National Geographic* magazine). And whoever originated the widely repeated statement that M'toto weighed 438 pounds must have had an urge for exaggeration, considering that such a poundage would be proportionate to an adult male gorilla nearly a foot taller than M'toto. As may be seen in the accompanying photograph of the gorilla and her trainer, Tomas (Fig. 124), the ape's bulk is certainly no more than half again as great as that of Tomas's, who presumably was a man of ordinary size.

After Gargantua died, late in 1949, M'toto was kept on exhibition by the circus until 1956, when Mrs. Hoyt requested that the aging anthropoid be returned to her. But M'toto lived on with her owner until her death sometime in 1968, which would have made her between thirty-six and thirty-seven years of age—an all-time record in longevity for a female gorilla in captivity. Mrs. Hoyt herself died just a year later.

14. *Makoko* and 15. *Oka*

These two splendid gorillas of the Bronx Zoo could not, unfortunately, be referred to as a "pair." However, though they never mated, they were brought up together and so are being dealt with

Fig. 124. M'toto and her long-time trainer and caretaker, José Tomás, of Cuba, who was one of the few persons to gain M'toto's confidence and friendship. (Photo by J. Steinmetz.)

Fig. 125. Ill-fated Makoko, a magnificent lowland gorilla, who drowned in a water-filled moat at the Bronx Zoo on 13 May 1951. Makoko, who at the time of his death had been in captivity nearly ten years, was eleven years and eleven months old and weighed 448 pounds. (Photo courtesy New York Zoological Park.)

Table 22. Body Weights of the two Lowland Gorillas, Makoko and Oka, at Successive Ages.
( ) = Estimated Weight

Date	Makoko		Oka	
	Weight, pounds	Age, yrs. & mos.	Weight, pounds	Age, yrs. & mos.
Sept. 7, 1941	28.16	2 – 3	20.46	1 – 5
Feb. 9, 1942	49.7	2 – 8	29.5	1 – 10
May 4, "	58.4	2 – 11	(34.5)	2 – 1
June 18, "	(59.8)	3 – 0	37	2 – 2
Jan. 28, 1943	70.0	3 – 8	53	2 – 10
June 7, "	77.0	4 – 0	61	3 – 2
Jan. 4, 1944	97.5	4 – 7	76	3 – 9
June 4, "	(113.8)	5 – 0	92	4 – 2
June 6, "	114	5 – 0	(92.2)	4 – 2
Jan. 8, 1945	134	5 – 7	113	4 – 9
May 9, "	145	5 – 11	120.5	5 – 1
Jan. 8, 1946	(181)	6 – 7	142	5 – 9
June 24, "	(209)	7 – 1	158	6 – 3
June 26, "	210	7 – 1	(159)	6 – 3
Oct. 11, "	(235)	7 – 4	(178)	6 – 6
Jan. 23, 1947	(252)	7 – 7	184	6 – 9
June 25, "	(279)	8 – 0	198	7 – 2
August , "	(290)	8 – 2	(205)	7 – 4
Feb. 25, 1948	(334)	8 – 8	228	7 – 10
May 21, "	(356)	9 – 0	(236)	8 – 2
Nov. 1, "	366	9 – 5	246	8 – 7
Sept. 1, 1949	408	10 – 3	282	9 – 5
May 13, 1951	448	11 – 11	(330)	11 – 1

here as though they had been a pair. The two gorillas, as infants, had been brought to the zoo on 7 September 1941. Makoko, the male, was estimated to be about 2 years and 3 months of age; and Oka, the female, about 1 year and 5 months. Both had been captured in French equatorial Africa by Philip Carroll, the well-known collector of African primates. Thus, Makoko and Oka were western lowland gorillas.

The following record of the growth in weight of these two gorillas was kindly furnished to me by Lee S. Crandall, then (1951) general curator at the Bronx Zoo. The age estimations are by me; those of Makoko as listed by the zoo make him three months older.

Although the poundages listed in the above table show a more or less uniform increase with age, a comparison with the average or typical poundages plotted on the graph in Figure 42 indicate that the rate of weight increase, from 3 years of age onward, in Makoko was 50 percent greater than average; and in Oka, more than double the average rate. Another rather surprising comparison may be made with the figures in Table 21, where the weights of the San Diego Zoo's two mountain gorillas are listed. Although both Mbongo and Ngagi continued to gain in weight until each weighed over 600 pounds, at the age of 10 years and 3 months Mbongo weighed 307 pounds, Ngagi 305 pounds, and Mkoko 408 pounds! Does this mean that Makoko, had he continued to live to, say, 15 years of age, would likewise have gained until he weighed over 600 pounds?

As is further seen in Figure 42, few captive gorillas are over average weight at the age of 3 years (most, indeed, being under rather than over); yet by 7 years they can be markedly "overweight," especially in the case of females. And if the heavier individuals do not possess proportionately thicker bones (which might justify their greater body weights), the only conclusion that can be drawn is that these gorillas are either overfed, underexercised, or both.

In some adult male gorillas, the excess poundage can be so evenly distributed over the body that the animal appears in fine physical condition and is highly impressive as a zoo specimen. This was the case with Bushman (see No. 7, preceding), who evidently could have weighed a hundred pounds less than he did, and still presented an imposing figure. And the same could be said of many other male gorillas, whose value as zoo attractions are, unfortunately (for the gorilla's health), often judged by the poundage those apes can be said to have attained.

Concerning the "menu" served each day to Makoko and Oka, Makoko received approximately 21 pounds of food and 9 pounds (4½ quarts) of water. Oka received about two-thirds as much as Makoko, or 15 pounds of food and 6 to 7 pounds (3–3½ quarts) of water. The food included the customary vegetables (celery, beets, carrots, string beans, etc.) and fruits (bananas, oranges, apples, grapes, etc.), but also 1½ pounds of horse meat, chopped and cooked. Additionally, each gorilla was given five slices of whole-wheat bread with syrup or jelly. At 9:00 A.M. and 4:00 P.M. a "gruel" was served, which was composed of a gallon of water, six eggs, and various minerals and vitamins. Of this gruel, which weighed perhaps 10 pounds, Makoko received 3/5 and Oka 2/5. Thus, Makoko and Oka were given a diet that was well balanced, so far as variety of essential ingredients was concerned, but which could have been considerably less in quantity without either of the apes reducing to less than normal weight (see chapter 11).

However, all contemplation of the status of Makoko and Oka as a "gorilla couple" can be dispensed with in view of Makoko's accidental, premature death. In the fall of 1950, the two apes were moved into new and more spacious living quarters, which included an outdoor promenade surrounded by a moat 14 feet wide, filled with water 6 feet deep. This moat separated the gorillas from their almost everpresent human audience, and vice versa. Evidently Makoko was intimidated by the "openness" of his new home, which lacked the iron bars to which he had been accustomed. Accordingly, he rarely ventured outside of his den. But on 13 May 1951, he unexpectedly pleased about 1,200 Sunday visitors by taking a leisurely stroll to the moat's edge. Then, evidently giving all his attention to the crowd, he slipped and fell into the moat. Animal keeper George Scott, who happened to be working nearby, saw Makoko tumble into the water. Knowing the gorilla could not swim, he courageously dived into the water to rescue him.

> "It took 10 minutes before Scott, with the help of three other attendants, succeeded in dragging the limp ape ashore. The zoo veterinarian and 12 oxygen tanks were rushed to the scene, but two hours of artificial respiration failed to revive Makoko."[17]

Thus, a most valuable animal—one that the zoo personnel had hoped might sire the first baby gorilla in captivity—was lost because his species is one of the few that does not naturally know how to swim.

After the officials at the Bronx Zoo had recovered from the shock of Makoko's tragic passing, they looked about for another possible mate for Oka, who, as previously mentioned, did not mate with Makoko. However, it was not until 1963—some 12 years after Makoko had passed on—that it was decided to bring together Oka and a male gorilla in the zoo named Mambo, who at the time had reached 13 years of age. But Oka meanwhile had become 23 years of age—equal by human standards to about 42 years—and was considered by the ape keepers of the zoo as being "probably too old." Perhaps Mambo also considered her as such! In any event, Oka never became a gorilla mother. Indeed, according to the list of gorilla births in captivity compiled by Marvin L. Jones (and in-

Fig. 126. "Incompatibility" is not a feeling restricted to men and women! Here, twenty-three-year-old Oka, on the left, and the thirteen-year-old Mambo are about as indifferent to each other as could possibly be. Much to the regret of the Bronx Zoo officials, their plan of mating these two seemingly eligible gorillas was a dismal failure. Possibly the age difference between the two subjects was too great! (Photo courtesy New York Zoological Society.)

Fig. 127. Albert, of the San Diego Zoo, in 1965, at the age of sixteen years. The profuseness of the light grey hair on his body and limbs gives him not only the "silver-back" coloration of adult male gorillas, but extends over his hips and even well down his legs. Albert, at a height of perhaps 62 inches and a weight of perhaps 300 to 310 pounds, is possibly the smallest fully adult male gorilla in the United States today. However, he is in fine physical condition at twenty-seven years and should attain a ripe old age—for a gorilla. (Photo courtesy San Diego Zoo.)

cluded here in chapter 13), as of 1 November 1971 no gorilla had ever been born in the Bronx, New York Zoo.

(A detailed account, with excellent photographs, of the zoo's unsuccessful endeavor to mate Mambo and Oka is given by Joseph A. David, Jr. in *Animal Kingdom*, September–October 1963, pp. 176–179.)

16. *Albert*; 17. *Bouba*; and 18. *Bata*

After the deaths of the two big mountain gorillas, Mbongo and Ngagi, in 1942 and 1944, respectively, the San Diego Zoo had only one other gorilla: "Kenya," a female, also of the mountain species, in its collection. But Kenya also died, on 23 September 1946, at the age of about 9½ years.

After nearly three years, during which period the zoo had no gorillas at all, three good-looking and healthy lowland specimens—a male and two females—arrived in San Diego on 10 August 1949.

They were given the names Albert, Bouba, and Bata (the latter being pronounced with the accent on the last syllable), respectively.

At the time of their arrival, Albert, who was noticeably smaller than either of the two females, was estimated by the zoo personnel to be about 6 months old; Bouba, 8 months; and Bata, 10 months. My own estimates, based on a graphic plotting of the young apes' body weights up to 2 years of age, is that Albert upon arrival was 7.9 months old; Bouba, 8.6 months; and Bata, 10.4 months. Regardless of which of these estimations was most nearly correct, the graphic charting of the weights showed in all three gorillas a remarkably uniform increase in weight during the first two years. Albert, who upon arrival weighed 8 pounds, grew at the rate of 1.60 pounds per month; Bouba, who weighed 9 pounds, at the rate of 1.70 pounds per month; and Bata, who weighed 10½ pounds, at the rate of 1.90 pounds per month.* From these differing rates of growth, it can be understood why the two female gorillas—Bouba and Bata—steadily outgrew little Albert until he was left far behind. Not until he was well along in adolescence did Albert come to outweigh Bouba and Bata, although he had long since learned how to assert his natural male dominance. He had managed to live on reasonably agreeable terms with the two females, although there was still a lot of bickering between the three, as though they had all come from the same family, which was not the case. Finally, however, Bata, who had always been the chief troublemaker, became so overly bossy that the zoo people sent her to the Fort Worth, Texas, Zoo, in exchange for a young pair of gorillas, "Mimbo" and "Timbo," which came to San Diego on 9 January 1964.

The successive weights and measurements of Albert, Bouba, and Bata, up until 3½ to 4 years of age (when the measurings were discontinued), are given in metric units in Table 7. The original list, which was recorded in pounds and inches, was kindly furnished to me by Edalee Harwell and Ken Stott, Jr., who were caretakers of the three gorillas over the years.

Albert, like most young male gorillas, was a showoff. When my wife and I visited the "gorilla nursery" shortly after the arrival of the three small apes, Albert shifted our attention which at the moment we had been giving to the two "girl" gorillas, Bouba and Bata, by thumping his chest and climbing up a rope, meanwhile riveting his attention upon us to see if we were watching and admiring his actions! Bouba, in contrast, was generally likened to a demure maiden, while Bata was regarded as a fun-loving tomboy.

Albert was not only a fine specimen of a lowland gorilla himself, but in due course a capable sire. Although he had no interest in this direction in either Bouba or Bata, when a young female named Vila was brought into the zoo on 27 March 1959, he "took" to her at once. On 3 June 1965, when Vila was only 7 years and 9 months of age, she gave birth to a female, which the zoo officials appropriately named "Alvila."* The newborn infant weighed 4 pounds 11½ ounces. Albert, the father, was then 16 years and 3 months old.

Many other gorillas came to the San Diego Zoo following the encouraging, death-free era that had been ushered in by the arrival in 1949 of Albert, Bouba, and Bata. Trib, a male lowland gorilla who came to the zoo on 30 July 1960, at the age of about 12 months, by the age of 13 years (in late 1972) was transferred from the San Diego quarters, where he and Albert did not "get along," to the zoo's new Wild Animal Park, which is located some thirty miles north of San Diego and comprises about 1,800 acres, 500 of which simulate the natural environments of the animals living in the park. Now, in the park, as well as in the San Diego Zoo, there is a thriving colony of gorillas—all of the western lowland species. (As of 1 April 1974, there was one adult male and three adult females in the zoo, and three adult males, two adult females, and two immature males in the park.)

Albert, whose pictorial likeness graces the jacket of this book, is shown also, as a six-month-old infant, in Figure 49. Today, as he approaches the age of 28 years, Albert is still decidedly small for a mature male lowland gorilla and weighs probably little if any over 300 pounds. Trib, in contrast, at 14 years weighed over 400 pounds. A year later, Trib mated with a ten-year-old charmer named Dolly; and on 15 October 1973, a male weighing a heavy 6 pounds and 4 ounces was born, which was named "Jim." And on 2 October 1974, Dolly gave birth to another infant, a female named "Binti" (see Fig. 148). Thus the former feeling—that it would be a miracle to ever expect the birth of a gorilla in captivity—is decidedly not felt at the San Diego Wild Animal Park, where the sim-

---

* Also, in the three infant gorillas (two males, one female) at the Jersey (Channel Islands) Zoo, the average rate of growth was about 1.90 pounds per month during the first few months. (See under "Jambo," No. 26, following.)

* In some places it has been stated that Alvila was the seventh gorilla to be born in captivity, "anywhere in the world." More correctly, Alvila was the eleventh in the world, although the sixth in the United States (see Table 26).

Fig. 128. Bata at the age of eleven years (1959). Her "frown lines" make her look perpetually worried; cover them, and she looks merely curious or inquisitive. (Photo courtesy San Diego Zoo.)

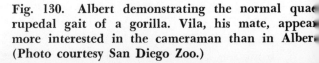

Fig. 130. Albert demonstrating the normal quadrupedal gait of a gorilla. Vila, his mate, appears more interested in the cameraman than in Albert. (Photo courtesy San Diego Zoo.)

Fig. 129. At the right is Bouba, at the age of ten years and nine months; while Bata peeks in at the left. These two female gorillas, although coming from different locations and "families," got along quite well in captivity for about fifteen years. (Photo by Van Nostrand; courtesy San Diego Zoo.)

Fig. 131. Massa (left), aged eight years; Trib (fourteen years and over 400 pounds); and Dolly (in tree), ten years—at the San Diego Wild Animal Park in 1973. (Photo by F. D. Schmidt; courtesy San Diego Zoo.)

ulated and spacious natural environment reduces "captivity" to a minimum.

Some further comments, particularly concerning the diets of young gorillas as followed at the San Diego Zoo, are given in chapter 11.

For a delightful, detailed account of the living habits and behavior of the young San Diego gorillas—Albert, Bouba, and Bata—see the series by Edalee Orcutt Harwell, the gorillas' dedicated nurse and assistant keeper, entitled "Gorilla Notes." These artcles ran in *Zoonooz*, the San Diego Zoological Society's official magazine, during 1950 and 1951.

19. *Phil*

One of the heaviest, though far from the tallest, captive lowland gorillas was "Phil," of the St. Louis Zoo, who died in 1958 at the age of 19 years. He gained wide prominence by being said to weigh, after death, no less than 776 pounds (!). Even though this alleged record weight was duly seen to be in error, it continued to be circulated as a fact, until there were even hunters who believed that giant, 800-pound gorillas might occur in unexplored parts of the African forests! There is one certain and conclusive way of clearing up such megalomanic misconceptions, and that is to measure the skeleton of the animal after death.

Fortunately, the limb bones of Phil were available. From his articulated skeleton, as shown in Figure 134, these limb-bone lengths were obtained (all from the left side): humerus, 429 mm; radius, 375 mm; femur, 391 mm; tibia, 308 mm; clavicle, 178 mm.[18] These dry-bone dimensions yield the following probable measurements during life: arm length 1050 mm (41.34 in.); span, or horizontal arm stretch 2424 mm (95.43 in.); leg length 777 mm (30.59 in.); sitting height 1050 mm (41.34

199

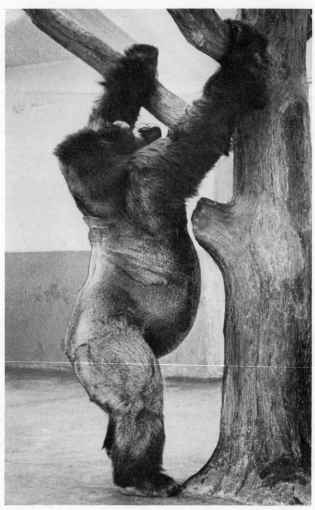

Fig. 132. Phil, of the St. Louis Zoo from 1941 to 1958, was one of the largest captive gorillas on record. His postmortem measurements: 5 feet 5½ inches tall; seventy-two-inch chest; sixty-four-inch waist (?); thirty-six-inch neck; fifteen-inch wrist. (Photo by Jack Zehrt; courtesy St. Louis Zoo; approximately one-twelfth natural size.)

in.); standing height 1664 (65.50 in). In view of Phil's height being publicized as 5 feet 11 inches (!), it is a mystery how such a measurement (?) was obtained.*

If Phil's posthumous chest girth of 72.25 inches can be accepted as having been taken correctly, and his sitting height to be the figure derived

---

* Another such instance is where the male mountain gorilla, Mbongo, of the San Diego Zoo, was measured after death by an inexperienced museum worker, who stated the gorilla's height to be 77 inches (!). This fantastic figure was derived by adding together head, trunk, and leg lengths, which actually overlapped. It should be added here, however, with reference again to Phil, that in an article by George McCue, in the St. Louis *Globe-Democrat* Sunday Magazine for 15 June 1958, (the year Phil died), this gorilla's height is reported as a more reasonable 5 feet 7 inches.

above (41.34 inches), his posthumous weight, as estimated by the procedure explained in chapter 5 (and Fig. 43) becomes: $72.25^2 \times 41.34 \times 0.00297$, or approximately 640 pounds. This weight compares closely with the weights recorded upon the deaths of the San Diego Zoo's two male mountain gorillas, Mbongo and Ngagi, which were 618 and 636 pounds, respectively. And since Mbongo was known to have reached at least 660 pounds in life, and Ngagi 683 pounds, it would appear reasonable to attribute to Phil a maximum living body weight of, say, 690 pounds. But it should be emphasized that the normal weights of these three significantly "overweight" (if judged by specimens living in the wild) gorillas should be within the range of 500 to 520 pounds. It would certainly seem reasonable to assume that overweight (with accompanying lessened physical activity) is as inimical to health and longevity in gorillas as it is known to be in human beings. A 640-pound gorilla "reducing" to 500 pounds would be comparable to a 200-pound, obese man reducing to a normal 156 pounds.

Phil's reported weight of 776 pounds was the result of his dead body being placed in the back of a jeep, after which the jeep (with Phil in it) was weighed on a public scales at an ice house. Later on, the jeep was weighed without Phil [and with more, or less, gasoline in the tank (?)], and this weight was deducted from the combined weight. The difference was assertedly 776 pounds. To repeat our opening statement, Phil was one of the heaviest captive gorillas on record; and his relatively short stature made it possible for him to have also what may have been the largest girths of chest, belly, neck, arms, and legs in either a lowland or a mountain gorilla.

Phil, as a youngster of about 2 years, had arrived at the St. Louis Zoological Park on 7 September 1941, weighing 26 pounds. He was one of eight young gorillas (a record shipment) brought from French equatorial Africa by the well-known animal collector Phillip Carroll. Two of the gorillas in this group were Makoko and Oka (see Nos. 14 and 15, preceding), which were acquired by the Bronx, New York Zoo. Three others, besides Phil, went assertedly to the St. Louis Zoo; however, in a "census" dated April 1944, only one, a 4½-year-old male named "Mattie," was listed; so presumably the other two specimens died sometime prior to that date.

Unlike some other gorillas, Phil was not afraid of water, and in the pool—which admittedly was shallow—in his living quarters he would fre-

Fig. 133. The mounted skeleton of Phil, in typical quadrupedal posture, compared with the skeleton of a man of average size in erect posture. (Photo courtesy St. Louis Museum of Science and Natural History; times 0.08 natural size.)

quently lie and enjoy himself. In 1954, when he was 15 years of age and fully adult, zoo officials hoped to mate him with some eligible lady gorilla. Regrettably, this hope never materialized, and Phil, like many other zoo gorillas, never became a parent.

Phil died sometime during the latter half of 1958, at the age of 19 years, after having had "heart disease" for many months previously. The disease caused massive edema, or fluid accumulation, in the body tissues. While this may have contributed to the largeness of Phil's girth measurements, he had, on the other hand, lost a considerable amount of weight by virtually "fasting" for several weeks prior to his death. His departure was a deep loss to the zoo and a source of sadness to the many visitors who came regularly to the zoo just to see Phil, the biggest gorilla in the country.

20. *Samson*

In decided contrast to the "lightweight" gorilla, Albert, and more in keeping with the heavyweight, Phil, is the huge and ponderous "Samson," of the Milwaukee Zoo. Samson is without a doubt one of the biggest and finest adult male gorillas to be seen in any zoo in the world today. He and a male companion named "Sambo" arrived at the Washington Park Zoological Garden on 10 October 1950. Samson, who then weighed 15 pounds, was estimated to be about 11 months old; and Sambo, who weighed 14¼ pounds, was about 12 months old. Sambo died in 1959 of tuberculosis.

Samson, in his spacious living quarters at the zoo, looks out at his human audiences through ¾-inch-thick laminated glass, which protects the people outside from Samson and protects Samson from possible contamination by the people!

As related by George Speidel, Director of the Milwaukee County Zoological Park:

I have been in this business for 42 years and

Fig. 134. "Samson," popular lowland gorilla of the Milwaukee Zoo, weighing 540 pounds on a two-thousand-pound scale on which visitors can check his weight on the outside of the heavily glassed cage in which he lives. Actually, his "cage" is a huge room, forty-two feet long, in which he is given every essential care to a gorilla's well-being. (Photo courtesy George Speidel, Director.)

have seen many animals in many zoos, but never had the opportunity to see one that has the appeal that Samson has. He is literally known all over the world. 14 percent of our visitors come just to see him, and he is the only animal that I have known that can hold a crowd while he is sleeping. Every move he makes is a dramatic gesture that seems to amuse or interest people.

Several years ago I was visiting a monkey research center near Tokyo, Japan and was amazed to see a life-size picture of Samson in their indoctrination center with many graphics around, explaining the animal, his weight, etc.

Samson eats about 28 pounds a day of fruits, vegetables, and a bit of meat. We have found that our gorillas are in need of Vitamin B^1, and he gets an ample dosage of this each day. He likes people, we believe. He knows all his keepers very well. I consider him one of my best friends. I have known him since purchasing him in New York when he weighed 12½ pounds.[19]

In the center of Samson's living room is a steel platform eight inches above the floor, where the big gorilla sits, sleeps occasionally, and dines regularly. The table is also a scale, which registers Samson's fluctuating weight on an illuminated dial outside the enclosure. His maximum weight to date has been 610 pounds, but more recently he has had his food allowance reduced somewhat, which has brought his weight down to 525 to 540 pounds. The dial of the scale, as shown in Figure 134, records up to 2000 pounds and at the time the photo was taken showed Samson to weigh 540 pounds.

When one of Samson's several daily meals is about to be served, a door at the left rear of his living room is electrically raised. Knowing what this means, Samson at once exits through that doorway down some stairs to another of his rooms. The door then closes, whereupon the keeper places Samson's meal on the platform, makes his departure, and double locks the door by which he entered. Then, the electrically operated rear door reopens. When Samson darts through it to get to his meal, his agility in coming through the small doorway makes him appear twice as huge as he really is.

In 1975, the zoo acquired an adult female gorilla named "Terra," with whom they hoped Samson would mate. After a year or more together, however, the two apes evidently have decided to be simply "friends."

A second gorilla named "Samson" was acquired by the Buffalo, New York, Zoo, on 14 March 1962. (No. 112 on the Chronological Record in chapter 13.)

21. King Tut

In contrast to some of the hoped-for (but in vain!) gorilla sires mentioned previously, "King Tut," of the Cincinnati Zoo, has distinguished himself, as has also his mate "Penelope," by fathering (as of 1974) no fewer than four offspring —three females and one male—all of which have survived. King Tut and Penelope, therefore, have certainly done their share towards increasing the gorilla population in this country. Names and other particulars of the four offspring—as well as of four other young ones born to gorillas at the Cincinnati Zoo up until 1974—are given in the Chronological Record in chapter 13.

A second "gorilla couple" at the Cincinnati Zoo has been as prolific as King Tut and Penelope. This pair is "Hatari" and "Mahari." Curiously, like King Tut and Penelope, Hatari and Mahari have produced three females and one male. Indeed, there would appear to be a numerical predominance in general of females among gorillas born in captivity, since out of the first thirty-three zoo births, thirteen were male and twenty female (!).

King Tut arrived at the zoo sometime in January 1951 (?) (other accounts say 1950 and 1952), when he was estimated to be 13 months old. Therefore, when he fathered his first offspring, a female named "Samantha," on 31 January 1970, he was somewhere between 20 and 22 years of age, depending upon which information as to his birth date is correct. Four years and 2½ months later, when he fathered another female, named "Tara," he would have been at least 24 years old, the comparable human age of which is 44 years. This gives King Tut another point of distinction: the oldest zoo gorilla on record to have become a father!

So popular has King Tut been at the Cincinnati Zoo that several years ago a realistic miniature model of him, carefully proportioned and cast (in solid plastic) about ⅛ life-size, was produced by the animal sculptor Francis W. Eustis. A reproduction of this statuette is shown in Figure 136.

22. Big Man

The pride of the Kansas City Zoological Gardens was "Big Man," who died just this year (1976) at the age of 19 years—the same age as did Phil, of the St. Louis Zoo. Big Man was captured by the well-known gorilla-and-chimpanzee collector Deets Pickett, D.V.M., of Kansas City. Big Man

Fig. 135. King Tut, 450-pound leader of the gorilla colony at the Cincinnati Zoo, is the proud father of three female and one male offspring. He is further distinguished by being the oldest (twenty-four years) zoo-gorilla sire. (Photo courtesy Zoological Society of Cincinnati.)

Fig. 136. A statuette of King Tut designed by the distinguished animal sculptor Francis W. Eustis, of Cincinnati. This photograph is about three-quarters the linear size of the statuette. (Photo courtesy Mr. Eustis.)

Fig. 137. Hatari, of the Cincinnati Zoo, at about the age of eight years. When only seven years and nine months (in January 1970), he became, by fathering a male named "Sam," one of the youngest gorilla sires on record. As of August 1974, he had fathered three more offspring—all females. His mate in all four cases was Mahari. (Photo by James F. Brown; courtesy Cincinnati Zoo.)

arrived at the zoo on 26 July 1958, at the age of about 12 months, weighing a relatively heavy 35 pounds. On 15 July 1968, he and his first mate, "Jungle Jeannie," became the parents of "Tiffany," a female weighing 4 pounds 2 ounces. And on 16 September 1969, with "Kribi Kate" as his mate, Big Man had a son, named "The Colonel," who weighed a hefty 6 pounds (that is, for a newborn gorilla). Dr. Pickett, who brought Big Man to the zoo, visited him throughout the years and was always remembered and welcomed, even when Big Man had grown to an imposing 450 pounds. The gorilla died of hepatitis and accompanying complications. Like the death of other big male gorillas who attracted thousands of visitors, those to the Kansas City Zoo were saddened by the loss of Big Man.

23. *Snowflake*

This unique "white" gorilla, which even by the natives who saw it was regarded as a phenomenon, was captured on 1 October 1966 in a rain forest in northern Rio Muni, Spanish equatorial Africa. The little white ape was found clinging to the body of its black-haired mother, who had been shot by a native farmer as she raided his banana patch. When notified of the rare specimen, the Spanish zoologist and animal collector Jorge Sabater Pi purchased "Snowflake" (who was called by the natives "Nfumu Ngi," or white gorilla), trained him, then shipped him to the zoo in Barcelona.

When first captured, the little gorilla, which was a male, weighed 19¼ pounds. This weight would normally presuppose an age of about 11 months. However, the stage of dental development indicated an age of about 2 years. Thus, in common with many other infant gorillas obtained under such circumstances, Snowflake was small for

Fig. 138. Big Man, being held by Deets Pickett, D.V.M., who captured the gorilla in the Cameroons, west Africa, when Big Man was about a year old, and brought him to the Kansas City Zoo. Here the gorilla, who was always happy to see Dr. Pickett, is shown at the age of about two years (1959). (Photo courtesy Dr. Pickett.)

long life in the congenial atmosphere of the Barcelona Zoo. Snowflake's living habits and behavior were essentially the same as those of other captive gorillas of his age, and this was true also of his intelligence.

The diet adopted for both Snowflake and Muni consisted—and may still consist—of a main meal, served about 1:30 P.M., in which the two gorillas are given bananas, apples or pears, bread and jelly, and a serving of boiled ham or roast chicken. In three smaller meals served at appropriate intervals they receive yogurt, hard-boiled eggs, rice, cookies, and some raw beef. The inclusion of a small portion of meat in the menu has been found to be relished by most captive gorillas, notwithstanding that in the wild state their diet is almost without exception vegetarian.

As to Snowflake's being an "albino" animal, strictly speaking this is not the case; for Snowflake has blue eyes (rather than pink), and flesh-colored

his apparent age. For a while after his arrival in Barcelona, during the cool winter months, Snowflake was cared for by the zoo's veterinarian, Dr. Roman Luera Carbo, and his wife, in their apartment. Señora Carbo, who had previously reared a score of baby apes, gave Snowflake the motherly care and attention so vital to young gorillas, particularly during the early stages of their captivity. Snowflake responded well and continued to grow and adapt himself to the conditions around him. Shortly, a normally colored, black male gorilla of the same age, named "Muni," was brought in as a companion for Snowflake, and the two young apes became close friends.

After four years (i.e., in 1970), when 6 years old, Snowflake, weighing about 100 pounds, was still small for his age—paralleling in this respect "Albert" of the San Diego Zoo—but was healthy and vigorous. Hopefully, he was destined for a

Fig. 139. Two-year-old Snowflake, the world's only white gorilla, enjoying an apple in his temporary home with Roman L. Carbó, zoo veterinarian, and his wife in their apartment in Barcelona, Spain. Later, the unique little gorilla was transferred to the Barcelona Zoo and has since become a parent, although his offspring was black. (Photo by Dr. Carbó.)

skin (rather than white). Curiously, while normally colored young gorillas have a tuft of white hair on the rump, in Snowflake this tuft is (or was) black.

In 1974, when he was about 10 years of age, Snowflake mated with a normally colored female gorilla of similar age, named Greta; and from that union came an infant that was black. However, the infant is carrying genes that are recessive for whiteness, so that in due course, by mating two of Snowflake's offspring, the Barcelona Zoo may be able to produce white gorillas.

Excellent detailed accounts of Snowflake, by the primatologist Dr. Arthur J. Riopelle, are given in two issues of the *National Geographic* magazine, and by Geoffrey H. Bourne and Maury Cohen in their book *The Gentle Giants*.[20]

24. Colo

While most of the gorillas discussed previously in this list have been adults, we shall now introduce several newborn gorillas, each of which gained widespread publicity and distinction from being one of the first gorillas ever born in captivity. Of these famous newborns, the very first was "Colo," a female of the typical western lowland species.

Colo derived her name from the place, Columbus, Ohio, where she was born on 22 December 1956. Her father was "The Baron" (c. 10 years old, 380 lbs.) and her mother was "Christina" (11 years and 4 months, 260 lbs.). Thus, The Baron and Christina were the first captive gorillas ever to become parents. Colo, whose gestation period had been 259 days, was appreciably under the average female birth weight, scaling only 3¼ pounds. She had been discovered in Christina's cage by Warren Thomas, a veterinary student, who started the tiny ape breathing by blowing his own breath into its lungs.

The unprecedented appearance of a newborn gorilla in captivity attracted a multitude of visitors, many of them zoo workers, to the Columbus venue. The little ape certainly made a fine, though perhaps unexpected, Christmas present for the zoo. Colo was not nursed by her mother and at first was placed in an improvised incubator. Privileged visitors were allowed to see the tiny gorilla, provided that each onlooker wore a germicidal breathing mask. Her diet consisted of a mixture of milk formula and boiled water, which had been prescribed by a local pediatrician. At three weeks of age Colo could gurgle, suck her thumb, and play with her toes. Shortly thereafter she was put on a diet of pabulum. By the age of 14 months, she had caught up with the average female weight for that age: 22 pounds.

Colo was distinguished further by being the first gorilla born in the United States—indeed, in the world—to give, in turn, birth to offspring. On 1 February 1968, when she was 12 years and 6 months of age (at last, the age of a captive gorilla was becoming definitely known!), she gave birth to a female who was named "Emmy," and who weighed three pounds ten ounces, being relatively small, as Colo had been at birth. The father (Colo's mate) was "Bongo," age 10 years 8 months. A second baby, by the same "gorilla couple," was born on 18 July 1969, when Colo was 12 years and 6 months of age, and Bongo 12 years and 1 month. This time the baby was a male, who was promptly named "Oscar" (to complement the movie name given to his sister "Emmy"). Oscar weighed a good 4 pounds 12 ounces.

Colo's debut into a field where ape keepers had been dubious about ever achieving a success provided a much-needed psychological "shot in the

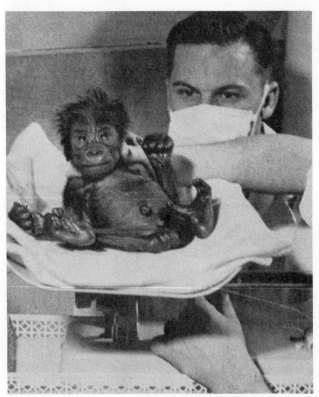

Fig. 140. Colo, the world's first gorilla born in captivity, being weighed by the zoo veterinarian, Robert Vesper, only whose arms appear here. In the rear is veterinarian student Warren Thomas, who first discovered Colo and saved her by applying artificial respiration. (Photo courtesy Columbus Zoo.)

Fig. 141. Goma, the first gorilla to be born in captivity in Europe (23 September 1959, at the Basel Zoo, Switzerland). She is shown here at the age of four days. For details, see text. (Photo by Dr. Ernst M. Lang; approximately one-half natural size.)

arm" to zoos everywhere that possessed gorillas. Now, while still far from an everyday occurrence, the actuality of the births in captivity of Colo and other "pioneer" gorilla infants has given zoologists and conservationists an assured means for saving the threatened gorilla species from extinction.

25. Goma

Like Colo, in the United States, this zoo-born gorilla, also a female, attained widespread publicity by being the first gorilla of either sex to be born in captivity in Europe—or, for that matter, in the Old World. She was not, however—as one writer of a book about gorillas has asserted—"the first to survive for any length of time." As we have seen, Colo in the United States survived long enough to become the mother successively of two offspring and, as far as I know, is still living, which would make her (in 1976) twenty years of age.

Goma—who was named after the village of that name (meaning "dancing festival") situated on the north shore of Lake Kivu in Ruanda, East Africa—has a name that comes within the habitat of the mountain gorilla, rather than one of the western lowland species, of which she is. Goma was born in the zoo at Basel, Switzerland, on 23 September 1959—2 years and 9 months after Colo had made her appearance in Columbus, Ohio. Goma had a gestation period of 289 days—a full month longer than that of Colo, yet weighed at birth only a tenth of a pound more (i.e., 4 pounds). Goma's father was "Stefi" (aged 12 years and 5 months, weight 350 pounds); and her mother was "Achilla" (aged 12 years, weight 165 pounds). Note the difference in the two parents' weights.

As in the case of Colo, the mother gorilla did not seem to know how to take care of her first-born baby. Colo's mother simply left her infant lying on the floor of the cage, where it would shortly have died had it not been for the timely aid of one of the zoo attendants who found her. Achilla actually held Goma in her arms, but the wrong way around, with the infant's back to her breast, so that the little one could not feed. On the following day, the zoo director, Dr. Ernest M. Lang, decided that Goma should be taken from her inept mother and brought up in the zoo's primate nursery.

About nineteen months after Goma's birth, her parents, Stefi and Achilla, had a second baby—this time a male, who was named "Jambo" and who is reviewed next on this list. A third offspring, another male, who was named "Migger," was born to Stefi and Achilla on 1 June 1964. And on 17 July 1968, a fourth infant—a female named, appropriately, "Quarta"—was born to the same pair. Thus, the Basel Zoo successfully administered the raising of four new-generation gorillas—all born to the same parents.

Goma, upon becoming of nubile age, is said by one author to have had a baby at the zoo in Frankfurt. However, according to the list of gorilla births compiled by Marvin L. Jones (and reproduced here in chapter 13), the only mother gorilla at the Frankfurt Zoo between 1965 and 1971 was "Makula" (not Goma).

For detailed information on the birth and upbringing of Goma, see the several accounts by Ernst Lang.[21]

26. Jambo*

As just mentioned in connection with No. 25, "Jambo" had the same parents as Goma (namely, Steffi and Achilla). Accordingly, he was born at the same zoo: Basel, Switzerland. The date of

---

* To avoid confusion, it should be mentioned that an infant lowland male of this same name came to the Highland Park Zoo in Pittsburgh in June 1951, at the estimated age of twenty-one months and a weight of 25¾ pounds. Thus, the United States Jambo is (or was) about 11½ years older than the Switzerland Jambo.

Fig. 142. Jambo, the third living gorilla ever born in captivity (17 April 1961), is here being affectionately cradled by his fourteen-year-old mother, Achilla, at the zoo in Basel, Switzerland. Nineteen months later, Achilla had become the mother of a female infant, Goma; but since she did not at that time exhibit the care she later gave to Jambo, Goma had to be taken from her and raised by human means. Fortunately, the little ape thereby did well. (Photo by E. M. Lang.)

Jambo's birth was 17 April 1961. He thus became the fourth (living) gorilla to be born in captivity in the world; the second to be thus born in Europe (Steffi and Achilla's first offspring, a female born on 29 March 1958, had been aborted); and the first male gorilla to be born in captivity either in Europe or perhaps the world (see the listing of births by Marvin Jones in chapter 13). Jambo had the additional distinction of being the first gorilla of either sex whose mother took care of him.

Jambo's name comes from the native Swahili word meaning "good morning." At birth he weighed the same as his sister Goma had weighed nineteen months earlier, namely, 4 pounds. At the age of 2 years, weighing 31 pounds, he was somewhat light for that age. However, he later made up this slight deficiency and turned out to be a splendid example of his species (western lowland).

Attendants at the Basel Zoo had unending entertainment watching mother Achilla take care of infant Jambo. But although Achilla carried her baby around with her, she made no effort to help him find one of her nipples. Fortunately, he soon learned to do this for himself. Thereafter he fed whenever he was hungry, and Achilla seemed to enjoy having him do so. Dr. Lang, director of the zoo, who kept almost constant watch on Achilla and her infant during the early stages of his development, reported many amusing as well as enlightening incidents. One of these was as follows. Achilla, who had learned to drink soup with a spoon, tried to feed Jambo in the same manner, but without much success. "Once she did manage to poke the loaded spoon at his mouth, but the next time she tried it he turned his head and got an earful."[22]

Much credit for the successful upbringing of Jambo is due to Carl Stemmler, head keeper at the Basel Zoo. By exercising endless patience, Stemmler, some six weeks after Jambo was born, succeeded in allaying Achilla's natural fears for the safety of her infant to the point where she freely handed the little ape over to Stemmler (!). Thereafter, assured of the keeper's good intentions, the mother gorilla would sit and watch Stemmler hold and play with Jambo just as though the baby gorilla were a human one. At the age of two years, Jambo was no longer breast-fed and was separated from his mother. His life at the Basel Zoo for the next several years was exemplary, so far as being a successfully raised, healthy young gorilla was concerned. On 27 April 1972, Dr. Lang took Jambo, who had just passed his eleventh birthday and was nearing the adult stage, to the Jersey (Channel Islands) Zoo. Jambo, just a year previously, had mated in Basel with Goma, his sister, so the zoo people in Jersey were happy to receive a potential mate for their two adult females, "N'Pongo" (born 1957 in the French Cameroons) and "Nandi" (born 1959 in the French Cameroons). By Nandi was born (14 July 1973) a male named "Assumbo" and by N'Pongo, on 11 September 1973, another male, named "Mamfe"—the father in each case being, of course, Jambo. A third baby gorilla, a female named "Zaire," was born on 23 October 1974 to Nandi, the father again being Jambo. According to attendants at the Jersey Zoo, "Jambo is of a friendly disposition and enjoys playing games, such as chasing or games with the water hose, but in line with the majority of mature male gorillas he cannot be trusted."[23]

* * *

From the foregoing series of condensed reviews —the number of which could readily be doubled or trebled—it is seen that gorillas range as widely in physique, intelligence, longevity, adaptability, behavior, and general individuality as do human beings, at least within a population of the same number. And while general averages may be drawn from each and all of these characteristics, it is the individual animal that provides the most interest. Hence the enthusiastic attention given by zoo-goers to individual gorillas, which may even come to be looked upon as personal friends. Not even the chimpanzee nor the orangutan engenders this degree of sympathetic relationship.

The vast gain in knowledge by zoologists and animal keepers during recent years concerning these most interesting subhuman primates is an assurance that progress in the rearing of gorillas in captivity will save the species from the almost certain extinction, which today faces it in the wild.

Notes

1. William T. Hornaday, in *Zoological Society Bulletin* (New York) 18, no. 1 (January 1918):1184; Henry W. Sheak, "Anthropoid Apes I Have Known," *Natural History*, January–February 1923, p. 48.
2. Sheak, "Anthropoid Apes I Have Known," p. 48.
3. Adolph H. Schultz, "Notes on the Growth of Anthropoid Apes, with Especial Reference to the Deciduous Dentition," *Report, Laboratory and Museum of Comparative Pathology, Zoological Society of Philadelphia*, 1930, pp. 34–35.
4. Roger Conant, "Meet the Champions," *Fauna* (Philadelphia

Zoological Society) 3, no. 2 (June 1941):48.
5. William M. Mann, *Wild Animals In and Out of the Zoo*, Smithsonian Scientific Series, vol. 6 (Washington, D.C., 1943).
6. Paul E. Steiner, "Anatomical Observations in a *Gorilla gorilla*," *American Journal of Physical Anthropology*, n.s. 12 (1954):145–65.
7. Frederick A. Ulmer, Jr., "Massa—Dean of Captive Gorillas," *America's First Zoo* (Philadelphia) 18, no. 4 (December 1966): 27–30.
8. Colin P. Groves, "Ecology and Taxonomy of the Gorilla," *Nature* (London) 213 (1967):890–93.
9. Kenheim Stott, Jr., personal communication, September 30, 1975.
10. C. R. Carpenter, "An Observational Study of Two Captive Mountain Gorillas (*Gorilla beringei*)," *Human Biology* 9, no. 2 (May 1937):175–96.
11. Belle J. Benchley, *My Friends, the Apes* (Boston: Little, Brown & Co., 1942).
12. Belle J. Benchley, "Mbongo and Ngagi," *Touring Topics*, October 1932, pp. 15, 17, 48; idem, "Mbongo and Ngagi," *Nature Magazine* 21, no. 5 (May 1933):217–22.
13. Belle J. Benchley, *Mountain Gorillas in the Zoological Garden, 1931 to 1940* (San Diego, Calif.: Publications of the San Diego Zoological Society, 1940), pp. 1–24.
14. Frank D. McKenney, J. Traum, and Aileen E. Bonestell, "Acute Coccidiomycosis in a Mountain Gorilla (*Gorilla beringei*), with Anatomical Notes," *Journal of the American Veterinarian Medical Association* 104, no. 804 (March 1944): 136–41.
15. Ivan Dmitri, "UGH! UGH! UGH!", *Saturday Evening Post*, January 4, 1941 (includes five full-color photographs of M'toto); see also A. Maria Hoyt, *Toto and I* (New York: J. B. Lippencott Co., 1941); B. H. Mc Court, "The Gorilla who lived like a Lord," *Animal Life* (Worcester, England), December 1965, pp. 6–9.
16. G. H. Bourne and M. Cohen, *The Gentle Giants* (New York: G. P. Putnam's Sons, 1975), pp. 65–75; Vernon Reynolds, *The Apes* (New York: Harper & Row, 1971).
17. In *Life Magazine*, May 28, 1951, p. 30.
18. Dianne K. Risser, personal communication, October 6, 1975.
19. George Speidel, personal communication, February 2, 1976.
20. Arthur J. Riopelle, " 'Snowflake,' the World's First White Gorilla," *National Geographic* 131, no. 3 (March 1967): 442–48; idem, "Growing Up with Snowflake," ibid. 138, no. 4 (October 1970):490–503; Bourne and Cohen, *The Gentle Giants*, pp. 199–203.
21. Ernst M. Lang, "Goma, das Basler Gorillakind," *Documenta Geigy*, 1960–61, nos. 1, 3, 8; Ernst M. Lang and R. Schenkel, "Goma, das Basler Gorillakind," ibid., 1961, nos. 6, 7; Ernst M. Lang, *Goma the Baby Gorilla* (London: Gollancz, 1962; New York: Doubleday, 1963).
22. Ernst M. Lang, "Jambo—First Gorilla Raised by its Mother in Captivity," *National Geographic* 125, no. 3 (March 1964): 446–53; idem, "Jambo, the Second Gorilla Born at Basel Zoo," *International Zoo Yearbook*, 1961.
23. Jeremy J. C. Mallinson, Phillip Coffey, and Jeremy Usher-Smith, "Maintenance, Breeding, and Hand-rearing of Lowland Gorilla at the Jersey Zoological Park," *Annual Report, The Jersey Wildlife Preservation Trust*, 1973, pp. 15–28.

11

Care and Feeding of Gorillas in Captivity

Breeding, Pregnancy, and Birth

Obviously, the first step in breeding a baby gorilla is to bring together two potential gorilla parents. Evidently in the wild such pairing-off takes place through "natural selection." But in a zoo, or other domicile of captive gorillas, those who supervise such breeding are fortunate if they have a single eligible pair of gorillas, or at the most two such pairs. Accordingly, the "selection" has to be artificial rather than natural and may prove either a success or a failure. As has been mentioned in chapter 10, Makoko and Oka did not mate; neither did Mambo and Oka, nor Gargantua and M'toto. Then there was, and possibly still is, Bobo, a 560-pound male gorilla at the Seattle Zoo, who wanted nothing to do with a dainty (350-pound) enchantress named Fifi, for the purchase of whom the citizenry of Seattle had raised $4000. And Bobo's indifference was not because he might have been run-down, for each day he was given a massive dose of vitamin E, a swig of pure wheat-germ oil, eight (yes, 8!) raw egg yolks, a jigger of thiamine, and a 20 mg injection (?) of male hormones. The cause of Bobo's sexual reluctance, according to the zoo's veterinarian, Dr. Kenneth Binkley, was: "Single male primates raised from babyhood in human homes [as was Bobo] are highly neurotic. Bobo has human inhibitions—he simply will not make an exhibition of himself in public."[1]

Then there was Chicago's popular male gorilla, Bushman, of Lincoln Park Zoo. When Bushman had reached nubile age, there were two or three eligible female gorillas in other zoos whose keepers were eager to mate them to the apparently lonely male. But Marlin Perkins, then director of the Lincoln Park Zoo, considered Bushman such an exceedingly valuable animal that he declined to take any chance—such as an attack from an antagonistic female—of endangering Bushman's status quo.

The lesson to be learned from these random examples is that the successful breeding in captivity of a viable baby gorilla is still a far-from-routine procedure. The potential parent gorillas must be compatible and willing; and once the probable date of conception has been recorded, a close watch must be kept on the pregnant mother and maintained for the anticipated arrival of the baby gorilla. The likely date of arrival cannot be established to the day and hour any more than in the case of a human birth. Among the approximate 100 gorillas born in captivity to date, the gestation period has ranged from 236 to 290 days, the average being 265. And as the number of births of gorillas increases, so should the probability of a wider range in the gestation period. As an indication of this probability, the extreme range in man (among many millions of births) would appear to be 135 to 375 days, the average being 280.

Up to August 1969, of 16 lowland female gorillas that had given birth, all of their first-born offspring had to be hand-reared, except one that died before this could be undertaken. According to Von R. Kirchshofer, of 32 living and still-births, 22 took place in the warm months between April and September. These births had occurred four times more frequently by day than by night.[2] However, as mentioned later in chapter 13, it would appear that, in general, gorillas in captivity as well as in the wild are born in approximately equal numbers in every month of the year. Accordingly, they can mate at any time during the year.

The birth in a gorilla of twins occurs only rarely. On 3 May 1967, at the zoo in Frankfurt, Germany, twin female lowland gorillas were born to "Abraham" and "Makula," but only one of the twins survived. On 18 September 1966, at the Kansas City Zoo, "Big Man" had fathered twins that were aborted, and of which the sex and the mother gorilla's name were not stated. Triplets have been born to chimpanzees, but never, so far as I know, to gorillas.

Because of the natural abdominal rotundity of the female gorilla, it is difficult or impossible to determine from the ape's appearance whether or not she is pregnant (see Fig. 104). At the St. Louis Zoo the only indication of pregnancy in the gorilla "Trudy" was a swelling of the breasts four months prior to her giving birth.[3] A similar enlargement of the breasts about 3½ months prior to giving birth occurred in 1970 in one of the two adult gorillas then at the Cincinnati Zoo.[4] A general gain in weight of the pregnant gorilla "Helen" of the Lincoln Park, Chicago, Zoo, was observed four months prior to her giving birth. Additional reports on this subject indicate that some time between the fourth and the sixth month, a state of pregnancy may be inferred from the swelling of the breasts that then occurs.

The first course of action, once a gorilla has been born in captivity, is to see whether or not the mother shows any inclination to hold to her body and to nurse her new offspring. However, from the meager evidence thus far recorded it would appear that newborn gorillas require a considerable amount of time to adapt to their mothers. "Jambo," for example, was not observed suckling from his mother until he was 35 hours old.[5] "Daniel," the first gorilla born at the Bristol (England) Zoo, similarly started after 32 hours. "Goma," the first gorilla born at the Basel Zoo, was left with her mother, "Achilla," for 36 hours. When, after that length of time, she had been unable to locate a nipple (and had not been assisted in this endeavor by the mother ape), she was removed from the cage for hand-rearing.[6] "Assumbo," the first baby born to "Nandi," at the Jersey Zoo, had not, even after 40 hours, been nursed by his mother. Accordingly, he, like Goma had been, was removed for hand-rearing. It should be added that unless the mother gorilla shows no interest in the removal of her offspring, and keeps at a distance, she should be put under sedation for a period sufficient to enable the baby gorilla to be removed safely. The means commonly employed for the sedation is an injection (usually of about forty-five mg) of Sernylan via a dart fired from a "Capchur" pistol.

The nursery in which an infant gorilla is to be hand-reared must be equipped with what, to an "outsider," would appear to be an incredible number of items. In short, the gorilla nursery must provide all the essentials required in a nursery for human infants. As a brief indication, there should be one, or preferably two, incubators, playpens, laundry baskets, cots and mattresses, and a pair of baby scales. Then there should be quantities of towels, sheets, blankets, face masks, surgical gowns, disposable diapers, plastic pants, etc. The feeding equipment will include numerous bottles and bottle warmers, measuring beakers, disposable syringes, etc.; while a large quantity of items will be needed for bathing, anointing, and keeping clean both the baby and its surroundings. In addition, medical and surgical equipment must be available for clamping and ligaturing, if necessary, the newborn's umbilical cord, and for giving essential vaccinations.

The development of various behavioral actions between the ages of 3 and 18 weeks in two male lowland infant gorillas—"Assumbo" and "Mamfe" —at the Jersey Zoo is shown in the following table, which is reproduced from an excellent paper on the hand-rearing of infant gorillas.[7] This table illustrates the wide variation in the respective number of days required for two different gorillas to learn and perform the same acts.

Table 23. Comparative Behavior in the Male Gorillas "Assumbo" and "Mamfe" between 3 and 18 weeks of Age (after Mallinson, Coffey, and Usher-Smith, of the Jersey Zoological Park. Jersey Wildlife Conservation Trust Annual Report, 1973).

"Assumbo" Age (days)	"Mamfe" Age (days)	Event
22	24	Finger and thumb sucking becoming pronounced.
30	24	First in the basket.
45	66	Lifting with hands pulling on horizontal pole over basket.
54	46	Interest in rattle.
63		Rolling in basket.
63	33	Watching own hand movements.
64	59	First in the playpen.
69	25	Observation of first interest in each other.
70	78	Crawling using arms only.
70	84	Rolling in the playpen.
76	85	Crawling using both arms and legs.
77	69	Pulling himself around playpen using the vertical bars.
86	105	Mouthing own feet.
93	94	Standing supported by vertical bars.
120	98	Climbing using hands and feet.
122	69	Observation of first interaction between them in playpen.
122	128	Biting when tired or annoyed, using threat 'cough'.
124	93	Sitting up independently.
127	151	Holding feed bottle.
140	129	Walking on all fours.
143	111	Threw first tantrum.
144	94	Attention seeking from foster parent.
154	97	Observation of first real 'play' together.
156		Swinging from horizontal with feet.
162	166	Tooth grinding.
170	133	Fighting other infant (over a towel).
180	145	Smelling objects rather than immediately mouthing them.
183		Undoing bolts on playpen.
185		Inventing a game.
195	170	Great interest in foster parent's mouth and face.

Although, as mentioned earlier, the hand-rearing of a baby gorilla deprives the young ape of the benefits of association and rapport with its parents and its "family" troop—which it would experience under natural conditions in the wild —it has the compensatory advantages, at least, of enabling the mother gorilla to have a baby every year, as do many humans, rather than two or three years apart, as is the case where the infant gorilla is nursed by its mother. Too, although the hand-reared infant is separated from its parents, it can be, and generally is, provided with a playmate of its own species and age, the provision of which goes a long way towards supplying the companionship and sense of security so vitally needed by gorillas of all ages. And where a gorilla companion is not available, a chimpanzee of similar age often proves a satisfactory substitute (see Fig. 61).

In a number of studies made at different zoos, observations of the behavior and treatment of the mother gorilla prior to parturition and of the baby gorilla during the first few weeks of its life in a nursery have been recorded in great detail. In one such report, for example, over 260 observations were listed by various members of the zoo's staff merely of a single event: a 9½-hour period of labor. Even the authors had to summarize this extensive report.[8] Readers (probably zoo workers, principally) interested in these detailed accounts are directed here in this chapter to notes 7, 8, 9 and 10.

In various comparisons that have been made between gorilla and human infants it has been demonstrated that in most forms of physical capability the ape babies surpass the human babies. This is only to be expected where the tests involve brachiation (hanging from a bar by the hands),

Fig. 143. Dolly, of the San Diego Zoo, with her infant female, Binti, here only a few days old. Binti was born on 2 October 1974. Dolly was then ten years and ten months old; while the father of Binti, Trib, was about fifteen years and two months. As in the case of Achilla, the mother gorilla in Basel, Switzerland, Dolly neglected her first-born, a male named Jim, but took proper care of her second-born, a female. (Photo by F. D. Schmidt; courtesy San Diego Zoo.)

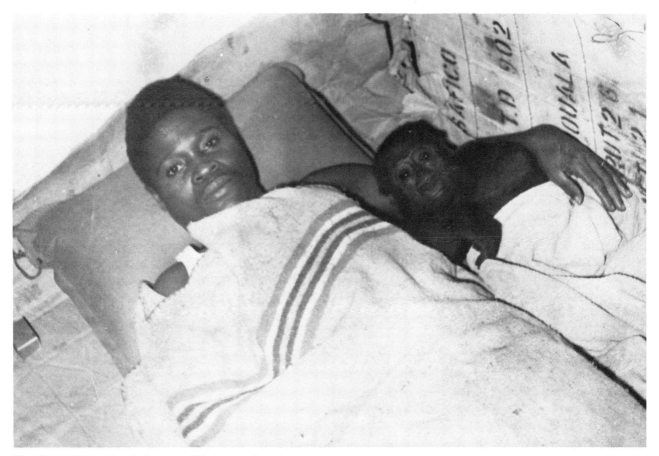

Fig. 144. Here, an infant gorilla being hand-reared in Africa is provided with a "round-the-clock" native nurse, whose continuous presence gives the emotionally insecure little ape the companionship and friendship so vital to its health and survival. (Photo courtesy Dr. Deets Pickett.)

since the apes' hooklike fingers and light lower limbs adapt them naturally for such activities. And, of course, the apes' handlike feet are endowed with far greater grasping power than are the feet of children, which are adapted primarily for supporting and walking, rather than grasping. Baby apes can even hang by their feet with ease —a feat that is totally beyond the limited grasping power of human toes, in which the big toe is not apposable to the other toes, as is oppositely the case among quadrumanous ("four-handed") apes and monkeys. So, in such tests, baby gorillas (and chimpanzees, orangutans, and gibbons) are "more advanced" at a given age than human infants, largely because they are better adapted anatomically for these tests. It may be added, however, that recently born gorillas do not have the ability to hang onto their mothers' hair to the extent shown by chimpanzees and orangutans, and they often have to be helped by their mothers when their grip starts to weaken. To sum up, at a given chronological age (postnatally), infant apes are ahead of infant humans in certain respects by reason of their possessing natural structural advantages. Another factor in the apes' favor is that they grow at a much faster rate. This is clearly indicated by the following comparison.

If, as extensive statistics show, the optimum white male infant weighs 7.7 pounds at birth and 23.2 pounds at 12 months, he has during that period increased in weight by 7.7 | 23.2, or $\times$ 3.013. Comparable weights for a healthy, lowland male gorilla are 4.7 pounds at birth and 20 pounds at 12 months, or $\times$ 4.255. The latter ratio, divided by that for the human infant (3.013), shows that the average male gorilla grows $\times$ 1.41, or over 40 percent, faster than a well-reared human baby boy. Indeed, since adopting, some years ago, a weight of 20 pounds as being typical for a year-old gorilla, it would appear that today a better goal should be 25 pounds. This for the reason that in first adopting 20 pounds, I was perhaps influenced unduly by the large number of malnourished infant gorillas then in captivity. Today, the typical growth of gorillas during the first year indicates that the optimum weight at the end of 12 months should be 20 pounds more than the weight at birth; that is, a gain of about 1.70 pounds per month (in either sex). This being the case, it makes the gorilla's rate of growth 76 percent faster than that of the human infant. And at ages beyond one year, the difference in the growth rate is even greater.

Thus, in physical tests—such as lifting or holding the head up while lying; turning the body over; crawling, standing, or sitting—it would seem that a gorilla infant should be compared with a human infant at least 50 percent older (rather than one of the same age), in order to properly take into account the more rapid maturation of the gorilla.

Ape Houses in Today's Zoos

One of the major factors in the success attained during recent years in the raising and accommodating of gorillas has been the modernizing of their living quarters—the "Ape House." Quite apart from the ape house itself are the spacious "grounds" surrounding it, in which an entire troop of gorillas can move about with almost as much freedom as a troop in the wild would enjoy. In many of the larger zoos, in addition to the terraced levels and runways, there are water-filled moats that separate the gorillas from the public, and in which—if the water is not too deep—the apes like to splash about, especially in warm weather. In view of the fact that a number of apes —chimpanzees as well as gorillas—have been drowned by falling into deep-water moats, precautions are now taken to prevent further accidents of this kind.

As an example of "modern" gorilla accommodations, the Ibadan Zoo in Nigeria has an enclosure measuring 2870 square feet, or equal to an area of over 53 by 53 feet. The same size and type of enclosure is used for chimpanzees. An overhanging roof shelters portions of the enclosures from sun and rain. The moat has a sloping floor and is divided partway across its entire length by a fence topped with electrified wires. This gives the gorillas a "swimming pool" 4½ feet wide by perhaps 50 feet in length, in which the maximum depth of the water is 3½ feet. Beyond the gorillas' area, the water continues on the other side of the electric barrier for a width of 10 feet and reaches a maximum depth of 6½ feet. It ends at a high wall topped by a railing, behind which the spectators watch the gorillas, and vice versa. Curiously, the chimpanzees do not like the water, so keep out of it; but the gorillas enjoy it and have even learned to swim for short distances, although they avoid putting their faces in the water. The gorillas, while disturbed if a person other than their keeper comes within close range, do not seem to be bothered by the crowds of visitors who view them from the far side of the moat.[11]

Each zoo, of course, must adapt its grounds and runways, moats or pools, and ape houses to the amount of space available and to such existing obstacles that may not be alterable or removable. With few exceptions, however, todays' gorillas are given ample room, both inside and outside their enclosures, to live comfortably and healthily.

As to the construction of the gorillas' indoor living and sleeping quarters, in most present-day zoos it is essentially as follows. First, the former, conventional iron-barred cage has been replaced by a wall of ¾-inch-thick laminated glass, which is installed in large sections. This glass has several important advantages over a barred cage: (1) it keeps the gorillas free from germs carried by spectators—in particular, germs connected with respiratory diseases, to which gorillas are highly susceptible; (2) it enables the spectators to see the apes more clearly; (3) the ventilation, humidity, and temperature within the glassed room can be controlled and kept at an optimal level; (4) the room, being enclosed, keeps disagreeable animal odors from reaching the spectators.

In some zoos, too, a scale on which the gorilla, or gorillas, may be weighed is arranged so that whenever one of the apes sits on a platform inside the cage, the weight is shown on a dial located outside the cage, where spectators may read the poundage (see Fig. 134). Also, most such "gorilla quarters" are equipped with ropes, bars, and other apparatus that can be used for exercise and play. A favorite object in the cage of a big male gorilla is a large auto or truck tire, which he can pick up and bend together. In the sleeping quarters, heaps of hay or straw are provided, from which each gorilla can arrange his, or her, own bed for the night. The beds, which take the place of the "nests" that gorillas in the wild construct out of the surrounding vegetation, are generally located on solid shelves placed some distance above the floor.

In warm weather, some zoo gorillas may prefer to remain outdoors at night, in which event they can make their bed or nest on a high shelf or terrace from whatever straw or leaves is lying about. Some gorillas like to stay outside even if

Fig. 145. Baby gorillas, like baby boys and girls, need frequent baths. This infant gorilla follows the usual pattern in not relishing the soapsuds in his (or her?) eyes. (Photo courtesy Dr. Deets Pickett.)

it is raining. The two male mountain gorillas "Mbongo" and "Ngagi" of the San Diego Zoo were of this nature and evidently got great pleasure out of cavorting in the rain—even if it lasted all night.

It may be added that the construction of present-day gorilla houses is so arranged that the attendants can look after each cage from a service corridor. Whenever it is necessary to enter the cages, the various doors involved in shunting the ape, or apes, from one room to another can in many zoos be operated electrically, as described in connection with "Samson" (No. 20, chapter 10). But even under these circumstances the gorillas get to know and favor their regular keepers. Some individual apes may become so attached to a particular attendant that they will sulk and refuse to eat if that person is kept away for any length of time.

Which brings us to the vital subject of how best to feed gorillas to keep them strong and healthy and enable them to live long.

Feeding Zoo Gorillas

The exacting task of feeding a captive gorilla often starts in the infant's crib, where he (or she) is given a baby's "formula." For the Jersey Zoo's gorillas, "Assumbo" and "Mamfe," the first feed-

Fig. 146. Not all gorillas like water, but seven-year-old Massa, of the San Diego Wild Animal Park, enjoys a dive and dip into the moat, where the water is only eighteen inches deep. A circular pool, twelve feet in diameter and two feet deep, is also supplied. (Photo courtesy San Diego Zoo.)

ings consisted of powdered milk to which was added a 5-percent Dextrose solution. Feedings were given every 3 hours and were rapidly increased in volume from the first day (c. 90 ml) to the seventh day (c. 380 ml).

From the foregoing period to six to eight months of age, Assumbo and Mamfe were fed mainly an increasing amount of the powdered milk, along with mashed bananas (usually 3 per day) and strained fruit, the latter being increased steadily in quantity. A detailed table of the foods given these two infant gorillas, including the quantity of each and the number of feedings per day of each, is given in the paper by Mallinson, Coffey, and Usher-Smith referred to previously (note 7; pp. 14–17).

Although Assumbo was given solid food when he was only a week old, and Mamfe when he was three weeks old, the female infant Zaire was not given such food until she was two months of age. She was also given smaller amounts because of her smaller size (see note 10; p. 46). Thus, it is seen that the feeding of newly born gorillas is largely an individual matter, each one having to be fed what, by experience, seems to be best tolerated and assimilated. Frequently, constipation develops, whereupon brown sugar is customarily added to the feedings, although there are other means of combating the condition.

Notwithstanding individual differences as to the kinds of food best tolerated in infancy, a more or less fixed total volume of food (solid plus liquid) is generally given in accordance with the age and weight of the gorillas. At the San Diego Zoo, the following schedule was adopted for the feeding of the three lowland gorillas—"Albert," "Bouba," and "Bata"—as they grew between the lowest and the highest body weights listed.[12] To the table as originally published I have here added the first column, in which is given the approximate average age of the three gorillas conforming with the weight shown in the second column.

Note that the quantities listed in the above chart are for all three gorillas combined. Accordingly, each gorilla received approximately one-third of the amounts listed, with Albert, who was the smallest, probably getting the least. But on these rations all three grew satisfactorily.

In regard to the kinds and amounts of food comprising the diets of adult or subadult gorillas, the following "menus" of some well-known zoo specimens should provide a general idea. "Bamboo," for example, who lived at the Philadelphia Zoo from 1927 until his death in 1961, when he had attained the record age of 34 years and 4½ months, was given daily amounts of food not exceeding a total of 4000 calories. The following details are of the diet he was following in 1957, when he was 30 years old. Bamboo was fed only twice a day. His first, and largest, meal included a liberal portion of a highly nutritious mixture called "Monkey Cake." This was composed of one part boiled ground beef and nine parts dry mixture. The meat and the dry mixture were covered with a sufficient amount of meat broth to make a stiff mash. This was pressed into shallow pans and placed in a refrigerator (45 degrees) overnight to harden, after which it was cut by the keepers into sizes appropriate to the size of the gorilla, chimpanzee, or other primate to which it was fed. The dry mixture consisted of the following amounts of ingredients: 20 percent whole wheat meal; 20 percent soybean meal; 20 percent rolled oats; 10 percent yellow cornmeal; 10 percent peanut meal; 5 percent dry skim milk; 5 percent brewer's yeast; 5 percent alfalfa-leaf meal; 2 percent oyster-shell flour; 2 percent iodized salt; 1 percent concentrated cod-liver oil. All of these items were mixed well together and then made into the "cake" described.

Bamboo was generally given about a one-pound serving of the cake. Some hours later, in the afternoon, he was served a lighter meal consisting of oranges and some leafy vegetable, such as let-

Feeding Chart—Daily Menu for Three Young Gorillas
(Supplemented by vitamin preparations)

Age	Approximate Weight of Gorilla	Fruits & Vegetables	Milk	Cereal	Hard-boiled Eggs
3 mos.	8–10 lbs.		3 oz. formula each 4 times a day	1 cup baby cereal	
6 mos.	12–15 lbs.	3–4 cans baby food, grated apple, mashed banana, orange juice	8 oz. formula each 3 times a day	2 cups baby cereal or pablum	
9 mos.	15–20 lbs.	6 cans Jr. food; 1 apple, 1 banana, 1 sweet potato each, oranges	8 oz. bottles each 3 times a day	2 cups pablum	
1 yr.	20–30 lbs.	8 cans Jr. food, bananas, tomatoes, pears etc.	8 oz. bottles each 3 times a day	2–3 cups pablum	
2 yrs.	30–40 lbs.	6 lbs. raw	3 qts.	3 cups	3
3 yrs.	50–70 lbs.	12 lbs.	3–4 cans	5–6 cups	3
4 yrs.	75– lbs.	15–18 lbs.	4 cans	8 cups	6

tuce or kale. Dr. Herbert L. Ratcliffe, the zoo's pathologist, who had worked out the "monkey-cake" mixture (which was low in cholesterol and fats of all kinds), felt that it was a major factor in keeping Bamboo healthy and enabling him to attain, for a captive gorilla, an unprecedented longevity.[13] In contrast to the many adult gorillas that had been allowed to overeat and so attain an impressive, but unhealthy, overweight condition, Bamboo at no time weighed much, if any, over 300 pounds, which included very little excess fat. A typical day's ration for Bamboo was 18 oranges (c. 7 lbs.); lettuce or greens equivalent to 4 heads (c. 5 lbs.); a one-pound slice of "Monkey Cake"; and 6 quarts of water (12 lbs). Thus, his total daily food intake was 13 pounds of solids and 12 pounds of water.

At a somewhat earlier date (1951), the Bronx Zoo's two adult lowland gorillas, "Makoko" the male and "Oka" the female, were fed a combined quantity of food consisting of the following items. Of the total quantity, Makoko received approximately 3/5 and Oka 2/5.

This dietary list was passed on to me by Lee S. Crandall, then general curator at the Bronx (New York) Zoological Park.[14] It was emphasized by him that the food quantities listed were those given to the two gorillas on a particular day, not necessarily every day. There was also a daily variation in the amount of food consumed, since on some days it was all eaten, while on other days some was left. The list should, however, provide a good idea of the variety of foods, minerals, and vitamins required in the daily dietary of two full-grown gorillas, one a male and the other a female. Note that three meals a day were served, and that some meat was included.

9:00 A.M. Gruel:

(Makoko to receive 3/5, Oka 2/5).

T = tablespoonful
t = teaspoonful

10 T. powdered milk
 8 T. sugar
 1 t. salt
60 T. Cerevim
 2 T. powdered malted milk
 2 T. cod liver oil
 1 T. honey
20 drops Zymadrops
 1/2 t. calcium phosphate
 1 gallon water
 6 eggs

2:00 P.M. Fruit and Vegetables:

(All food washed and picked over. Makoko to receive slightly over half)

3 bunches celery
2 bunches beets
2 bunches carrots
1 scallion
3 lbs. string beans
1½ lbs. horse meat, chopped and cooked
12 oranges
20 apples
6 lbs. bananas
1 lb. grapes
Other fruit in season, such as peaches, pears, cherries, cantaloupe, etc.
3 cans pineapple bits, per week

4:00 P.M. Gruel:

Same as at 9 A.M., except omitting cod liver oil, calcium phosphate, and Zymadrops. (Makoko ⅗, Oka ⅖)

10 slices whole wheat bread with syrup or jelly—5 to Makoko, 5 to Oka

As a rough estimate, it would appear from the above menu that Makoko, who weighed 468 pounds, was given each day about 21 pounds of food and 9 pounds of water; and Oka, who weighed 290 pounds, was given about 15 pounds of food and 6 or 7 pounds of water.

If the amount of solid food consumed daily is compared with the body weight of the gorilla, it develops that Makoko's intake amounts to about 4½ percent, and Oka's to a little over 5 percent. In Bamboo the ratio was only 4⅓ percent. On the basis of Bamboo's size (surface area, not weight), Oka's food-intake ratio (to body weight) should have been not more than 4½ percent, and Makoko's about 3¾ percent.

The food intake of the two huge mountain gorillas "Mbongo" and "Ngagi," former residents of the San Diego Zoo, consisted, during the period 28 April to 5 May 1940, of the following average amounts per ape per day (all except the milk and the eggs are listed in pounds):

Oranges 3–4, grapefruit 2, bananas 6–10, carrots 2–6, sweet potatoes 1, apples 3½–6, pears 2–4, tomatoes 1–2, celery 2–4, lettuce 2–3, bread ½. On two days of the week, cherries were given (¾–1 lb.), and on one day, green corn (3 lbs). The total daily amount of food, exclusive of milk and eggs, for Mbongo (who at the time weighed about 600 pounds) ranged from 23½ to 34½ pounds, and for Ngagi (who weighed about 530 pounds) from 29½ to 39½ pounds. In addition to the foregoing fruit and vegetables, each gorilla was given 2 quarts of milk a day. Mbongo was given 5 eggs a week, and Ngagi 12 eggs a week. Possibly it was felt that Ngagi was "underweight," since a year earlier, at 13½ years of age, he had weighed about 470 pounds to Mbongo's 450 pounds, Mbongo then being about 13. Note that no meat was given; nor is there any mention in this food list of vitamins, minerals, or other food supplements. In this connection the date (1940) must be taken into account, since at that time the "scientific" feeding of anthropoid apes was just getting under way.

While most of the food items in the above list were served only once a day, pears and apples were given twice a day, and bananas usually three times a day. All in all, during the aforementioned period of eight days Mbongo was fed about 37 pounds of food a day, and Ngagi about 40 pounds a day.[15] The amount of water they drank was not listed.

Although the food requirements of gorillas are not identical to those of the smaller anthropoid apes, such as chimpanzees (which are more frugivorous and do not need such vast quantities of food that develop the huge bellies characteristic of gorillas), nevertheless they should conform, as far as calories are concerned, to the general principle that the amount of food required is in ratio to the relative surface area of the body. The amount of muscular work performed, of course, influences this determination and therefore must be assumed to be equal in the animals being compared. It should be safe to assume that the physical activity of adult captive gorillas is no greater than that of average men and women in everyday life. Certainly it is markedly less than the almost ceaseless activity of chimpanzees.

Which brings us to the question of how much food anthropoid apes require on the basis of their surface area, which corresponds relatively with their muscular cross-section or muscular strength. Samuel Brody, an expert in the field of bioenergetics, derives the surface area (also the basal metabolism) by making it proportionate to the body weight raised (actually, lowered) to the 0.73 power. From this he finds that the daily food requirement in pounds in anthropoid apes is equal to the body weight in pounds $^{0.73} \times 0.184$.[16] This requirement, for various body weights, is shown in Table 24. Bringing the body weight to the 0.73

Table 24. Daily Food Requirements in Anthropoid Apes on the Basis of Body Weight and Ordinary Physical Activity.

Body Weight, lbs.	Food Requirement (relative) on basis of $BW^{0.73}$	Daily Food Ration, lbs., (excluding water)	Daily Water Ration, quarts	Approx. Calories (ordinary activity)
10	0.186	1.0	0.5	350
20	0.309	1.6	0.8	550
30	0.415	2.2	1.1	740
40	0.512	2.7	1.4	920
50	0.603	3.2	1.6	1090
60	0.689	3.7	1.8	1250
70	0.771	4.1	2.0	1400
80	0.850	4.5	2.2	1550
90	0.926	4.9	2.4	1700
100	1.000	5.3	2.6	1850
120	1.146	6.1	3.0	2100
140	1.284	6.8	3.4	2350
160	1.415	7.5	3.8	2600
180	1.539	8.2	4.1	2800
200	1.659	8.8	4.4	3000
220	1.777	9.4	4.7	3200
240	1.893	10.0	5.0	3400
260	2.006	10.6	5.3	3600
280	2.119	11.2	5.6	3800
300	2.230	11.8	5.9	4000
350	2.504	13.3	6.6	4500
400	2.768	14.7	7.3	5000
450	3.022	16.0	8.0	5500
500	3.267	17.3	8.7	6000
550	3.504	18.6	9.3	6450

power causes a doubling of the weight to bring an increase in the surface area of about × 1⅔ (rather than × 2). It may be noted that the same type of formula is employed in calculating feeds for horses, the body weight being raised to the 0.75 power for mature horses and to the 0.52 power for growing foals.[17]

Unfortunately, it is not possible to say just how much an adult gorilla should weigh by determining what it actually does weigh. What the proper, or normal, weight should be is governed by the stature (standing height) of the ape and the thickness of its bones (the latter measurement indicating the proportionate size of the associated muscles). While the bony measurements, and the implied normal weight, of human subjects may readily be determined,[18] no such determination (or cooperation!) can be expected from gorillas—especially adult males, to whom the appearance of a tape measure or calipers might suggest a snake or some other object repellent or fearsome. Although the tamed and trained adult female gorilla "Susie," of the Cincinnati Zoo, allowed her trainer to enter her cage and measure her, this was an exceptional instance and one that might be dangerous to attempt duplicating with a full-grown male. In view of these difficulties, the most practical way of determining the dimensions of a gorilla's frame, and so of its proper weight, is to measure the long bones (humerus, radius, femur, tibia), as is explained in chapter 5, and take in addition the circumference of the humerus at the midpoint of its length. The appropriate body weight is then derived by squaring the circumference of the humerus, multiplying by the standing height, and then multiplying by one of the following factors. The procedure is simplified by first dividing both the humerus circumference (which is recorded in millimeters) and the standing height (also taken in millimeters) by 100, which will yield a figure of 8 to 30, usually, instead of 8,000,000 to 30,000,000. The 8 to 30 (whatever measurement is determined) is then multiplied by one of the following factors to obtain the body weight in kilograms: for adult male lowland gorillas, × 8.63; adult female lowland gorillas, × 8.83; adult male mountain gorillas, × 8.32; adult female mountain gorillas, × 8.94.

By this procedure, the body weights obtained for both Mbongo and Ngagi are about 230 or 231 kg (510 pounds), and for Phil, of the Cincinnati Zoo, 234 kg (516 pounds). These weights are a far cry from the actual weights attained by these two mountain gorillas and one lowland gorilla, all of which at one stage exceeded 600 pounds. They illustrate the inimical effects of excessive weight on longevity, to say nothing of the needless extra expense of providing prodigious quantities of food. In a number of instances, zoo keepers have become aware of this situation and have wisely cut down on the amount of food supplied to their gorillas, thereby contributing to the apes' health and longevity if not to their massiveness of appearance.

Table 24, on page 221, shows the interrelationships of body weight, relative surface area, daily food requirement, daily water requirement, and approximate need of calories (assuming a medium amount of physical activity), respectively. While specifically intended for gorillas, the relationship between weight and food required is approximately the same for other animals within the weight range listed; for example, a man weighing 160 pounds and engaging in ordinary physical activity requires food of an amount and nature that will produce about 2600 calories daily. Again, a chimpanzee weighing 100 pounds should have about 5.3 pounds of solid food, including milk, per day, along with 2 to 3 quarts of water. And since even a one thousand-pound horse at "hard work" requires only about 25 pounds of dry food a day, it should be evident that a 500-pound gorilla, taking life relatively easy, should need no more than the 17.3 pounds of food listed in the table. This is less than half the weight of food served even today to certain adult male captive gorillas whose body weight has been allowed to reach or surpass 600 pounds. Female gorillas, too, have in some cases been grossly overfed. Roly-poly "Susie" finally reached a weight of 458 pounds, which was clearly over 50 percent greater than it should have been. Even at 300 pounds she was a record-sized specimen of her sex and amply impressive weight-wise.

It should be added that Table 24 is presented, not as a rigid, infallible schedule, but rather as a guide to rational feeding that should be useful for reference in the preparation of diets for gorillas.

Notes

1. Kenneth Binkley, in *Time*, August 14, 1964, p. 57.
2. Von R. Kirchshofer, *Gorillazucht in Zoologischen Garten und Forsshungsstationen* (Frankfurt: Frankfurt Zoological Society, 1970).
3. R. J. Frueh, "A Captive-born Gorilla . . . at St. Louis Zoo," *International Zoo Yearbook*, vol. 8, 1968, pp. 128–32.
4. R. Lotshaw, "Births of Two Lowland Gorillas . . . at Cincinnati Zoo," *International Zoo Yearbook*, vol. 11, 1971, pp. 84–87.
5. Ernst M. Lang, "Jambo, the Second Gorilla born at Basel Zoo," *International Zoo Yearbook*, vol. 3, 1964, p. 84.
6. Ernst M. Lang, *Goma, The Baby Gorilla* (London: Golancz, 1962).
7. Jeremy J. C. Mallinson, Phillip Coffey, and Jeremy Usher-Smith, "Maintenance, Breeding, and Hand-rearing of Lowland Gorilla at the Jersey Zoological Park," *Annual Report, The Jersey Wildlife Preservation Trust*, 1973, Table 12, p. 26.
8. Ibid., pp. 10–11.
9. Duane M. Rumbaugh, "The Behavior and Growth of a Lowland Gorilla and Gibbon," *Zoonooz* (San Diego) 39, no. 7 (July 1966):8–17 (for the female gorilla "Alvila"). See also *Zoonooz* (San Diego) 38, no. 9 (September 1965):12–20, for the birth of "Alvila."
10. P. Coffey and J. Pook, "Breeding, Hand-Rearing, and Development of the Third Lowland Gorilla . . . at the Jersey Zoological Park," *Eleventh Annual Report, The Jersey Wildlife Preservation Trust*, 1974, pp. 45–52 (the female gorilla "Zaire").
11. Robert R. Golding, "Why Zoos in Africa?", in *Our Magnificent Wildlife* (New York: Reader's Digest Press, 1975), pp. 212–13.
12. Joan Morton Kelly, "The First 100 Pounds Are the Hardest," *Zoonooz* 26, no. 3 (March 1953):4.
13. Herbert L. Ratcliffe, "New Diets for the Zoo," *Fauna* (Philadelphia Zoological Society) 2, no. 3 (September 1940): 62–65; idem, "Diet Keeps Oldest Gorilla Healthy," *Science Digest*, February 1957 (back cover).
14. Lee S. Crandall, personal communication, December 17, 1951. A diet list followed in November 1950 for Makoko and Oka, before Makoko died from drowning.
15. Belle J. Benchley, *Mountain Gorillas in the Zoological Garden, 1931–1940* (San Diego, Calif.: San Diego Zoological Society Publications, 1941).
16. Samuel Brody, *Bioenergetics and Growth* (New York: Hafner Publishing Co., Inc., 1964), p. 478.
17. David P. Willoughby, *Growth and Nutrition in the Horse* (Cranbury, N.J.: A. S. Barnes & Co., Inc., 1975) p. 136.
18. David P. Willoughby, "An Anthropometric Method for Arriving at the Optimal Proportions of the Body in any Adult Individual," *Research Quarterly* (American Physical Education Association) 3, no. 1 (March 1932):48–77.

12

Gorillas in Fiction and Fantasy

One of the delightful features of science fiction, particularly the type that may be called "science fantasy," is that the reader may temporarily find release from the world of facts and logic and let his imagination run riot. However, if the reader happens to be a serious scholar as well as a sci-fi addict, he cannot help from observing, and sometimes even objecting to, certain parts of a story that violate all rules of both actuality and possibility.

Recently, quite by chance, I came across an early issue of a science-fiction magazine in which, among "letters from readers," was one from Isaac Asimov, who at the time was only 17 years of age. Yet even at that stage he knew enough about science in general to protest, in a highly knowledgeable manner, against sci-fi authors who created such biological (and geometrical) impossibilities as "giant ants."[1]

The same objection can be levelled against 50-foot gorillas, or even 10-foot human giants: they simply do not occur. In the motion picture *1,000,000 Years B.C.*, which presents a reasonably realistic tyrannosaurus rex in action, my "rapport" with the picture was shattered when a spider as big as a dinosaur also briefly appeared. But one must expect such things every now and then, even though motion pictures are steadily improving in their depiction of biological fantasies. The realistic facemasks worn by the "gorillas" and "chimpanzees" in *Planet of the Apes* was one such notable advance. But often some unwarranted conceptions are disseminated by a persuasive author who attempts to deal with an unfamiliar subject.

To elaborate on my statement that there are no human giants even 10 feet in height, it can be pointed out that the tallest man in medical history was Robert Wadlow, of Alton, Illinois, who attained an accurately ascertained height of 8 feet 11.1 inches. As Wadlow was just past 22 years of age at the time of his death, in 1940, and was still growing at an appreciable rate, he probably would have attained a full 9 feet in height if he had lived even 7 or 8 months longer. However, a height of 10 feet is something else. The probability of such a height may be compared to the likelihood of a financier becoming at least a trillionaire, if not a quadrillionaire! The possibility is there, but the probability is practically nil. And among women, the attainment of a height of even 8 feet is questionable. At the present time there is a woman basketball player named Iuliyaka Semenova, in Soviet Russia, who stands 6 feet 11 inches and is rightly regarded as a phenomenon, even though such a stature among male basketball players is not uncommon.

Be all this as it may, in the field of fantasy even a 10-foot giant is hardly worth mentioning. Lemuel Gulliver, the hero of *Gulliver's Travels*, by Jonathan Swift (1726), was only 1/12 as tall as the giants who lived in a country called Brobdingnag. So, if Gulliver was six feet tall and weighed, say, 180 pounds, the Brobdingnagians, if built in the same proportion, would be 72 feet tall and would weigh $12 \times 12 \times 12 \times 180$, or 311,040 pounds each! However, the strength of the giant's muscles would be in proportion to their cross-sectional area, and this would be only 12×12, or 144 times Gulliver's cross-section and strength. Accordingly, the muscular strength of each Brobdingnagian would be only 144/311,040, or 1/2160, that of Gulliver in proportion to his weight. The giant's relative strength, therefore, would be as though Gulliver had only the strength of a creature weighing 1/2160 of 180 pounds, or about 1/12 of a pound; that is, the size of a mouse. So Gulliver would have nothing to fear from the Brobdingnagians, since, even if they were lying flat on their backs, they would lack the strength to move one of their arms. Indeed, like a large whale that is stranded, they would probably have suffocated under their own weight!*

From a geometrical standpoint, therefore, there can never be—at least on Earth—a 72-foot human giant, nor even a 50-foot King Kong. But that is no reason why such giants, in fantasy, cannot run and jump and climb just as readily as normal-sized men and apes.

The latter is just what happens in films such as *King Kong* (1933), where the colossal ape moves his 300,000-pound body with the same facility as though he weighed only 1/1000 as much. The mechanical and engineering problems that had to be overcome in the designing and making of the various-sized models used in creating the illusion of a realistic, 50-foot gorilla reflect great credit on the producers.** One of the main points in-

---

* A number of scholarly papers on what could be termed the geometry of size have been published. Here are a few references on the subject.[2]

** A detailed, illustrated explanation of the laborious and ingenious methods employed to give apparent life to King Kong and the prehistoric creatures inhabiting his domain is given in *The Illustrated London News* for May 13, 1933, pp. 684–685. A preview of the 1976, twenty-four-million-dollar production of *King Kong* by Dino de Laurentiis, in which a forty-foot "gorilla," highly automated and animated by a team of twenty operators is used, is given in *Time*, October 25, 1976, pp. 64–70.

Fig. 147. "King Kong" atop the Empire State Building, crushing an airplane in his gigantic paw. One of the many still shots released by Radio Pictures Corporation in their 1933 production of this perennially entertaining motion picture for young and old alike.

variably ignored in such fanciful motion pictures, however, is that there is only one of each kind of creature represented. No troop of apes, nor even the parents of the simian hero, are ever even suggested. Apparently the featured animal was the only one of his kind, devoid equally of parents or offspring! Too, the locus is generally on some remote island, usually of limited extent, where in reality, if dinosaurs were involved, nothing short of a continent teeming with lush vegetation would meet the demand for food. But again, in fantasy, such considerations must be put to one side. Possibly the fruits, nuts, and vegetables on the tiny island grew anew as rapidly as King Kong consumed them!

As this is being written, a new and more elaborate production of *King Kong* is being filmed. Concurrently, in England, a "rip-off" film entitled *Queen Kong* is being produced. Possibly the logical sequel to these epic motion pictures would be to bring the two principals together in a *Wedding of the Kongs*!

Fig. 148. The "apes" (not once are they called gorillas) in the "Tarzan" stories were—as one commentator put it—"strictly from Burroughs' imagination . . . they were unlike any apes in the world." Here is shown one that is evidently not bluffing!

Over the years, beginning in 1908 with a one-reel film by Gaumont Studios entitled *The Doctor's Experiment, or Reversing Darwin's Theory*, over sixty movies have been made in which improvised apes, mostly gorillas, were the subjects. A list of these films has been assembled in the book *The Gentle Giants—the Gorilla Story*.[3] In the present chapter my purpose is not to go into the details of these motion pictures, but rather to give the reader an idea of how various authors and illustrators over the years have each contributed their ideas of what gorillas are (or might be) like. The range of imagination exhibited by these purveyors of gorilla lore has certainly been extensive!

In connection with fiction dealing simply with "apes" (rather than ape giants), the *Tarzan* stories by Edgar Rice Burroughs (1875–1950) are doubtless the best known. In fact, few books on any subject have enjoyed the immense popularity and sales that Burroughs's writings did over a span of some thirty years, and which in revived editions they continue to do. But in the *Tarzan* series, not once did Burroughs use the words *gorillas*, or *chimpanzees*. It was always simply "apes," and Burroughs left it to his readers to figure out what species of anthropoid Tarzan was adopted by. Too, ERB never explained who taught Tarzan to swim, since certainly it wasn't one of the "apes"; nor how the apes swung through the treetops with the greatest of ease, even though they were the size of full-grown gorillas. Notwithstanding such omissions, as a youth I was entranced by the writings of Burroughs, especially where, as in his Martian stories and *At the Earth's Core*, he introduced all kinds of improbable men and equally weird animals.

As to the Tarzan movie films, beginning with the first, in 1918, which starred Elmo Lincoln, there have been at least forty made in the United States and perhaps a half-dozen in Europe. Fourteen different actors have starred as Tarzan in the United States films. The first, with Elmo Lincoln in the title role and Enid Markey as "Jane," opened at the Broadway Theater, New York City, on Sunday, 27 January 1918. It was modestly announced as "The most stupendous, amazing, startling film production in the world's history. Produced in the wildest jungles of Brazil at a cost of $300,000. Staged with wild lions, tigers, elephants, baboons, apes, cannibals." Note the inclusion of tigers and baboons. Obviously any conformity with the established geographical distribution of animals was ignored. In another part of the theater announcement it says, "battle between an ape and gorilla." Since in his first book of the Tarzan series, *Tarzan of the Apes*, Burroughs describes the "ape" that killed Tarzan's father as "weighing probably three hundred pounds," he doubtless had in mind a gorilla. In that case, what species was the "ape" mentioned in the announcement as doing battle with a gorilla? Surely not a poor little chimpanzee!

Again, as in the case of *King Kong*, there is not much one can hope for in analyzing such film interpretations for zoological accuracy. Burroughs was a gifted, imaginative novelist, not a zoologist; but it would appear that in some instances his nescience of (or disregard for) biological facts was even transcended in some of the Hollywood interpretations of his writings.

Despite all this, I recall the thrills I experienced, as a teenager in 1918, in viewing the first showing in San Francisco of *Tarzan of the Apes*. I recall even the music that was being played on the pipe organ in the theater. In the first lonely scenes, where Lord Greystoke (Tarzan's father-to-be) and his pregnant wife were set ashore on the jungle-clad coast of west Africa, the music that rumbled from the organ was Franz Schubert's *The Erlking*. Later in the picture, Elmo Lincoln, with his immense (48- to 53-inch) chest made a

Fig. 149. Here is a rare old poster advertising the first *Tarzan* series, in which the first movie Tarzan, Elmo Lincoln, was the star. In those days, a tamed gorilla simply wasn't available. Accordingly, we see here not a gorilla, nor even a chimpanzee—as was later used by Johnny Weismuller—but an orangutan (from Borneo or Sumatra, not Africa), riding with Elmo on an Indian elephant! What's the difference, as long as it was in a jungle!?

Fig. 150. Ki-Gor, a jungle-living white superman like Tarzan, was conceived by author John Peter Drummond. Here, armed only with a knife, he is about to defend himself and his lady friend from a charging "bull" gorilla, evidently bent on annihilating them. The name of the illustrator was not given.

clout caught tight about his lean, hard loins. With a toss of his head, he threw his blonde mane of hair back out of smouldering grey eyes. His right hand twisted with restless strength along the shaft of his war spear.[4]

Naturally, located as he is in an area swarming with gorillas, Ki-Gor has frequent run-ins with the apes, as is thrillingly narrated by Drummond in his stories.

Another fanciful story involving apes is *The White Gorilla,* by Elmer Brown Mason. This tale appeared in *Fantastic Novels* magazine for November 1949. It centers on a huge, white gorilla, which was controlled in a somewhat hypnotic manner by a native tribes-girl, and which was kept on hand to slay enemies of the tribe. A conception of the "white gorilla" being directed by the girl who controlled him is presented by artist Virgil Finlay in Figure 151. Finlay was one of those rare illustrators who could realistically portray animals of all kinds, as well as human beings.

Way back in 1889, the famous circus tycoon P.T. Barnum, in one of the books published under his name, told about "A Strange Battle" between

believable Tarzan, although, of course, the acting in those early days of cinematography simply did not compare with what we see today.

But Edgar Rice Burroughs was not the only novelist to bring together the plot of "apes," a lone white "lord of the jungle," and usually some beautiful white girl waiting to be rescued. In the series of magazine articles by John Peter Drummond, such a hero is "Ki-Gor," a white man of Tarzan's physique and prowess, only blond rather than dark-haired. Numerous episodes of Ki-Gor's adventures in the African jungle appeared in *Jungle Stories* in the 1940s. Figure 150 depicts one of Ki-Gor's hair-raising encounters with a charging male gorilla. Author Drummond describes Ki-Gor thus:

> Even among the tall, powerfully-built Masai, the great-thewed White Lord was a commanding figure. He wore only a leopard-skin breech-

Fig. 151. "The White Gorilla," after a pen drawing by Virgil Finlay in *Fantastic Novels* for November 1949.

a gorilla and a crocodile. Although the "battle" was wholly imaginary, it was told as though there were eyewitnesses to it! Part of the extraordinary account is quoted here, if only to demonstrate what a "showman" can do with words.

> Bending slightly forward, the gorilla indulged in some odd grimacing motions, much like those of the ordinary monkey, and which were meant to tantalize the crocodile into coming forth and assailing him.
>
> Uttering his cry in a half-suppressed voice, the gorilla made a leap forward, as if to alight on the snout of the other. Instantly those jaws opened like a vast steel trap, and, had the gorilla made the bound that he really appeared to have started upon, he would have been caught in a vise from which ten times his power would not have extricated him.
>
> But with inimitable dexterity, the animal [gorilla] turned himself to one side and leaped backward, eluding the mouth, which snapped shut with a sound that startled the spectators.
>
> Hardly had the gorilla jumped when the crocodile doubled himself sideways, and his great tail made a terrific sweep, like that of a scythe in the hands of a giant. It whizzed over the ground where the gorilla was standing, but did not hit him.
>
> He [the gorilla] bounded into the air with a nimbleness that could not have been surpassed, and the next moment did a thing so incredible that the hunters could hardly believe their eyes.
>
> The crocodile knew he was going to miss before his furious blow was delivered, and, with astonishing agility, he wheeled with open jaws to seize the exasperating enemy; but the same dexterity that had saved the latter an instant before did not fail him now. He darted like a flash to the left, then sprang directly upon the back of the saurian, and, bending over, grasped his forelegs.
>
> One was seized in either hand, and, summoning his Samson-like strength, he leaned backward and jerked with might and main.

Fig. 152. Imaginary combat between a gorilla and a crocodile, as related by P. T. Barnum. For details, see text.

Fig. 153. In the story "Gorilla Ghost," by Lawton Ford (*Jungle Stories*, Summer 1948), an infuriated male gorilla crushes a hapless native in his viselike grip, as illustrated here. The artist's name was not given.

Fig. 154. This spirited dry-brush drawing, by Lee Conrey, appeared in one of the Sunday newspaper supplements in the 1930s. Unfortunately, the accompanying story was not preserved. One wonders which of the antagonists won! (At least we know which one we are pulling for!)

Fig. 155. An imaginary combat in the jungle between a male gorilla and a pair of leopards. While a number of anatomical discrepancies appear in this painting, the artist has nevertheless portrayed with lively action a drama of the wilds that could possibly have taken place; although, normally, a leopard hunts alone and at night.

Fig. 156. "Abduction by a Gorilla," a conception by the French sculptor Emmanuel Fremiét (1824–1910). Since Fremiét completed this statue in 1859, it is doubtful that he ever saw an adult gorilla of either sex. The statue stands in the Musee des Beaux Arts, Nantes, France.

The spectators heard the bones crack, and they knew that both the crocodile's legs were broken like a couple of pipe-stems.[5]

And that was the end of the crocodile, whose fate the "spectators" had been privileged to witness presumably from just across the water! That the illustrator of this alleged encounter was no less capable than the writer of it is demonstrated in Figure 152, which shows the climax of the fight.

As is depicted in Figure 153, Lawton Ford, in his story "Gorilla Ghost," tells about an infuriated and vindictive male gorilla, "showing yellow fangs of finger length" (!), which crushed to death a young native tribesman in his irresistible embrace.

A number of illustrators have each given their idea of a fight between a gorilla and a big snake. This is interesting from the standpoint that practically all apes and monkeys have an instinctive dread and repugnance against snakes of any size, even foot-long ones. And in view of the gorilla's habitat, a really large snake would have to be a python, a serpent like the South American boa constrictor, only larger. Since the common African python (*P. sebae*) may attain a length of 20 feet, it can be appreciated what a formidable opponent it could be, even to an adult male gorilla. An idea of what such a combat might be like is illustrated in Figure 154. Artist Lee Conrey has endowed his python with the proportions of a true monster. One can only hope that, like *King Kong,* the gorilla succeeds in breaking loose from the serpent's constricting coils and emerges from the combat victorious. However, in no factual account of gorillas in the wild that I can recall is there any mention of an encounter between one of these giant apes and a python. Each probably has a healthy respect for the other!

In chapter 8, Figure 97, an illustration is given of an imaginary, although quite possible, fight between an adult male lowland gorilla and a leopard. As mentioned, however, no experienced gorilla would risk the devastating effects of a leopard's claws if it could keep the cat at arm's length and either strangle it or break its spine with a ponderous blow.

In Figure 155, the artist has put the gorilla in an even more vulnerable position by having the ape attacked by a pair of leopards. While such a possibility is conceivable, it would have to occur during the leopards' mating season, which is usually the only time when a pair of the cats goes about together. But it would be difficult to visualize a gorilla coming out of a close-contact combat with even a single leopard, let alone two, without being torn to ribbons. The terrific, knifelike gashes that can be inflicted by one of the large cats was shown in an actual battle that took place some years ago between a grizzly bear and a cougar, or puma. Although the much-larger bear killed the cougar, the cat before expiring had ripped so deeply into the bear's abdomen with its claws that the bear was disembowelled and shortly died also. Gorillas, although immensely powerful, are not constructed like meat-eating predators, and in particular lack the razorlike claws that make all cats, big or small, such efficient killers.

As a concluding illustration of once-supposed behavior in the gorilla, Figure 156 is presented. In this example of the work of the French animal sculptor Emmanuel Fremiet, the old yarn of a woman-abducting gorilla is shown as though actually taking place. One can only imagine how the gorilla held his struggling victim with one arm without crushing her, while he hobbled on three legs all the way back to his "lair." In early days almost any fearsome act could be attributed to the gorilla by the highly suggestible natives who lived in the same gloomy jungle environment. Happily, the introduction of modern means of exploration and enlightenment has removed most, if not all, of the once-held superstitious beliefs, including that of a woman-kidnaping ogre of the forest.

Notes

1. Isaac Asimov, in *Astounding Science-Fiction* 22, no. 3 (November 1938):152–54.
2. Gaylord Johnson, "Why Can't Men Become Giants or Jump Like Grasshoppers?", *Popular Science Monthly*, April 1931, pp. 60–61, 144; Florence Moog, "Gulliver Was a Bad Biologist," *Scientific American*, November 1948, pp. 52–55; Peter K. Weyl, *Men, Ants, and Elephants. Size in the Animal World* (New York: Viking Press, 1959); Isaac Asimov, "Just Right," *Fantasy and Science Fiction*, March 1969, pp. 89–98.
3. Geoffrey H. Bourne and M. Cohen, *The Gentle Giants* (New York: G. P. Putnam's Sons, 1975), pp. 44–45.
4. John Peter Drummond, "The Golden Claws of Raa," *Jungle Stories* 4, no. 4 (August–October 1948):4.
5. Phineas Taylor Barnum, *The Wild Beasts, Birds, and Reptiles of the World: The Story of their Capture* (Chicago, 1889), pp. 454–56.

13

A Census of Zoo Gorillas

My object in this chapter is, as the title indicates, to list the numerous gorillas that have been brought, over the years, into the United States. Wherever it is known, the name and the estimated age of the ape are recorded, along with the name of the zoo or other domicile to which the animal is being sent. The sex—while almost invariably known—in at least several instances was not definitely established until some years after the ape's arrival. As to the age being "estimated," this has been necessary in all cases where the gorilla was not born in captivity, which means every case listed in Table 25. The estimation of age is generally made by the zoo's veterinarian or medical doctor. The basis of the determination is mainly the stage of the animal's dental development, although the size and general behavior of the young ape may also be taken into consideration. So far as plotting the growth in weight of the gorilla is concerned, the age as estimated is generally sufficiently close to the probable true age to make no difference in interpreting the rate of growth.

I hasten to add that Table 25 is not only lacking in many places in detailed information, but also in probably omitting a number of individual gorillas that were brought into the United States during recent years. The reason for this incompleteness is that my latest correspondence with zoo directors, so far as taking a "gorilla census" was concerned, ended in 1971. I can only hope that readers who are interested in some particular gorilla, or gorillas, for which some of the desired information is here lacking, may obtain the desired facts by writing directly to the zoo involved. I shall be happy to add to Table 25 any additional information thus obtained, or sent to me by a zoo worker. But obviously it is impossible to keep any such list strictly "up to date," since the total population of gorillas, like that of humans, is changing constantly, in respect both to births and to deaths.

Since, then, the information in Table 25 is mainly no more recent than of the year 1971, in the column designated "Present estimated bodyweight" the weights (which in most cases are actual rather than estimated) should be considered as having been taken that year or earlier, if the age is not specifically listed. Many of these listings may be regarded as representing the full mature weights of the gorillas. One listing in the table (No. 69) should not be there. I learned later that "Lilly" was the same female mountain gorilla in the Bronx Zoo as "Sumaili" (No. 37). Too, No.

139 (Shamba) is the *same* as No. 120, the repetition resulting from the transference from one zoo to another.

A number of earlier "censuses" of gorillas have been made, both of specimens in the United States and abroad. The earliest such list that I have is of gorillas in the United States as of April 1944. This list names only 16 individuals—7 males and 9 females. The range in age is from 4 to 18 years, the oldest individuals being "Bamboo" (Philadelphia) and "Susie" (Cincinnati), both 18.

In a list presented by Bernard F. Riess, et al, in 1949, 22 gorillas are named—13 males and 9 females.[1] It should be realized that both this list and that of 1944 name only the gorillas currently living in U.S. zoos. Many specimens died in earlier years—at least 15 up to 1948, including 2 males and 2 females at the San Diego Zoo, which were of the comparatively rare mountain species.

Robert Yerkes, in recording the gorillas living in the United States in 1951, listed 40 specimens.[2] This number appears to correspond with that of the gorillas listed in Table 25. It is interesting to note that all three of the lists just mentioned contain the names of "Bamboo" and "Massa," both of the Philadelphia Zoo; and that Massa, although No. 15 in Table 25, has continued to survive while nearly 200 younger gorillas have made their appearance in American zoos.

The German zoologist, Dr. Bernhard Grzimek, in a review published in 1953, listed 44 captive gorillas in the United States and 12 in Europe.[3] Grzimek listed a male gorilla, "Bobo," at the St. Louis Zoo as well as in Seattle, whereas I have found no other listing of a "Bobo" in St. Louis.

In 1960, a census conducted by Honegger and Menichini listed 72 gorillas in the United States and 31 in Europe and Asia.[4]

In 1961, Edalee Harwell, of the San Diego Zoo, compiled a list of the gorillas then residing in 31 United States zoos and the Yerkes Primate Laboratory. Altogether there were 84 specimens—46 males and 38 females.[5]

Evidently the zoo-gorilla world population increased spectacularly between 1961 and 1962. In a report by James Fisher, he says: "In 1962 there were 148 lowland gorillas in 60 zoos and 12 mountain gorillas in six zoos."[6]

By 1965, there were 219 captive gorillas distributed throughout the zoos of the United States (118), Canada (6), Europe (75), Asia (15), and Africa (5). Of these, 101 were lowland males, 103 lowland females, 7 mountain males, 7 mountain females, and one a mountain (?) gorilla of unknown sex.[7]

Between 1965 and the present time, so many zoo gorillas have had offspring that the total gorilla population has appreciably increased, even though few if any of these now rigorously protected apes are allowed to be taken out of Africa. Accordingly, it is impossible to say how many captive gorillas there are today throughout the world. A fair guess would be in the neighborhood of 300 living individuals.

Turning now to Table 26, we present a list, as of 1 November 1971, of gorillas born in captivity the world over. This list is the product of independent surveys made by Gary Clarke of the Topeka, Kansas Zoo; Marvin L. Jones of the San Francisco Zoo; and the present author.* My thanks go to Mr. Clarke and Mr. Jones for their kind permission to republish these statistics here in Table 26.

In reference to Table 25, one of the reasons for its incompleteness is that a number of zoos and animal dealers did not respond to the writer's request for information. In view of this and other factors, it would appear safe to say that from 1897 to the end of 1976 at least 200 gorillas, mostly captured young ones, have been brought into the United States. Of this number, probably over one-half are living today. In addition to the gorillas in American zoos and primate collections, there are perhaps in the neighborhood of a hundred captive specimens in Europe, Asia, Africa, and elsewhere in the Old World. All of which would indicate a total captive-gorilla population of at least 250 animals and possibly as many as 300.

Table 26 is similarly incomplete; yet a number of interesting and possibly significant facts may be drawn from the list as it stands. An average of the known or estimated ages of the parent gorillas in this tabulation yields figures of approximately 10½ years for the fathers and 10¾ years for the mothers. These ages refer to the parents at the time of their first-born offspring. When the ages for repeated births are averaged, that for the fathers becomes 12 years 10 months, and for the mothers 12 years. The latter figures are almost exactly the average ages of the two respective sexes at maturity. The age range of the father gorillas is from 6 years 10 months (Mighty Omega) to 24 years 4 months (King

---

* Marvin Jones' presentation of this material was published in the January 1972 issue of the *Newsletter* of the American Association of Zoological Parks and Aquariums under the title, "Breeding of the Gorilla in Captivity." To that list I have added in Table 26 the columns on Weight at Birth and Ages of Parents.

Table 25. Chronological Record of Living Gorillas Brought to the United States.

NOTE: This list admittedly is incomplete, especially as regards gorillas which arrived later than 1968. However, the information it contains should provide a useful supplement to later lists.

No.	Name	Sex	Species	Estimated Birthdate	Place Received	Date of Arrival	Estimated Age on Arrival
1	?	♂	Lowland	?	Boston	May 2, 1897	(infant)
2	Madame Ningo	♀	Lowland	?	N.Y. Zool. Park	Sept. 23, 1911	2–3 yrs.
3	Dinah	♀	Lowland	Mid-1911	N.Y. Zool. Park	August 21, 1914	3 yrs.
4	John Daniel	♂	Lowland	Oct. 1916	Ringling Bros. Circus, N.Y. City	March, 1921	nearly 4½ yrs.
5	Sultan (or John Daniel II)	♂	Lowland	Early 1920	Kept in London by Miss Cunningham	May, 1924	4–4½ yrs.
6	Congo	♀	Mountain	Oct. 1919	near Jacksonville, Fla. (later to Ringlings at Sarasota)	Oct. 1925	6 yrs.
7	Bamboo	♂	Lowland	Sept. 1926	Philadelphia Zoo	Aug. 5, 1927	10¾ mos.
8	Janet Penserosa	♀	Lowland	Spring, 1927	N.Y. Zool. Park	Oct. 30, 1928	17–19 mos.
9	N'Gi	♂	Lowland	Sept. 1926	Nat. Zool. Park, Washington, D.C.	Dec. 5, 1928	2 yrs. 3 mos.
10	Jimmie	♂	Lowland	June, 1927	N.Y. Zool. Park	Nov. 1929	2 yrs. 5 mos.
11	Bimbo	♂	Lowland	1928	? (by H. Bartel)	1929	12 mos.
12	Bushman	♂	Lowland	1928	Lincoln Park Zoo, Chicago	1930	2 yrs. +
13	In Europe (1927) "Congo" In U. S., Susie	♀	Lowland	Mid-1926	Cincinnati Zool. Gardens	U.S.A. 1929 Zoo, June 11, 1931	3 yrs. 5 yrs.
14	?	?	?	?	St. Louis Zool. Gardens		
15	Massa	♂	Lowland	Dec. 30, 1930	Brooklyn, N.Y. Philadelphia Zoo	Mrs. Lintz, Sept. 1931 Zoo, Dec. 30, 1935	9 mos. 5 yrs.
16	Okero (or "Snowball")	♂	Mountain	Mid-1929	National Zool. Park, Washington, D.C.	Sept. 1931	2 yrs. +
17	Mbongo	♂	Mountain	Oct. 1926	San Diego Zoo	Oct. 5, 1931	5 yrs.
18	Ngagi	♂	Mountain	April, 1926	San Diego Zoo	Oct. 5, 1931	5½ yrs.
19	Gargantua (first, "Buddy")	♂	Lowland	June, 1931	Brooklyn, N.Y. Ringling Bros. Circus	Mrs. Lintz, Dec. 1932 Dec. 1937	18 mos. 6½ yrs.
20	Congo II (or Miss Congo)	♀	Lowland	March, 1935	Chicago Zool. Park, Brookfield, Ill.	June 22, 1936	15 mos.
21	Suzette	♀	Lowland	March, 1935	" "	Sept. 25, 1936	18 mos.
22	Sultan (II)	♂	Lowland	July, 1936	" "	Oct. 22, 1937	15 mos.
23	M'Toto (or Toto)	♀	Lowland	Dec. 1931	Mrs. E. K. Hoyt, Havana Ringlings, at Sarasota	Nov. 1932 Feb. 1941	11 mos. 9 yrs. 2 mos.
24	Kenya	♀	Mountain	Mar. 1937	San Diego Zoo	Sept. 18, 1941	4½ yrs.
25	Kivu	♀	Mountain	?	San Diego Zoo	Sept. 18, 1941	
26	Makoko	♂	Lowland	June, 1939	N.Y. Zool. Park	Sept. 7, 1941	2 yrs. 3 mos.
27	Oka	♀	Lowland	April, 1940	N.Y. Zool. Park	Sept. 7, 1941	1 yr. 5 mos.
28	Mattie	♀	Lowland	March, 1937	St. Louis Zool. Park	Sept. 7, 1941	4½ yrs.

Bodyweight on Arrival	Present Estimated Bodyweight	Date of Death	Estimated Age at Death	Length of Time in Captivity (U.S.A.)	Remarks
15½ lbs	———	May 7, 1897	18 mos.	5 days	Height 26 in. Brought from Liverpool, Eng.
25 lbs. (malnutrition)	———	Oct. 5, 1911	2–3 yrs.	12 days	Height 34 in., span 47 in. Brought from Africa by R. L. Garner.
40½ lbs.	———	July 31, 1915	4 yrs.	11 mos. 10 days (1 yr. prior in Africa)	On Sept. 1, 1914: height 36 in., span 50½ in. Brought from Africa by R. L. Garner.
112 lbs.	———	last week in April, 1921	4½ yrs.	c. 1 month (after c. 1½ yrs. in England)	In March, 1921: height 40½ in.
80 lbs.	———	1927 (in London)	7½ yrs.	———	In May, 1924: height 41¾ in.
65 lbs.	at 8¼ yrs. 160 lbs.	April 23, 1928	8½ yrs.	2½ yrs.	At 6¼ years; height 38 in., weight 65 lbs. At 7¼ years; height 47 in., weight 128 lbs. At 8¼ years; height 51 in., weight 160 lbs.
11 lbs. (?)	at 13½ yrs. 435 lbs. (?)	Jan. 21, 1961 (weight 281 lbs.)	34 yrs. 4½ mos.	33 yrs. 5½ mos.	
17¼ lbs.	at death 165 lbs.	Aug. 23, 1940 (killed)	13–13½ yrs.	11 yrs. 10 mos.	In November, 1929: weight 35½ lbs.
30 lbs.	at 3½ yrs. 56 lbs.	March 10, 1932 (weight 62 lbs.)	5½ yrs.	3 yrs. 3 mos. (after 1 yr. prior in Africa)	Captured on Jan. 17, 1928 by J. L. Buck, when 16 mos. old. Weight at 3½ years, 56 lbs.
35½ lbs.					
17 lbs.					At 16 months, weighed 19½ lbs.
37 lbs.	at 20 yrs. 550 lbs.	Dec. 31, 1950	22 yrs. +	20 yrs. +	Weight at 8 years, 210 lbs.; at 13 years, 461 lbs.; at 14½ years, 505 lbs.; at 16 years, 530 lbs.
45 lbs.	at 18½ yrs. 420 lbs.	Oct. 29, 1947	c. 21 yrs.	c. 18 yrs.	Height at 17 years, 62 inches (a ♀ record).
90 lbs.					
18 lbs.	at 10 yrs.				Still living (1976) at age of 46 years.
140 lbs.	360 lbs. (?)				
35 lbs.	———	Oct. 1932	3 yrs. +	13 mos.	Obtained as an infant by Martin Johnson from natives.
	Maximum, in early 1942, 618 at death, 660 lbs. (!)	Mar. 15, 1942	15½ yrs.	c. 10½ yrs.	Height at death, 67½ inches. Captured by Martin Johnson in December 1930.
125 lbs.	Max., in 1943 683 lbs. (!) at death 636	Jan. 12, 1944	17 yrs. 9 mos.	12 yrs. 3 mos.	" " " " " "
147 lbs.	at death 312 lbs.	Nov. 25, 1949	18½ yrs.	17 yrs.	Height at death, 67¾ inches.
22 lbs.	at 7 yrs. 4 mos. 250 lbs.	Sept. 22, 1949	14½ yrs.	13 yrs. 3 mos.	Height at death, 66 inches.
12⅝ lbs.	at 7 yrs. 4 mos. 240 lbs.	1948	13 yrs. +	c. 12 yrs.	
24½ lbs.	at 6 yrs. 125 lbs.	Oct. 27, 1942	6¼ yrs.	5 yrs.	Height on arrival, 30 inches.
18¾ lbs. at 3 mos.	at 9 yrs.				Captured in February, 1932. Died in 1968 at age of 37 years. Height, 57 in.
9 lbs.	438 (?) lbs. at 7 yrs.				
60 lbs.	148 lbs.	Sept. 23, 1946	9½ yrs.	5 yrs.	
		Jan. 10, 1942	?	−4 mos.	
28 lbs.	at death 448 lbs.	May 13, 1951	11 yrs. 11 mos.	9 yrs. 8 mos.	Weight at 4 years, 9 months, 104 lbs.
20 lbs.	at 13 yrs. 8 mos. 280 lbs.				Weight at 4 years, 86 lbs.; at 27 years, 245 lbs.
	at 7 yrs. 180 lbs.				Weight (est.) at 7 years, 180 lbs.

(Continued)

Table 25, cont'd

No.	Name	Sex	Species	Estimated Birthdate	Place Received	Date of Arrival	Estimated Age on Arrival
29	Phil	♂	Lowland	Sept. 1939	St. Louis Zool. Park	Sept. 7, 1941	2 yrs.
30	Carolyn	♀	"	Jan. 1940	Central Park Zoo, N.Y. City	July 1, 1943	3½ yrs.
31	Joanne	♀	"	"	"	"	3½ yrs.
32	Artie	♂	"	Oct. 1939	"	"	3¾ yrs.
33	Sinbad	♂	"	March, 1948	Lincoln Park Zoo, Chicago	Oct. 1, 1948	8 mos.
34	Rajah	♂	"	April, 1947	"	"	18 mos.
35	Irvin Young	♂	"	Dec. 1946	"	"	22 mos.
36	Lotus	♀	"	Feb. 1946	"	"	2 yrs. 8 mos.
37	Sumaili	♀	Mountain	June, 1948	N.Y. Zool. Park	June 15, 1949	12 mos.
38	Albert	♂	Lowland	Feb. 1949	San Diego Zoo	Aug. 10, 1949	6 mos.
39	Bouba	♀	"	Dec. 1948	"	"	8 mos.
40	Bata	♀	"	Oct. 1948	"	"	10 mos.
41	(M'Golo) Big Boy	♂	"	Jan. 1948	Cincinnati Zool. Soc.	July 23, 1949	18 mos.
42	Mr. Bud	♂	"	Nov. 1947	Colorado Springs, Cheyenne Mtn. Zoo	Sept. 1949	22 mos.
43	Toto (II)	♀	"	Aug. 1949	Sarasota, Florida Ringling Bros. & Barnum & Bailey Circus	Nov. 1949	3 mos.
44	Gargantua II	♂	"	Sept. 1948		"	14 mos.
45	Bobby	♂	"	Sept. 1947	St. Louis Zool. Park	Jan. 1950	2 yrs. 4 mos.
46	Arno	♂	"	?	New York City	June, 1950	
47	Goliath	♂	"	Oct. 1949	"	"	8 mos.
48	Gorgeous	♀	"	July, 1949	Colorado Springs, Cheyenne Mtn. Zoo	June 16, 1950	11 mos.
49	Yokadouma	♀	"	Nov. 1947	Cleveland Zool. Park	June, 1950	2 yrs. 7 mos.
50	Bubu	♂	"	Dec. 1947	"Monkey Jungle", Goulds, Florida	"	2 yrs. 6 mos.
51	Juba	♂	"	Oct. 1948	San Antonio, Texas Zool. Soc.	"	20 mos.
52	Nimo	♀	"	"	"	"	20 mos.
53	Eseka	♂	"	Nov. 1949	Cleveland Zool. Park	Sept. 1950	10 mos.
54	Samson	♂	"	Nov. 1949	Milwaukee, Washington Park Zool. Garden	Oct. 10, 1950	11 mos.
55	Sambo	♂	"	Oct. 1949	"	"	12 mos.
56	Sheba	♀	"	Dec. 1948	Cincinnati Zool. Soc.	Jan. 1951	2 yrs. 1 mo.
57	King Tut	♂	"	Dec. 1949	"	"	13 mos.
58	Macombo, or The Baron	♂	"	Jan. 1947	Columbus, Ohio Municipal Zoo	Jan. 1951	4 yrs.
59	Millie, or Christina	♀	"	Aug. 1945	"	"	17 mos.
60	Christopher	♂	"	July, 1945	"	March, 1951	18 mos.
61	Baby	♂	"	April, 1950	Chicago Zool. Park, Brookfield, Illinois	Apr. 27, 1951	12 mos.
62	Sappho	♀	"	April, 1949	"	"	2 yrs.
63	Mambo	♂	Lowland	June, 1950	N.Y. Zool. Park	May 27, 1951	11 mos.
64	Zulu	♂	"	Feb. 1951	Colorado Springs, Cheyenne Mtn. Zoo	Nov. 12, 1951	9 mos.

Bodyweight on Arrival	Present Estimated Bodyweight	Date of Death	Estimated Age at Death	Length of Time in Captivity (U.S.A.)	Remarks
26 lbs.	at death, c. 600 lbs. (!)	1958	19 yrs.	17 yrs.	Weight at 4½ years, 110 lbs.; at 15 years, 500 lbs.
c. 50 lbs.	at 30 yrs. 250–275 lbs.				Weight at 4½ years, 75 lbs.
c. 55 lbs.	at 30 yrs. 275–300 lbs.				
c. 73 lbs.					Weight at 4 years, 3 months, 93 lbs. Traded to St. Louis Zoo, March 5, 1949.
12 lbs. ?	at 27 yrs. 530 lbs.				Weight at 2 years, 39½ lbs.
18 lbs.					
28 lbs.		(died) Sept. 15, 1961	15 yrs. 7 mos.	12 yrs. 11 mos.	
42½ lbs.					
16½ lbs.	at 11 yrs. 260 lbs. at 18 yrs. 292 lbs.				Weight at 17 years, 9 months, 292 lbs.
8 lbs.					Weight at 6½ years, 119 lbs.
9 lbs.	at 6 yrs. 8 mos. 158 lbs.				
10½ lbs.	at 6 yrs. 10 mos. 190 lbs.				Traded to Fort Worth Zoo, Jan. 3, 1964. (See Nos. 124 and 125).
17½ lbs.		early 1968	c. 20 yrs.	c. 18½ yrs.	
34½ lbs.					
39½ lbs.					
17½ lbs.	at 19½ yrs. c. 205 lbs.				
27¾ lbs.					
29¾ lbs.					
11 lbs.					
15 lbs.	at 20 yrs. 600 lbs. (!)				
14¼ lbs.		1959	c. 10 yrs.	c. 9 yrs.	
29¾ lbs.					
13 lbs.					
71 lbs.	at 23½ yrs. 320 lbs.				
18 lbs.	at 21 yrs. 291 lbs.				
21 lbs.					
17½ lbs.	c. 500 lbs.				
36 lbs.		Sept. 5, 1965	16 yrs. 5 mos.	14 yrs. 5 mos.	
16¼ lbs.	at 19 yrs. 345 lbs.				
14¾ lbs.					

(Continued)

Table 25, cont'd

No.	Name	Sex	Species	Estimated Birthdate	Place Received	Date of Arrival	Estimated Age on Arrival
65	Mitzi	♀	Lowland	Oct. 1949	"Monkey Jungle", Goulds, Florida	Nov. 1951	2 yrs. 1 mo.
66	M'Jingo	♂	"	Feb. 1949	Bedford, Va. Noell's Ark Gorilla Show	"	2 yrs. 9 mos.
67	Robert	♂	"	Aug. 1950?	Baltimore, Md., Druid Hill Park	Nov. 1951 ?	15 mos.
68	Bobo	♂	"	July 26, 1951	Seattle, Wash., Woodland Park Zoo	Dec. 1951 (family) Dec. 1953 (zoo)	5 mos. 2 yrs. 5 mos.
69	Lilly (same as No. 37)	♀	Mountain	June, 1948	N.Y. Zool. Park	June 15, 1949	
70	Jambo	♂	Lowland	June, 1951	Pittsburgh, Highland Park Zoo	March, 1953	21 mos.
71	Hugo	♂	"	Jan. 1950	Houston, Texas Zoo	"	3 yrs. 2 mos.
72	Nikumba	♂	"	Nov. 1953	Nat. Zool. Park, Washington, D.C.	Feb. 24, 1955	15 mos.
73	Moka	♀	"	June, 1953	"	"	20 mos.
74	Jim-Jim	♂	"	June, 1954	Detroit Zool. Park	May 6, 1955	11 mos.
75	Maximo	♂	"	June, 1954	"	July 16, 1955	13 mos.
76	Mesou	♀	"	Feb. 1955	"	"	5 mos.
77	Mike	♂	"	Sept. 1955	Fort Worth Zool. Park	May 5, 1956	8 mos.
78	Abe	♂	"	Jan. 1956	Colorado Springs, Cheyenne Mtn. Zoo	July 17, 1956	6 mos.
79	Hercules	♂	"	Sept. 1955	"	"	10 mos.
80	Fifi	♀	"	c. Sept. 1954	Seattle, Wash., Woodland Park Zoo	Sept. 26, 1956	c. 2 yrs.
81	Becky	♀	"	Feb. 1956	Colorado Springs, Cheyenne Mtn. Zoo	Oct. 13, 1956	8 mos.
82	Bathsheba	♀	"	Mar. 1956	"	"	7 mos.
83	Jimmy	♂	"	May, 1956	Dallas Zoo	May 8, 1957	12 mos.
84	Togo	♂	"	Nov. 1954	Toledo, Ohio Zool. Soc.	May 29, 1957	2 yrs. 6 mos.
85	Jenny	♀	"	July, 1955	Dallas Zoo	July 28, 1957	2 yrs.
86	Penelope	♀	"	Spring?1954	Cincinnati Zool. Gardens	Aug. 8, 1957	3 yrs. + ?
87	Porta	♀	"	April, 1954	Toledo, Ohio, Zool. Soc.	Oct. 1, 1957	3½ yrs.
88	Scoop	♂	"	Oct. 1956	San Diego Zoo	Apr. 25, 1958	18 mos.
89	Big Man	♂	"	July, 1957?	Kansas City Zool. Gardens	July 26, 1958	12 mos. ?
90	Bongo	♂	"	June, 1957	Columbus, Ohio Municipal Zoo	Oct. 1, 1958	16 mos.
91	Mighty Mr. Moore	♂	"	c. May, 1954	San Diego Zoo	Nov. 20, 1958	c. 4½ yrs.
92	Vila	♀	"	Sept. 1957	"	Mar. 27, 1959	18 mos.
93	Kribi Kate	♀	"	June, 1958	Kansas City Zool. Gardens	June 21, 1959	12 mos.
94	Jungle Jeannie	♀	"	June, 1958	"	"	12 mos.
95	Bwana	♂	"	April, 1958	San Francisco Zoo	Oct. 21, 1959	18 mos.
96	Missus	♀	"	April, 1957	"	"	2 yrs. 6 mos.
97	Copy	♂	"	June, 1958	San Diego Zoo	Dec. 6, 1959	18 mos.
98	Tanga	♂	"	April, 1959	Milwaukee County Zool. Park	Apr. 24, 1960	12 mos.
99	Terra	♀	"	July, 1959	"	"	9 mos.
100	Trib	♂	"	July, 1959	San Diego Zoo	July 30, 1960	12 mos.
101	Congo	♂	"	June, 1959?	Honolulu, Hawaii, Zoo	Sept. 28, 1960	c. 15 mos. ?

Bodyweight on Arrival	Present Estimated Bodyweight	Date of Death	Estimated Age at Death	Length of Time in Captivity (U.S.A.)	Remarks
27½ lbs.					
49½ lbs.					
21 lbs.					
(zoo) 60 lbs.	at 13 yrs. 560 lbs.	Feb. 22, 1968	17 yrs. 7 mos.	15 yrs. 2 mos. (zoo)	
25¾ lbs.					
60 lbs.					
17 lbs.	at 16 yrs. 450 lbs.				
22 lbs.		Aug. 30, 1968	15 yrs. 2 mos.	13½ yrs.	
18 lbs.	400 lbs.				
22 lbs.	520 lbs.				
10 lbs.	330 lbs.				
11 lbs.	450 lbs.				
11 lbs.	350 lbs.				
14 lbs.	400 lbs.				
c. 30 lbs. ?	350 lbs.				
14 lbs.	225 lbs.				
12½ lbs.	200 lbs.				
25 lbs.	500 lbs.	1973	c. 17 yrs.	c. 16 yrs.	
35 lbs.	400 lbs.				
48 lbs.	250 lbs.				
45 lbs. ?	135 lbs.				
45 lbs.	170 lbs.				
		Aug. 13, 1959	2 yrs. 10 mos.	16 mos.	
35 lbs.	at death, 425 lbs.	1976	c. 19 yrs.	c. 18 yrs.	
	500 lbs.				
	at 11¾ yrs. 390 lbs.				
		Nov. 21, 1958	c. 4½ yrs.	1 day	
23½ lbs.					
20 lbs.	300 lbs.				
18 lbs.	300 lbs.				
26 lbs.	at 10 yrs. 7 mos. 286 lbs.				
54 lbs.	at 5 yrs. 125 lbs.				
		June 2, 1960	2 yrs.	6 mos.	
17 lbs.	430 lbs.				
12 lbs.	240 lbs.				
19¼ lbs.	at 14 yrs., over 400 lbs.				
20 lbs.	375–400 lbs.				

(Continued)

Table 25, cont'd

No.	Name	Sex	Species	Estimated Birthdate	Place Received	Date of Arrival	Estimated Age on Arrival
102	Solomon	♂	Lowland	Dec. 1959	Oklahoma City Lincoln Park Zoo	Sept. 30, 1960	9 mos.
103	Sheba II	♀	"	Jan. 1959	"	"	20 mos.
104	Cameroun	♂	"	Sept. 1959?	Honolulu, Hawaii, Zoo	Feb. 25, 1961	c. 17 mos. ?
105	Fern	♀	"	June, 1960?	Philadelphia Zool. Garden	May 11, 1961	11 mos. ?
106	Pogo	♀	"	Feb. 1958	San Francisco Zoo	July 21, 1961	3 yrs. 5 mos.
107	Hazel	♀	"	July, 1960	Phoenix Zoo	Oct. 9, 1961	15 mos.
108	Bahati	♀	"	Jan. 1960	San Diego Zoo	Jan. 10, 1962	2 yrs.
109	Kukuma	♂	"	March, 1960	"	"	22 mos.
110	Ella	♀	"	Jan. 1960	"	"	2 yrs.
111	Yula	♀	"	Jan. 1961	"	Jan. 22, 1962	12 mos.
112	Samson II	♂	"	March, 1960	Buffalo, N.Y. Zool. Gardens	Mar. 14, 1962	2 yrs.
113	Moemba	♀	"	Aug. 1960	Oklahoma City Lincoln Park Zoo	Nov. 28, 1962	27 mos.
114	Boma	♀	"	Mar. 1959	"	Apr. 6, 1963	4 yrs. 1 mo.
115	Jonesie	♀	"	c. Feb. 1961	Buffalo, N.Y. Zool. Gardens	Apr. 21, 1963	2 yrs. +
116	Vivante	♀	"	July, 1962	San Diego Zoo	Apr. 23, 1963	9 mos.
117	Luka	♀	Mountain	Jan. 1956	Oklahoma City Lincoln Park Zoo	Apr. 26, 1963	7 yrs. 3 mos.
118	Makuba	♂	"	Oct. 1954	"	"	8 yrs. 6 mos.
119	Pilipili	♂	Mountain	1957	N.Y. Zool. Park	May 15, 1963	c. 6 yrs.
120	Shamba	♀	Lowland	June, 1959	Oklahoma City Lincoln Park Zoo	Oct. 25, 1963	4 yrs. 4 mos.
121	Alpha	♀	"	c. Feb. 1961	Chicago Zool. Park, Brookfield, Illinois	Aug. 2, 1963	c. 2 yrs. 6 mos.
122	Beta (II)	♀	"	c. Feb. 1961	"	"	c. 2 yrs. 6 mos.
123	Junior	♂	"	Jan. 1963	San Diego Zoo	Oct. 29, 1963	9 mos.
124	Beta	♀	"	Oct. 1948	Fort Worth Zool. Park	Jan. 4, 1964	15 yrs. 3 mos.
125	Mimbo	♂	"	July, 1962	San Diego Zoo	Jan. 9, 1964	18 mos.
126	Timbo	♀	"	Oct. 1962	"	"	15 mos.
127	Jacob	♀ (!)	"	c. May, 1956	San Francisco Zoo	Apr. 8, 1964	8 yrs. 1 mo.
128	Kathryn	♀	"	c. Mar. 1963?	Philadelphia Zool. Garden	June 24, 1964	c. 15 mos. ?
129	Dolly	♀	Lowland	Dec. 1963	San Diego Zoo	Sept. 9, 1964	9 mos.
130	Femelle	♀	"	c. Jan. 1959?	Nat. Zool. Park, Washington, D.C.	Jan. 21, 1965	c. 6 yrs. ?
131	Mighty Omega	♂	"	Jan. 1963	Chicago Zool. Park, Brookfield, Illinois	Jan. 21, 1965	2 yrs.
132	Hatari	♂	"	Apr. 1962?	Cincinnati Zool. Soc.	"	2 yrs. 9 mos. ?
133	Mahari	♀	"	Jan. 1963?	"	"	2 yrs. ?
134	Cloudy	♂	"	June, 1964	Los Angeles Zoo	Mar. 2, 1965	9 mos.
135	Betsy	♀	"	June, 1964	"	"	9 mos.
136	Bendera	♂	"	c. Oct. 1963?	N.Y. Zool. Park	Apr. 6, 1965	18 mos. ?
137	Tunuka	♀	"	c. Nov. 1963?	"	"	17 mos. ?
138	M'Wasi	♀	"	c. July, 1963?	"	"	21 mos. ?
139	Shamba	♀	"	June, 1959	Dallas Zoo	May 22, 1965	c. 6 yrs.

Bodyweight on Arrival	Present Estimated Bodyweight	Date of Death	Estimated Age at Death	Length of Time in Captivity (U.S.A.)	Remarks
18 lbs.	at 8 yrs. 11 mos. 45 lbs. (!)				
21 lbs.	at 3 yrs. 5 mos. 85 lbs.				
23 lbs.	375–400 lbs.				
13 lbs.	c. 150 lbs. ?				
c. 70 lbs.	at 10 yrs. 9 mos. 200 lbs.				
30 lbs.	250 lbs.				
34 lbs.					
31 lbs.		Mar. 23, 1962	2 yrs. 2 mos.	2 mos.	
14 lbs.		Jan. 21, 1969	8 yrs.	7 yrs.	
38 lbs.	300 lbs.				
24 lbs.	at 3 yrs. 3 mos. 40 lbs.				
38 lbs. (emaciated)	at 4 yrs. 8 mos. 80 lbs.				
36 lbs.	200 lbs.				
		Oct. 27, 1963	15 mos.	6 mos.	
135 lbs.	at 7 yrs. 10 mos. 150 lbs.				
180 lbs.	at 9 yrs. 1 mo. 215 lbs.				
156 lbs.	at c. 12 yrs. 465 lbs.				
40 lbs.	at 2 yrs. 9 mos. 45 lbs.	(Sent to Dallas Zoo, May 22, 1965)			
c. 30 lbs.					
25 lbs.					
	350 lbs.	(*Bata,* from San Diego Zoo)			
	at 9 yrs. 9 mos. 343 lbs.	(from Ft. Worth Zoo, in exchange for *Bata*)			
c. 170 lbs.	at 12 yrs. 3 mos. 240 lbs.	(was until 1968 thought to be a *male*)			
20 lbs.	100 lbs. +				
	Received from Brazzaville (Congo) Zoo				
c. 105 lbs. ?	at c. 10 yrs. 7 mos., 180 lbs.				
30 lbs.					
39 lbs.	at c. 8 yrs. 185 lbs.				
26 lbs.	at c. 7 yrs. 115 lbs.				
12 lbs.	at c. 6 yrs. 135 lbs.				
12 lbs.	at c. 6 yrs. 125 lbs.				
23 lbs.	at c. 4 yrs. 90 lbs.				
c. 19 lbs.	at c. 3 yrs. 11 mos., 75 lbs.				
25 lbs.	at c. 4 yrs. 3 mos. 74 lbs.				
142 lbs.	250 lbs.	(Same as No. 120)			

(Continued)

Table 25, concluded

	No.	Name	Sex	Species	Estimated Birthdate	Place Received	Date of Arrival	Estimated Age on Arrival
	140	Capt. Jack	♂	Lowland	c. Nov. 1964	Los Angeles Zoo	Nov. 17, 1965	c. 12 mos.
	141	Kay	♀	″	″	″	″	″
	142	(unnamed)	♀	″	″	″	″	″
	143	″	♀	″	″	″	″	″
	144	Chuma	♂	″	c. Aug. 1964?	N.Y. Zool. Park	May 10, 1966	21 mos. ?
	145	Sukari	♀	″	c. Nov. 1964?	″	″	18 mos. ?
	146	Kongo	♂	″	Aug. 1965	Central Park Zoo New York City	May 11, 1966	9 mos.
	147	Lulu	♀	″	″	″	″	9 mos.
	148	Muke	♀	″	?	St. Louis Zool. Park	Oct. 24, 1966	?
(c.80?)	149	Om Bom	♂	″	Mar. (?) 1955	″	Sept. (?) 1956	c. 18 mos.
	150	Massa (II)	♂	″	Jan. 1966	San Diego Zoo	July 28, 1968	2 yrs. 6 mos.
	151	Little John	♂	″	Nov. 1967	″	Aug. 8, 1968	9 mos.
	152	Cinnamon	♀	″	Sept. 1967?	Seattle, Wash. Woodland Park Zoo	Nov. 1968	14 mos. ?
	153	Nutmeg	♀	″	Nov. 1967?	″	″	12 mos. ?
	154	Ginger	♀	″	Oct. 1967?	″	″	13 mos. ?
	155	Sasha (or "Vanilla")	♀	″	May, 1968	Houston Zool. Gardens	Nov. 11, 1968	6 mos.
	156	Kiki	♂	″	Jan. 1968?	Seattle, Wash. Woodland Park Zoo	June, 1969	17 mos. ?
	157	Pierrot	♂	″	July, 1968?	″	″	11 mos. ?
	158	Bomba (or "Otto")	♂	″	c. Mar. 1967	Houston Zool. Gardens	June 26, 1969	c. 2 yrs. 3 mos.
	159	Westy	♂	″	Aug. 1967?	Philadelphia Zool. Garden	July 11, 1969	23 mos. ?
	160	(unnamed)	♂	″	″	″	″	″
	161	″	♀	″	″	″	″	″
Some previously overlooked zoo arrivals (incomplete list)		Rudy	♂	″	Apr. 1957?	St. Louis Zool. Park	June 18, 1958	14 mos. ?
		Trudy	♀	″	″	″	″	14 mos. ?
		Casey	♂	″	c. Nov. 15, 1956	Como Zoo, St. Paul Omaha Zoo	May 14, 1959 ?	2½ yrs.
		Bwana	♂	″	c. Apr. 1958	San Francisco Zoo	Oct. 21, 1959	c. 18 mos.
		Missus	♀	″	c. Apr. 1957	″	″	c. 2½ yrs.
		Don	♂	″	c. Nov. 1, 1968	Como Zoo, St. Paul	May 1, 1969	c. 6 mos.
		Donna	♀	″	″	″	″	c. 6 mos.
		Brigitte	♀	″	?	Omaha Zoo	?	?
		Kisoro	♂	″	?	Chicago (Lincoln Park)	?	?
		Mumbi	♀	″	?	″	?	?
		Jim	♂	″	?	Kyoto, Japan	?	?
		Be-Be	♀	″	?	″	?	?
		Freddy	♂	″	?	Chicago (Lincoln Park)	?	?
		Helen	♀	″	?	″	?	?
		Benoit	♀	″	?	Cincinnati Zoo	?	?

Bodyweight on Arrival	Present Estimated Bodyweight	Date of Death	Estimated Age at Death	Length of Time in Captivity (U.S.A.)	Remarks
15 lbs.	at c. 5 yrs. 130 lbs.				
"	at c. 5 yrs. 120 lbs.				
"	at c. 5 yrs. 120 lbs.				
"	at c. 5 yrs. 120 lbs.				
c. 26 lbs.	at c. 3 yrs. 2 mos. 67 lbs.				
c. 23 lbs.	at c. 3 yrs. 5 mos. 63 lbs.				
c. 22 lbs.					
c. 18 lbs.					
?					
(to Dallas Zoo on Mar. 20, 1973) 42 lbs.	Transferred to Wild Animal Park on Apr. 16, 1972				
17 lbs.					
19 lbs.					
16 lbs.					
17 lbs.					
9 lbs.	at 21 mos. 36 lbs.				
22 lbs.					
14 lbs.					
36 lbs.	at c. 3 yrs. 50 lbs.				
c. 28 lbs. ?	43 lbs.				
"	40 lbs.				
"	40 lbs.				

Bodyweight on Arrival	Present Estimated Bodyweight	Date of Death	Estimated Age at Death	Length of Time in Captivity (U.S.A.)	Remarks
	at 14 yrs. 545 lbs.				
26 lbs.					
54 lbs.	at 5 yrs. 125 lbs.				
12 lbs. 1 oz.	at c. 18 mos. 36 lbs.				
12 lbs. 11 oz.	at c. 18 mos. 41 lbs.				
?					
?					
?					
?					
?					
?					
?					
?					

Table 26. Chronological List of Gorillas Born in Captivity.

All gorillas are of the *lowland* species, except where the ranking number is preceded by an asterisk.

Ranking Fetus	Viable	Date	Zoo	Sex	Name of Young	Weight at Birth, lbs. and oz.	Names of Parents Father	Names of Parents Mother	Ages of Parents and remarks concerning young Father	Ages of Parents and remarks concerning young Mother
1	1	Dec. 22, 1956	Columbus	♀	Colo	3 – 4	The Baron (380 lbs.)	Christina (260 lbs.)	c. 10 yrs.	11 yrs. 4 mos.
2	—	Mar. 29, 1958	Basel	♀	—	(aborted)	Stefi	Achilla	10 yrs. 11 mos.	10 yrs. 6 mos.
3	2	Sept. 23, 1959	Basel	♀	Goma	4 – 0	Stefi (350 lbs.)	Achilla	12 yrs. 5 mos.	12 yrs.
*4	3	Oct. 26, 1959	IRSAC-Lwiro (Bukavu, Zaire)	?	—	4 – 5¼	?	?	(killed by mother at birth)	
5	4	Apr. 17, 1961	Basel	♂	Jambo	4 – 0	Stefi (400 lbs.)	Achilla (165 lbs.)	14 yrs.	13 yrs. 7 mos.
*6	5	June 1, 1961	IRSAC-Lwiro (Bukavu, Zaire)	?	—	?	?	?	(died June 24, 1961)	
7	6	Sept. 9, 1961	Washington	♂	Tomoka	5 – 4¾	Nikumba	Moka	7 yrs. 10 mos.	8 yrs. 3 mos.
8	7	Jan. 10, 1964	Washington	♂	Leonard	5 – 7¾	Nikumba	Moka	10 yrs. 2 mos.	10 yrs. 7 mos. (died at Toronto Zoo)
9	8	June 1, 1964	Basel	♂	Migger	?	Stefi	Achilla	17 yrs. 1½ mos.	16 yrs. 8½ mos.
10	—	Oct. 6, 1964	Colo. Springs	♂	—	?	Hercules	Becky	9 yrs. 1 mo.	8 yrs. 8 mos. (4 to 6 weeks premature)
11	9	Feb. 13, 1965	Kansas City, Missouri	♂	Abe	(stillborn)	Big Man	Kribi Kate	c. 7 yrs. 7 mos.	6 yrs. 8 mos. (died)
*12	10	May 6, 1965	IRSAC-Lwiro (Bukavu, Zaire)	?	—	?	?	?		(lived)
13	11	June 3, 1965	San Diego	♀	Alvila	4 – 11½	Albert	Vila	16 yrs. 3 mos.	7 yrs. 9 mos.
14	12	June 22, 1965	Frankfurt	♂	Max	4 – 10	Abraham	Makula	c. 10 yrs.	c. 9 yrs.
15	—	Aug. 17, 1965	Toledo	♂	—	5 – 12	Togo	Porta	8 yrs. 9 mos.	11 yrs. 4 mos.
16	13	Sept. 2, 1965	Colo. Springs	♀	Roberta	(stillborn)	Hercules	Becky	10 yrs.	9 yrs. 7 mos. (died Sept. 6, 1965)
17	14	Dec. 18, 1965	Dallas	♀	Vicki	4 – 0	Jimmy	Jenny	9 yrs. 7 mos.	10 yrs. 5 mos. (sent to Alberta Game Farm)
18	—	April, 1966	Frankfurt	?	—	(aborted)	Abraham	Makula	c. 10 yrs. 10 mos.	c. 9 yrs. 10 mos.
19	15	Aug. 19, 1966	Dallas	♂	Micki	4 – 8½	Jimmy	Shamba	10 yrs. 3 mos.	7 yrs. 2 mos.
20	—	Sept. 18, 1966	Kansas City, Missouri	?	—	?	Big Man	?	c. 9 yrs. 2 mos.	(aborted twins)
21	—	" " "	"	?	—	?	"	"	"	"
22	16	Oct. 16, 1966	St. Louis	♂	Mzuri	4 – 10	Rudy	Trudy	c. 9 yrs. 6 mos.	c. 9 yrs. 6 mos.
23	17	Apr. 8, 1967	Washington	♀	Inaki	4 – 0	Nikumba	Moka	13 yrs. 5 mos.	13 yrs. 10 mos.
24	18	May 3, 1967	Frankfurt	♀	Ellen	4 – 3	Abraham	Makula	c. 11 yrs. 11 mos.	c. 10 yrs. 11 mos. (Fraternal twins) (lived)
25	19	" " "	"	♀	Alice	3 – 12	"	"	"	" (died June 7, 1969)

(Continued)

Table 26. Chronological List of Gorillas Born in Captivity. (Continued)

Ranking Fetus	Viable	Date	Zoo	Sex	Name of Young	Weight at Birth, lbs. and oz.	Names of Parents Father	Mother	Ages of Parents and remarks concerning young Father	Mother
26	—	June 24, 1967	Kansas City, Missouri	♂	—	? (stillborn)	Big Man	Jungle Jeannie	c. 9 yrs. 11 mos.	c. 9 yrs.
27	20	Sept. 19, 1967	Kansas City, Missouri	♀	—	? (stillborn)	Big Man	Kribi Kate	c. 10 yrs. 2 mos.	9 yrs. 3 mos. (died Sept. 23, 1967)
*28	—	Oct. 7, 1967	Antwerp	?	—	? (premature)	Kisubi	Pega	c. 9 yrs.	?
29	21	Nov. 14, 1967	Toledo	♂	Figoro	5 – 0 (est.)	Togo	Porta	11 yrs.	13 yrs. 7 mos. (died Jan. 7, 1968)
30	—	Dec. 23, 1967	Cincinnati	♀	—	? (stillborn)	King Tut	Penelope	18 yrs.	c. 13 yrs. 8 mos.
31	22	Feb. 1, 1968	Columbus	♀	Emmy	3 – 10	Bongo	Colo	10 yrs. 8 mos.	11 yrs. 1 mo.
32	—	Feb. 15, 1968	Dallas	♂	—	? (stillborn)	Jimmy	Shamba	11 yrs. 9 mos.	3 yrs. 8 mos.
*33	23	June 9, 1968	Antwerp	♀	Victoria	4 – 2	Kisubi	Quivu	?	?
34	24	July 15, 1968	Kansas City, Missouri	♀	Tiffany	4 – 2	Big Man	Jungle Jeannie	(now in Topeka Zoo) c. 11 yrs.	c. 10 yrs. 1 mo.
35	25	July 17, 1968	Basel	♀	Quarta	?	Stefi	Achilla	21 yrs. 3 mos.	20 yrs. 10 mos.
36	—	Mar. 6, 1969	Omaha	?	—	? (aborted)	Casey	Brigitte	c. 12 yrs. 4 mos.	?
37	26	Mar. 7, 1969	Dallas	♂	Max	4 – 6	Jimmy	Shamba	12 yrs. 10 mos.	3 yrs. 9 mos.
38	27	Apr. 15, 1969	Frankfurt	♀	Salome	?	Solomon	Makula	?	(now in Topeka Zoo) c. 12 yrs. 10 mos.
39	28	July 10, 1969	Columbus	♂	Oscar	4 – 12	Bongo (390 lbs.)	Colo	12 yrs. 1 mo.	2 yrs. 6 mos.
40	29	Aug. 26, 1969	Toledo	♀	Happy	5 – 0 (est.)	Togo	Porta	12 yrs. 9 mos.	15 yrs. 4 mos.
41	30	Aug. 27, 1969	Omaha	♂	—	?	Casey	Benoit	c. 12 yrs. 9 mos.	?
42	31	Sept. 16, 1969	Kansas City, Missouri	♂	The Colonel	6 – 0	Big Man	Kribi Kate	c. 12 yrs. 2 mos.	11 yrs. 3 mos. (died Aug. 28, 1969)
43	32	Nov. 15, 1969	Chicago Zoo (Brookfield, Ill.)	♀	—	?	Mighty Omega	Alpha	6 yrs. 10 mos.	3 yrs. 9 mos. 1969 (died Nov. 24, 1969)
44	33	Nov. 15, 1969	Colo. Springs	♂	Chey-Gor	c. 4 – 0	Hercules	Becky	14 yrs. 2 mos.	13 yrs. 9 mos.
45	34	Nov. 30, 1969	Kansas City, Missouri	♀	Violet	?	Big Man	Jungle Jeannie	c. 12 yrs. 4 mos.	c. 11 yrs. 5 mos. (died June 2, 1970)
46	35	Jan. 23, 1970	Cincinnati	♂	Sam	4 – 9½	Hatari	Mahari	7 yrs. 9 mos.	c. 7 yrs.
47	36	Jan. 31, 1970	Cincinnati	♀	Samantha	4 – 15	King Tut	Penelope	21 yrs. 1 mo.	c. 15 yrs. 9 mos.
48	37	Mar. 1, 1970	Omaha	♀	Miss Vicky	?	Casey	Brigitte	c. 13 yrs. 4 mos.	? (died June 3, 1970)
49	38	May 26, 1970	San Francisco	♂	Kwanza	?	Bwana	Missus	12 yrs. 1 mo.	13 yrs. 1 mo.
50	39	July 10, 1970	Dallas	♀	Moja Demba	?	Jimmy	Shamba	14 yrs. 2 mos.	11 yrs. 1 mo.
51	40	July 22, 1970	Chicago (Lincoln Park)	♀	Kumba	?	Kisoro	Mumbi	c. 12 yrs.	?
52	41	Aug. 29, 1970	Omaha	♀	?	?	Casey (545 lbs.)	Benoit	c. 13 yrs. 9 mos.	?

(Continued)

Table 26. Chronological List of Gorillas Born in Captivity. (Concluded)

Ranking Fetus	Viable	Date	Zoo	Sex	Name of Young	Weight at Birth, lbs. and oz.	Name of Parents Father	Name of Parents Mother	Ages of Parents and remarks concerning young Father	Mother
53	42	Oct. 29, 1970	Kyoto, Japan	♂	Makk	?	Jim	Be-Be	?	?
54	43	Nov. 24, 1970	Basel	♀	Souanke	?	Stefi (?)	Kathy	23 yrs. 7 mos. (died Nov. 28, 1970)	?
55	44	Feb. 2, 1971	Takamatsu, Japan	♂	Fujio	?	Ricky	Rinko	?	?
56	—	Feb. 12, 1971	Colo. Springs	♀	—	(stillborn)	Hercules	Bathsheba	15 yrs. 5 mos.	14 yrs. 11 mos.
57	45	Mar. 6, 1971	Bristol	♂	—	?	Samson	Caroline	?	?
58	46	Apr. 10, 1971	Bristol	♂	Daniel	?	Samson	Delilah	(died Mar. 16, 1971)	?
59	47	Apr. 12, 1971	Cincinnati	♂	Ramses I	?	King Tut	Penelope	22 yrs. 4 mos.	c. 17 yrs.
60	48	Apr. 21, 1971	Chicago (Lincoln Park)	♀		?	Freddy	Helen	?	?
61	49	May 2, 1971	Basel	♂	Tamtam	?	Jambo	Goma	10 yrs. 1 mo.	11 yrs. 7 mos.
62	50	May 17, 1971	Chicago Zoo (Brookfield, Ill.)	♂	Weaver	?	Mighty Omega	Alpha	8 yrs. 4 mos.	10 yrs. 3 mos.
63	51	May 27, 1971	Frankfurt	♀	Dorle	?	Matze	Makula	?	c. 14 yrs. 11 mos.
64	52	July 4, 1971	San Francisco	♀	Hanabi-Ko	?	Bwana	Jacqueline	13 yrs. 2 mos.	?
65	53	Sept. 12, 1971	Cincinnati	♀	Kamari	?	Hatari	Mahari	9 yrs. 5 mos.	?
66	54	Sept. 22, 1971	Omaha	♀		?	Casey	Benoit	c. 14 yrs. 10 mos.	8 yrs. 8 mos.
		May 29, 1972	Washington	♂	Mseni	5 – 10	Nikumba	Femelle	18 yrs. 6 mos.	c. 13 yrs. 4 mos.
		July 13, 1972	Cincinnati	♀	Mopaya	?	King Tut	Penelope	22 yrs. 7 mos.	c. 18 yrs. 3 mos.
		Jan. 1, 1973	Cincinnati	♀	Gigi	?	Hatari	Mahari	10 yrs. 8 mos.	9 yrs. 11 mos.
		July 14, 1973	Jersey (Channel Is.)	♂	Amani Assumbo	4 – 14	Jambo	Nandi	12 yrs. 3 mos.	c. 14 yrs.
		Sept. 11, 1973	Jersey (Channel Is.)	♂	Mamfe	5 – 10	Jambo	N'Pongo	12 yrs. 5 mos.	c. 16 yrs.
		Oct. 15, 1973	San Diego (Wild Animal Park)	♂	Jim	6 – 4	Trib	Dolly	14 yrs. 3 mos.	9 yrs. 10 mos.
		Apr. 15, 1974	Cincinnati	♀	Tara	?	King Tut	Penelope	24 yrs. 4 mos.	c. 20 yrs.
		Aug. 16, 1974	Cincinnati	♀	Mata Hari	?	Hatari	Mahari	12 yrs. 3 mos.	11 yrs. 6 mos.
		Sept. 8, 1974	Lincoln Park, Chicago	?		?	Frank	Mary	?	?
		Oct. 2, 1974	San Diego (Wild Animal Park)	♀	Binti	4 – 0 (?)	Trib	Dolly	15 yrs. 3 mos.	10 yrs. 10 mos.
		Oct. 23, 1974	Jersey (Channel Is.)	♀	Zaire	4 – 14	Jambo	Nandi	13 yrs. 6 mos.	c. 15 yrs. 3 mos.
		Jan. 28, 1975	Phoenix	♂	Fabayo	4 – 8 (est.)	Trib	Hazel	15 yrs. 6 mos.	14 yrs. 6 mos.
		Jan. 29, 1975	Jersey (Channel Is.)	♂	Tatu	?	Jambo	N'Pongo	13 yrs. 9 mos.	c. 17 yrs. 4 mos.
		Dec. 1975	Kansas City, Missouri	♀	?	?	Trib	Kribi Kate (?)	16 yrs. 5 mos.	17 yrs. 6 mos.

14 Births occurring after Sept. 22, 1971 (incomplete)

Tut). For the mothers it is from 6 years 8 months (Kribi Kate) to approximately 20 years (Penelope).

The longest period of productivity in a father gorilla was that of Stefi, of the Basel, Switzerland, Zoo, provided he sired Souanke in 1970. His span in that case was from 10 years 11 months to 23 years 7 months, or 12 years 8 months. Otherwise it would have been from 10 years 11 months to 21 years 7 months, or 10 years 4 months. But the latter span is surpassed by that of Nikumba, of the National Zoo, Washington, D.C., with a span of from 7 years 10 months to 18 years 6 months, or 10 years 8 months. The longest productive period among gorilla mothers was that of Kribi Kate, of the Kansas City, Missouri, Zoo, with from 6 years 8 months to 17 years 6 months, or 10 years 10 months. Converted to human ages, the gorilla sires would be productive from about 12 years to 40 years, and the gorilla dams from about 10 years to 30 years.

The most times a gorilla has become a father is 6, possibly 7. 6 sirings, including one of twins, were recorded of Big Man. Stefi sired 5 times, possibly 6. Jambo, at least 5. Trib, 4 or 5. Among females, Makula became a mother 5 times, including one of twins; Penelope, 5 times; Achilla and Kribi Kate, each 4 times.

As to the sex ratio, among the 71 newborn gorillas in Table 26 of which the sex was recorded, 33 were males and 38 were females.

Among the 79 births listed, 10 occurred in September; 8 each in April, May, July, and October; 6 each in January, June, and August; 5 each in March and November; and 4 in December. From this it is evident that among gorillas mating and parturition may take place at any time of the year. However, from the present data it is clear that more births occur during warm weather than during the winter months. Whether this has any significance remains to be determined. The births took place much more frequently by day than by night.

Among the 80 births listed in Table 26 of captive gorillas during the last 20 years, it is interesting and encouraging to note that while it took 12 years 9 months for the first 40 births to take place, only 7 years 3 months were required for the second 40. From this it is clear that the birth rate is markedly accelerating. And if the present rate acceleration continues, it would appear that another 40 births should take place within only the next 5 years. This possibility would certainly boost the lowland gorilla population. But the population of captive mountain gorillas is something else; and steps should be taken to improve their exceedingly low and disappointing rate of births. Among the 68 surviving gorilla births here recorded, 65 were of lowland gorillas and only 3 of the mountain species.

Where do the names for zoo gorillas come from? As Tables 25 and 26 show, the name for a particular gorilla can be almost anything, so long as the name is in keeping with the sex of the animal. Sometimes, however, a name may be given on the assumption that the ape's sex is known, only to turn out later to be an error. This happened to "Jacob" (No. 127 in Table 25), who proved in due course to have been a female all along! "Massa" also was at first, and for many years thereafter, thought to be a female, as was even "Ngagi," of the San Diego Zoo, who turned out to be a male mountain gorilla of record-sized dimensions.

Here is a sampling of names of gorillas mentioned in this book, along with the sources of the names (as presumed by the writer):

Achilla (female, Basel): the feminine equivalent of *Achilles* (?).
Albert (male, San Diego): after King Albert of Belgium, who in 1925 established Albert National Park as a sanctuary for the mountain gorilla. "Albert," however, is a western lowland gorilla.
Bata (female, San Diego): a town on the coast of equatorial Guinea.
Benoit (female, Omaha): a town in west-central Mississippi (?).

Fig. 157. Two gorilla mothers at the San Diego Wild Animal Park, April 1976. At the left, "Aunt" Vila is carrying 2½-year-old Jim, son of Trib and Dolly; while Dolly, following her, carries 1½-year-old daughter Binti. (Photo courtesy San Diego Zoo.)

Fig. 158. "Missus," a three-year-old female lowland gorilla at the San Francisco Zoo (1960). The little girl is showing that Missus enjoys being tickled (as indeed most young gorillas do), and she is bringing out in Missus about the nearest thing to laughter that a gorilla can come up with. (Photo courtesy Dr. Deets Pickett.)

Binti (female, Jersey Is.): after Mt. Bintimani in Sierra Leone (?).
Bongo (male, Columbus): a village in northwestern Zaire (formerly Belgian Congo); and another in southwestern Gabon.
Bwana (male, San Francisco): a colloquial native African name for "Master."
Chey-Gor (male, Colorado Springs): after (Chey)enne Mountain Zoological Park in Colorado Springs and (Gor)illa.
Colo (female, Columbus): after the city in Ohio where she was born.
Gargantua (male, circus): after the giant in Rabelais' story, *Gargantua and Pantagruel*.
Goma (female, Basel): a village on the north shore of Lake Kivu, in Ruanda.
Inaki (female, Washington): evidently a modification of the native word *Ingagi*, meaning "gorilla."

Kribi Kate (female, Kansas City, Mo.): Kribi is a town on the southwest coast of Cameroon.
Makoko (male, Bronx, N.Y.): a town (Makokou) in Gabon.
Mamfe (male, Jersey Is.): a native village in northwestern Cameroon.
Massa (male, Philadelphia): a native corruption of the English word "Master."
Mbongo (male, San Diego): after the Alu(mbongo) mountains in easternmost Zaire, where he and Ngagi were captured.
Mzuri (male, St. Louis): "Missouri."
Nandi (female, Jersey Is.): a hypothetical hyenalike beast ("Nandi bear") of the East African forests. Also the name of a tribe of natives of that region.
Ngagi (male, San Diego): a native term (N'Gagi, or Ingagi) for "gorilla."
N'Pongo (female, Jersey Is.): one of the local African names for "gorilla."
Trib (male, San Diego): so-named from having been presented to the San Diego Zoo by James S. Copley, publisher of the San Diego Union and Evening (Trib)une. The name "Trib" was selected as the result of a contest conducted by the San Diego Union.
Zaire (female, Jersey Is.): after the African state, Zaire (formerly Belgian Congo).

Some gorilla names appear to be after towns that are not even in Africa. Either that, or the names of such towns are simply coincidental to names of gorillas derived from some other sources. While the colossal fictional ape *King Kong* is generally associated with some remote island in the South Pacific (to which place no gorilla is likely to have migrated!), there is a town named Kong in the northern part of Ivory Coast, Africa.

Sometimes, names for baby gorillas can be adopted that indicate the names of the parents, which helps to identify both the offspring and its parents. This occurred in the name "Alvila," a female of the San Diego Zoo, which name was compounded from that of the father (Al)bert and the mother, Vila. Apparently such naming occurred also in "Tomoka," a male of the National Zoo in Washington, D.C., although here only the mother gorilla's name (Moka) was involved.

In any case, it is evident that there is no limitation to the names that have, and can be, bestowed on captive gorillas. In the foregoing list, African-derived names are purposely interpreted; but there are many other names that need no interpretation and simply follow those given to human infants:

Jim, Mike, Solomon, Abraham, Jacob, Phil, et al, among boys; and Penelope, Carolyn, Dolly, Susie, Betsy, Hazel, et al, among girls (girl gorillas, that is!).

Notes

1. B. F. Riess, Sherman Ross, S. B. Lyerly, and H. G. Birch, "The Behavior of Two Captive Specimens of the Lowland Gorilla," *Zoologica* (New York Zoological Society) 34, pt. 3 (November 30, 1949):111–18 (2 plates—Makoko and Oka).
2. Robert Yerkes, "Gorilla Census and Study," *Journal of Mammalogy* 32, no. 4 (1951):429–36.
3. Bernhard Grzimek, "Die Gorillas ausserhalb Afrikas," *Der Zoologische Garten* (NF) 20 (2/3) (1953):173–85.
4. R. E. Honegger and P. Menichini, "A Census of Captive Gorillas with Notes on Diet and Longevity," *Der Zoologische Garten* (NF) 26 (3/6) (1962):203–14.
5. Edalee Orcutt Harwell, personal communication, April 5, 1962.
6. James Fisher, *Zoos of the World* (London: Aldus Books, 1966), p. 91.
7. George H. Pournelle, "The Gorilla . . . its Status in Captivity," *Zoonooz* (San Diego).

See also the studbook of gorillas compiled by Von R. Kirschshofer in the *International Zoo Yearbook* for 1973, London.

14

The Gorilla's Chances for Survival

Animals become extinct for various reasons. In the case of the dinosaurs, which ruled the earth for a period of over 100,000,000 years, changing climate and food supply, along with accelerating competition from mammals (some of which ate the dinosaur's eggs), in due course led to the extinction of the giant reptiles. But within historic times the chief exterminator of the lower animals has been man. His geometric increase in numbers, over a living area of limited extent, has caused him to alter and to put to his own use a steadily increasing amount of space that previously he had shared with other living creatures. Hence, as the inevitable result of the human population increase, most forms of animal life, particularly land-living mammals, are doomed to extinction in the wild (which will no longer be "wild"), possibly within the next two or three generations. Here, let me quote from one of my earlier articles on the extermination of various forms of wildlife:

> Paradoxically, the more numerous and conspicuous an animal species becomes, the sooner it may disappear. A classic illustration is the Passenger Pigeon, which in the middle 1800s populated the middle and eastern United States in millions, yet was virtually extinct before the year 1900. The American bison, or buffalo, which now exists in only a few controlled herds, once swarmed over the prairies in numbers estimated up to sixty million. The pronghorn antelope probably was equally numerous. Fortunately, both these picturesque mammals were saved from extinction by legislation. In Africa, many splendid game animals have experienced a similar decimation. The springbok, a beautiful antelope now seen only in small groups, was described by early travelers as migrating in immense *trekbokens* containing from ten thousand to fifty thousand individuals.
>
> The quagga probably never existed in such large numbers, but its relatively sudden disappearance from its age-old homeland, the veld, constitutes a sad but enlightening instance of what may happen when human foresight is nonexistent.[1]

While I was writing this, I noticed an account in the day's newspaper that stated that the California condor, which once roamed the entire Pacific Coast from Canada to Baja California, was now reduced to a colony of about forty-five individuals. The account was captioned: "CAPTIVE BREEDING SEEN AS CONDOR'S LAST HOPE." Only the same hope now prevails for the gorilla, particularly the Kivu or mountain species. This possibility has been well expressed by Dr. Deets Pickett, the gorilla collector, now retired, who recently wrote: "The only possible survival for these wonderful animals is by importation into this country and breeding as many as possible."[2]

Dr. Pickett, who made a number of trips into the Cameroons, each time bringing back with him captured specimens of gorillas, mostly of infantile or juvenile age, for American zoos, made additional remarks about gorillas in the wild and in captivity which I think should be of interest to readers here. These comments, which were written in 1960, apply in particular to the western lowland gorillas Pickett encountered, or brought back with him, from the Cameroon country.

> Two or three days ago, driving down the trail, we heard a tremendous "uuuuuugh, uuuuuugh, uuuuuugh," just like the sound Big Man* makes when he is displeased—only multiplied a hundred times. Amid a great shaking of leaves and branches, there appeared a dark and tremendous body, topped by a gargoyle-like head. After we slammed on the brakes, this huge gorilla looked us over disapprovingly for an instant, then faded away into the jungle. Truly a thrilling sight.

> With the number of gorillas we have glimpsed, and the hundreds of signs which they have left engraved into the earth, as well as the beds and nests, I am beginning to think that rather than the gorilla being rare, it is the hunter of the gorilla who is rare. There are few who care enough about seeing the gorilla in his native jungle to go to the necessary trouble. It is true that the gorilla inhabits only a relatively small, circumscribed area of the earth's surface and only pockets within that area. Many people believe that there are not more than a few hundred gorillas on earth, but there must be several thousand. The reputation for extreme rarity is probably due in part to the fact that the gorilla lives in the most outlying districts of wildest jungle. Their preference is for the thickest of bush, where, many times, you cannot see more than two or three yards in any direction. One is more apt to hear than to see them. That has occurred with me a number of times. Then, too, miles of turning, weaving and twisting one's body through vines, brush, and over fallen trees requires more than a few days of physical training. Men with the best of physiques are tired at the end of a few miles of such.

---

* A male gorilla delivered by Dr. Pickett to the Kansas City Zoo in 1958, when the little ape was about a year old. He grew to be a five hundred-pound giant, sired seven offspring, and died at the age of about 19 years in 1976 (see Fig. 138).

Another reason for the belief that gorillas are so rare is that there are relatively few in captivity.* Not over ten percent of those captured have lived to be displayed in zoological gardens. Gorillas, in their remarkable resemblance to man, have a great susceptibility to his infections. They have had no opportunity to build up a racial immunity; so, many of the diseases which barely affect man, prove rapidly fatal to gorillas. Another even more important condition affecting the longevity of gorillas in captivity is their extreme emotionalism. To people who are not familiar with the gorilla, his expression seems somewhat placid, but underneath that placid exterior he is a seething, writhing mass of emotions. Many adult gorillas have died

Fig. 159. This sad-faced little western lowland gorilla, photographed after having been taken away from his mother, may be seeking companionship with any friendly person or animal that he can approach. The tiny antelope he is stroking appears to be a young duiker. (Photo courtesy Dr. Deets Pickett.)

---

* This was written in early 1960, before the marked success that has been attained in the zoo-raising of gorillas since then.

Fig. 160. An adult male mountain gorilla killed (speared) by natives in the area immediately west of Lake Kivu (el. 7200 feet), in 1944. Mercifully, the animal appears to have received a spear-thrust directly in the heart. (Photo courtesy F. L. Hendricks, then director of the Kivu Experimental Station.)

Fig. 161. The propped-up figure of an adult male lowland gorilla, shot near Yaoundé, in the Cameroons. Note the characteristic barrel-shaped torso, tapering fingers, thick wrists and ankles, lack of "calf," etc. (Photo from Brehm's *Tierleben*.)

in captivity, and on autopsy no visible organic lesions have been found. The pathologists have been at a loss to explain the deaths. Many times these animals are those who have remained alone. Having no desire to live, they finally die from extreme melancholy.

It is for that reason that gorillas need tender, loving care from the very start. They have been harshly jerked out of their normal environment and away from their mother and father. Unlike most captured animals, they cannot be tucked away in a box and fed two or three times a day. The diet must be exact, and hygiene of the highest type used. Vaccinations against standard childhood diseases are indispensable.
Gorillas are normally very affectionate animals. When one of the children or a mate has been killed, the others will frequently remain in the vicinity for weeks, crying with low, moaning wails.

In discussing the merits of gorillas, it should not be inferred that the gorilla hasn't earned his reputation for ferocity. However, he is an animal who believes in tending strictly to his own business; and, as long as you leave him alone and don't try to interfere with him or his family, he will leave you alone. Of course, what he thinks of as interference may differ somewhat from your own definition. You may be just wandering down a jungle trail, and, if you are moving too silently and come upon him too suddenly, he may think that you are doing it with malice aforethought and give you a little trouble—and just a little trouble from a male gorilla is more than enough! However, even if you should come upon his family in such a manner, if you don't annoy him further, the chances are that he will charge you with a horrifying roar, but will stop short of where you are, beat his chest a few times, and when his family have moved on will then melt away himself. He actually wants no trouble with you —which is a good thing for you![3]

As to the gorilla's precarious position today in the wild, it can only be said that a number of factors are altering the ape's up-until-now natural habitat, and so threatening its survival—indeed, contributing to its extinction. One of these factors is the destruction of the forests wherein the gorillas have always lived. The felling of trees may take place either for the purpose of supplying lumber, or for the clearing of areas in which farming may be carried on. With the trees goes, of course, a great deal of the plants on which the gorillas feed. In west Africa, however, this does not seem to be the major threat to the gorillas'— that is, the western lowland gorillas'—survival. Rather, it is the hunting of these gorillas by natives, who not only have an age-old penchant for gorilla flesh, but who also use the gorilla meat to feed their dogs (which are used on the gorilla hunts). While the western, as well as the eastern

Fig. 162. This fine example of the taxidermist's art is of a male mountain gorilla, mounted by the famous Rowland Ward Studios of London, way back in 1914 (before laws were passed prohibiting the killing of their species). The standing height of this specimen was 62 inches, and the chest girth was 55¾ inches. These measurements indicate a living body weight of about 360 pounds, or a rather small adult example of the mountain species. (Photo courtesy British Museum of Natural History.)

Fig. 163. Two adult male lowland gorilla heads, drawn nearly a hundred years apart. Above is a wood engraving by the famous old-time German animal illustrator Friedrich Specht (1839–1909). Below is a recent pen drawing of "Guy," of the London Zoo, by Dr. Adolph H. Schultz, Emeritus Professor of Physical Anthropology at the University of Zurich, through whose courtesy the drawing is here reproduced. Note the benign countenance of the captive gorilla Guy, in contrast to the rather fearsome aspect of the jungle gorilla pictured above by artist Specht.

mountain, gorillas are supposed to be protected by law, it appears impossible to enforce that law. As an example of the flagrant disregard of the law, Dr. Pickett relates that "When I was last in Africa there was one 100-mile strip of jungle road on which I learned that at least a hundred gorillas were killed in a 30-day period of time" (see note 2, this chapter).

What with law-ignoring natives equipped with modern rifles and ammunition, and activated by their need for food—by which they may justify their securing of gorilla meat—there is not much need for stressing the destruction of jungle forests as the prime threat to the lowland gorillas' survival, as some writers on the subject continue to do. Also, while the actual amount of food taken by gorillas from native plantations is relatively slight, any at all would be sufficient from the owner's point of view to warrant the killing of the trespassers.

With the mountain gorilla, which resides in east Africa in a different type of environment, the picture is somewhat different. While some of the native tribes of that region, such as the Bambuti pygmies, may have a craving for gorilla meat, evidently this is not the chief factor operating against the gorillas' survival. Rather, it is the encroachment on their habitat or territory by native farmers and their farm animals, principally cattle, which not only overrun the gorillas' supposed sanctuary, but also consume or trample much of the vegetation. Although total protection was granted the mountain gorillas of the Virunga volcanoes region in 1968, here again it has been difficult to enforce the legislation. The native officials assigned to the job, being poorly paid, are ready acceptors of bribes from poachers and cattle raisers. It is impossible to impress a hungry African native with the need for species conservation.

Thus, the mountain gorilla is the most endangered of the African apes. This is not only by reason of its limited numbers in the wild (there being almost certainly fewer than 1000 individuals remaining), but also because its breeding in captivity is not proceeding at a rate sufficient to ensure its survival—even with man's assistance. It can only be hoped that the latter sole chance for the species existence may be promoted successfully.

Although it is no consolation, the plight of the orangutan is even worse than that of the gorilla, since its capacity to breed in capacity is far more limited and uncertain; therefore, it will perhaps be the first of the great apes to suffer extinction. We can only hope that the gorilla, in both its lowland and mountain species, may by man's ingenuity, determination, and foresight be saved for the contemplation and education of future generations.

Notes

1. David P. Willoughby, "The Vanished Quagga," *Natural History* 75, no. 2 (February 1966):60–63.
2. Deets Pickett, personal communication, September 26, 1976.
3. Deets Pickett, included in a "news letter" from Youndé, Cameroons, dated January 14, 1960.

See also the excellent account of gorilla and orangutan conservation by Vernon Reynolds in his book *The Apes* (New York: Harper & Row, 1967), pp. 251–67.

Select Bibliography (General)*

* See also the *illustrated guidebooks* issued by various zoos (e.g., San Diego, St. Louis, Milwaukee, Chicago). See also under "Notes" at the end of each chapter in this book.

Akeley, Carl E. "Gorillas—Real and Mythical." *Natural History* 23 (September–October 1923):441. (See also *World's Work*, August 1922, p. 397.)

———. "Hunting Gorillas in Central Africa," *World's Work* 44 (June, July, August, September 1922):169, 307, 393, 525, respectively.

Barns, T. Alexander. *Across the Great Craterland to the Congo*. London, 1923.

———. "Hunting the Morose Gorilla." *Asia* (New York), February 1928, pp. 116, 154.

Beebe, B. F. *African Apes*. New York: David McCay Co., 1969.

Benchley, B. J. *My Friends, the Apes*. Boston: Little, Brown, &.Co., 1942.

Bingham, H. C. *Gorillas in a Native Habitat*. Carnegie Institute of Washington Publication No. 426, 1932.

Blancou, L. "The Lowland Gorilla." *Animal Kingdom* (New York Zoological Society) 58 (1955):162–69.

Booth, A. H. "The Zoogeography of West African Primates." *Bull. Inst. Franc. Afr. Noire* 20 (1958): 587–682.

Bourne, G. H., and Cohen, M. *The Gentle Giants—The Gorilla Story*. New York: G. P. Putnam's Sons, 1975.

Brehm, Alfred Edmund. *Brehm's Life of Animals* (English edition). Chicago, 1896. (See also the German 4th edition in 13 vols., vol. 4, Leipzig, 1922.)

Burbridge, Ben. *Gorilla*. New York: Century Co., 1928.

Carpenter, C. R. "An Observational Study of Two Captive Mountain Gorillas." *Human Biology* 9, no. 2 (May 1937):175–96. (Mbongo and Ngagi of the San Diego Zoo.)

Cook, David, and Hughes, Jill. *A Closer Look at Apes*. New York and London: Franklin Watts, 1976.

Coolidge, Harold J., Jr. "Notes on Four Gorillas from the Sanga River Region." *Proceedings of the Academy of Natural Sciences, Philadelphia* 88 (1936): 479–501.

Derscheid, J. M. "Notes on the Gorillas of the Kivu Volcanoes." *Ann. Soc. Royale Zool. Belgique* (Brussels) 58 (1927):149–59.

Donisthorpe, J. "A Pilot Study of the Mountain Gorilla." *South African Journal of Science* 54 (1958): 195–217.

Dorst, Jean. *A Field Guide to the Larger Mammals of Africa*. Boston: Houghton Mifflin Co., 1970.

Du Chaillu, Paul B. *Explorations and Adventures in Equatorial Africa*. New York, 1861.

Eimerl, S., and DeVore, I. *The Primates*. New York: Time/Life Books, Life Nature Library, 1965.

Elliot, Daniel G. *A Review of the Primates*. 3 vols. New York: American Museum of Natural History, 1913.

Emlen, J. T., and Schaller, G. B. "Distribution and Status of the Mountain Gorilla . . . 1959." *Zoologica* (New York) 45 (1960a):41–52.

———. "In the Home of the Mountain Gorilla." *Animal Kingdom* (New York Zoological Society) 63 (1960b):98–108.

Fossey, Dian, and Campbell, Robert M. "Making

Friends with Mountain Gorillas." *National Geographic* 137, no. 1 (January 1970):48–67.

———. "More Years with Mountain Gorillas." *National Geographic* 140, no. 4 (October 1971):574–85.

Garner, R. L. *Gorillas and Chimpanzees.* London: Osgood, McIlvaine & Co., 1896.

———. "Gorillas in their Jungle." *Bulletin of the New York Zoological Society* 17 (1914):1102–4.

Gatti, Attilio. *The King of the Gorillas.* New York: Doubleday, Doran & Co., 1932.

Geddis, H. *Gorilla.* London: Andrew Melrose, 1955.

Gregory, W. K., and Raven, H. C. *In Quest of Gorillas.* New York: The Darwin Press, 1937.

Groves, Colin P. *Gorillas.* New York: Arco Publishing Co.; London: Arthur Barker, Ltd., 1970.

Grzimek, Bernhard. *Among Animals of Africa.* New York: Stein & Day, 1970. Chap. 13, "The Much-maligned Gorilla."

———. "The Gorilla." In *Grzimek's Animal Life Encyclopedia.* 13 vols. New York: Van Nostrand-Rinehold, 1972. Chap. 22, 10:525–48.

Hooton, Ernest. *Man's Poor Relations.* New York, 1942.

———. *Up from the Ape.* New York: The Macmillan Co., 1961.

Hornaday, W. T. "Gorillas, Past and Present." *Bulletin of the New York Zoological Society* 18 (January 1915):1181–85.

Jenks, A. L. "Bulu Knowledge of the Gorilla and Chimpanzee." *American Anthropologist* 13 (1911): 56–64.

Johnson, Martin. *Congorilla.* New York: Blue Ribbon Books, 1931.

Kawai, M., and Mizuhara, H. "An Ecological Study of the Wild Mountain Gorilla." *Primates* 2 (1959), 3 (1961).

Keith, Arthur. "An Introduction to the Study of Anthropoid Apes. I. The Gorilla." *Natural Science* 9 (1896):26–37.

Merfield, F. G. *Gorillas Were My Neighbors.* London: Longmans, Green & Co., 1956.

Morris, Ramona and Desmond. *Men and Apes.* New York: McGraw-Hill, 1966.

Napier, J., and Napier, P. *A Handbook of Living Primates.* New York: Academic Press. 1967.

Nott, J. Fortune. *Wild Animals Photographed and Described.* London, 1886. "The Gorilla," pp. 526–38.

O'Reilly, John. "The Amiable Gorilla." *Sports Illustrated,* June 20, 1960, pp. 69–76.

Pitman, C. R. S. "The Gorillas of the Kayonsa Region, Western Kigezi, Southwest Uganda." *Smithsonian Institution Report for 1936,* pp. 253–75.

Raven, Henry C. "Gorilla: The Greatest of all Apes." *Natural History* 31 (May–June 1931):231–41.

Reade, W. Winwood. "The Habits of the Gorilla." *American Naturalist* 1 (1868).

Reynolds, Vernon. *The Apes.* New York: E. P. Dutton & Co., 1967.

Rosen, Stephen I. *Introduction to the Primates . . . Living and Fossil.* Englewood Cliffs, N.J.: Prentice-Hall, 1974.

Sanford, L. J. "The Gorilla; being a Sketch of its History, Anatomy, General Appearance and Habits." *American Journal of Science* (2) 33 (1862):48–64.

Savage, T. S., and Wyman, J. "Notice of the External Characters and Habits of *Troglodytes gorilla*, a New Species of Orang from the Gaboon River." *Boston Journal of Natural History* 5 (1847):417–43.

Schaller, George B. *The Mountain Gorilla.* Chicago: University of Chicago Press, 1963.

———. *The Year of the Gorilla.* Chicago: University of Chicago Press, 1964.

Schultz, Adolph H. *The Life of Primates.* New York: Universe Books, 1969.

Yerkes, Robert M. and Ada W. *The Great Apes.* New Haven, Conn.: Yale University Press, 1929.

Zahl, Paul A. "Face to Face with Gorillas in Central Africa." *National Geographic* 117, no. 1 (January 1960):114–37.

NOTE: For specialized or technical papers dealing with the gorilla, see under "Notes" at the end of each chapter in this book. Consult also the systematically arranged "References" at the end of the book *The Apes* by Vernon Reynolds, listed above. Other extended bibliographies of books and articles dealing with gorillas appear at the ends of the following two papers:

Sabater-Pi, Jorge, and Jones, Clyde. *Comparative Ecology of "Gorilla gorilla" and "Pan troglodytes" in Rio Muni.* Basel: S. Karger & Co., 1971. (68 references.)

Schultz, Adolph H. "Age Changes and Variability in Gibbons." *American Journal of Physical Anthropology,* n.s. 2, no. 1 (March 1944):124–29. (117 references.)

Indexes

Index of People

An asterisk (*) following a page number indicates citation in a note. A page number in parentheses indicates citation in a caption.

Akeley, Carl E., 41, 45*, 49, 53, (54), 61, 61*, 62, 105*
Akroyd, R., 53, 55*, 163, 165, 170*
Alexeev, Vasili, 154
Aschemeier, C. R., 49, 55*
Asimov, Isaac, 223, 231*

Barns, T. Alexander, 52, 55*, 64, (65), (66), (67), 105*, 170*, 204
Barnum, Phineas Taylor, 155, 227, (228), 231*
Bartel, H., 77
Bartlett, A. D., 147, 157*
Battell, Andrew, 36, 37, 37*, 45*
Beach, Frank A., 157*
Benchley, Belle J., 11, 183, 186, 210*, 222*
Bingham, Harold C., 52, 55*, 170*
Binkley, Kenneth, 211, 222*
Birch, H. G., 249*
Blancou, L., 54, 56*
Boldrick, E. H., (81)
Bonestell, Aileen E., 210*
Bourne, Geoffrey H., 169, 171*, 191, 206, 210*, 231*
Bowdich, Thomas Edward, 37, 57, 105*
Bowerman, Walter G., 155, 158*
Brabant, H., 34*
Brace, C. Loring, 34*
Bradley, H. E., 61, (61), 62
Bradley, Mary Hastings, (61), 105*
Brehm, Alfred Edmund, 11, 24*, 44, 45*, 60, 61, 105*, (147), 157*, (160), (252)
Brindamoor, Biruté G., 122, 127*
Brody, Samuel, 220, 222*
Brosse, M. de la, 37
Brown, James F., 204
Buck, J. L., 77, 154, 178
Budich, G., (74)
Bullock, Michael, 45*

Burbridge, Ben, 45, 52, 55*, (176)
Burbridge, Juanita Cassil, (176)
Burnet, James, 37
Burroughs, Edgar Rice, 152*, 225, (225)
Butler, Albert E., (54)
Byrne, Peter, 27, 27*, 34*
Byron, Lord, 145

Campbell, T. D., 157*
Carbo, Roman Luera, 205, (205)
Carpenter, C. R., 107*, 184, 210*
Carroll, Philip, 194, 200
Chamberlain, A. F., 155, 158*
Christy, Cuthbert, 49*
Clarke, Gary, 233
Coffey, Phillip, 104, 106*, 210*, 213, 218, 222*
Cohen, M., 171*, 206, 210*, 231*
Collins, Alfred C., (66)
Comfort, Alex, 155, 158*
Conant, Roger, 177, 209*
Connelly, C. J., 143, 157*
Conrey, Lee, (229), 230
Coolidge, Harold Jefferson, 46, 49, 55*, (64), 76, 78, 103, 105*, 115, 116, 126*, 165, 170*
Cornish, J., 60, 105*
Cousins, Don, 50, 55*
Cragg, John, 127*
Crandall, Lee S., 194, 219, 222*
Cromwell, Oliver, 145
Cronin, Edward W., Jr., 34*
Cunningham, Alyse, 45, 173, (174), (175)
Cuvier, Baron Georges, 145

Dandelot, Pierre, 55, 56*
David, Joseph A., Jr., 196
Delano, Frank E., 152, 153, 157*

Derscheid, Jean M., 53, 56*
Deschryver, Adrien, 53, 56*, (162)
Disney, Walt, 25*
Dmitri, Ivan, 191, 210*
Dorst, Jean, 55, 56*
Dressman, William, 181
Drummond, John Peter, 227, (227), 231*
Dubois, Eugen, 149, 157*
Du Chaillu, Paul Belloni, 38, 38*, 39, 40*, 41, 44, 49, 59, 105*, 110, 151, 167
Duckworth, W. L. H., 73, 105*

Eisenstadt, Alfred, (188)
Elliot, Daniel Giraud, 46, 55*, 115, 126*
Emlen, J., 53, 56*
Erikson, G. E., (135), 139, 141*
Eustis, Francis W., 202, (204)

Falkenstein, Julius, 42, 44, 160, 170*, 172
Finch, G., 154, 157*
Finlay, Virgil, (227)
Fisher, James, 233, 249*
Flower, Stanley S., 155, 157, 158*
Ford, H. A., 38
Ford, H. J., (123)
Ford, Lawton, (228), 230
Fossey, Dian, 52, 55*, 159, 167, 171*
Frayer, David W., 34*
Fremiét, Emmanuel, (122), 230, (230)
Frueh, R. J., 222*

Gaffikin, Philippa, 96*, (97)
Garn, S. M., 34*
Garner, R. L., 161, 170*
Garrison, Ron, (88), (103)
Gatti, Attilio, 53, 56*, 62, (63), 105*
Gavan, James A., 98, 105*, 138, 141*
Gleser, Goldine C., 171*
Golding, Robert R., 216, 222*
Good, A. I., 51, 55*
Goodall, Jane, 112, 113, 126*
Goose, Denys H., 34*
Grauer (hunter), 49
Gray, J. E., 59, 105*
Green, D. L., 34*
Green, John, 26, 27, 34*
Gregory, William King, (62)
"Greystoke, Lord," 225
Groves, Colin P., 49, 50, (51), 55*, 76, (162), 183, 210*
Grumley, Michael, 34*
Grzimek, Bernhard, 126*, 140*, 141*, 159, 160, (162), 168, 169, 170*, 171*, 233, 249*
Gulliver, Lemuel, 224

Hackett, C. J., 157*
Hahler, Paul E., 34*
Hall, G. Stanley, 158*
Hanno, 35, 45*
Harrisson, Barbara, 98, 106*
Harwell, Edalee Orcutt, 197, 199, 233, 249*
Hastings, N. S., 181
Hathaway, Millicent L., 87, 105*
Hermes, Dr. von, 42, 44
Heck, Heinz, 115, 126*
Heck, Ludwig, 60, 105*

Hendricks, F. L., (252)
Hess, Lilo, (148), (150)
Hill, W. C. O., 117, 127*
Hilzheimer, Max, 60, 105*
Hofer, Helmut, 19
Hollister, N., 101, 106*
Honegger, R. E., 233, 249*
Hornaday, William T., 41, 45*, 60, 68, 105*, 117, 127*, 173, 209*
Hoyt, Maria, 157, 191, 210*
Hrdlička, Aleš, 17, 24*, 34*, 143, 157*
Hughes, Jennifer, 105, 106*
Hurme, V. O., 98, 99*, 106*
Huxley, Thomas Henry, 138, 141*

"Jane," 225
Jay, P. C., 157*
Jerison, Harry J., 143, 157*
Johnson, Gaylord, 231*
Johnson, Martin, 53, 56*, (81), 183, (183), (185)
Jonas brothers, (64)
Jones, Clyde, 50, 54, 55*, 56*, 163*, 170*
Jones, Marvin, 155, 158*, 195, 208, 209, 233, 233*

Keith, Arthur, 54, 56*
Kelly, Joan Morton, 222*
"Kigor," 227, (227)
Kirchshofer, Von R., 212, 222*, 249*
Kirkpatrick, G. E., (88)
Kitahara-Frisch, J., 34*
Knight, Charles R., (95)
Koenigswald, G. H. R. von, 28, 30, 31*, 34*
Kohler, Wolfgang, 157*
Koppenfels, Hugo von, 44
Kortlandt, Adriaan, 112, 126*
Krantz, Grover S., 27, 34*
Kuenzel, Wilhelmine, 157*

Lang, Andrew, (123)
Lang, Ernst M., (207), 208, 209, 210*, 222*
Langley, Nina Scott, (151)
Lankester, E. Ray, 155, 158*
Laurentis, Dino de, 224*
Leigh, William R., (54)
Lewis, A. B., 34*
Lewis, E. S., 21
Liburnau, L. von, 71
Lincoln, Elmo, 225, (226)
Linnaeus, Carolus, 24
Lintz, Gertrude, 182, 189
Livingstone, David, 113, 126*
Lombroso, Cesare, 50
Lotshaw, R., 222*
Lydekker, Richard, 45*, 61, (95), 105*, 117, 127*
Lyerly, S. B., 249*

McCourt, B. H., 191, 210*
McCue, George, 200*
McGregor, J. H., (62)
McKenney, Frank D., 210*
Mallinson, Jeremy J. C., 210*, 213, 218, 222*
Mann, William M., 154, 157*, 178, 192, 210*
Markey, Enid, 225
Martin, Rudolf, 143, 157*
Mason, Elmer Brown, 227

259

Matschie, Paul, 49
Menard, Wilmon, 41*
Menichini, P., 233, 249*
Mollison, T., 103, 106*, 137, 138, 141*
Monboddo, Lord, 37
Moog, Florence, 231*
Mooney, Mark, (182)
Moore, Benson R., (179)
Moore, Mr., 42
Mydans, Carl, 123, 127*

Neuville, A. de, (124)
Nissen, H. W., 98, 105*, 112, 126*
Nostrand, Van, (198)
Nott, J. Fortune, 45*, 69

Oertzen, Jasper van, 50, 55*
Oppenheim, Stefanie, 103, 106*, 143, 157*
Owen, Richard, 38, 45*, 59, 105*, (146)

Packard, Vance, 148, 157*
Parker, Neave, (29), (126), (164)
Paschen, H., 59, 60, (60)
Patry, Maurice, 153
Patterson, Roger, 25, 25*, 31
Pearl, Raymond, 22, 24*, 143, 157*
Pechuel-Loesche, Dr., 44
Penny, E. C., 175
Penny, Major, 175
Penny, Mrs., 175
Perkins, R. Marlin, 180, 211
Phillips, Captain, 182, 189
Pickett, Deets, (169), 202, 204, (205), (215), 217, (248), 250, 251, (251), 251*, 254, 255*
Pitman, C. R. S., 53, 55*, 56*, 165, 170*
Pook, J., 104, 106*
Poulton, Peter, (72)
Pournelle, George H., 249*
Purchas, Samuel, 35, 45*

Quiring, Daniel P., 143, 144, 157*

Radcliffe, Herbert L., 219, 222*
Randall, Francis, 12, 34*, 50, 55*, 96, 97, 103, 106*, 145, 157*, 166, 170*
Raven, Henry C., 62, (62), 105*
Reade, William Winwood, 39, 41, 44, 45*
Redshaw, Margaret, 105, 106*
Reichenow, Edward, 161, 170*
Reiss, Bernard F., 233, 249*
Reynolds, Vernon, 112, 113, 126*, 191, 210*, 255*
Riessen, A. H., 98, 105*
Riopelle, Arthur J., 206, 210*
Risser, Dianne K., 199, 210*
Robinson, Edgar, 180
Rode, P., 116, 127*
Rosen, Stephen, 15*
Ross, Sherman, 249*
Rothschild, Walter, 49, 55*, 59, 60, 60*, 105*
Rumbaugh, Duane M., 104, 106*, 149, 151, 157*

Sabater-Pi, Jorge, 50, 54, 55*, 56*, 163*, 170*
Saint-Hillaire, Isadore Geoffroy, 38, 45*
Sanborn, Elwin R., (114)

Sanchez, Tony, 152, 157*
Sanderson, Ivan T., 27, 34*, 164, 170*
Savage, Thomas S., 38, 45*, 110-12, 126*
Schaller, George, 53, 56*, 159, 160, 161, 170*
Schmidt, F. D., (104), (138), (199)
Schubert, Franz, 225
Schultz, Adolph Hans, 11, 12, 22, 46, 50, 55*, 62, 76, 78, 81, 89, (95), 96, 98, 98*, 105*, 107*, 115, 127*, 132, (137), 140, 141*, 143, 145, 151, 157*, 165, 170*, 175, 177*, 183*, 189, (254)
Schwarz, Ernst, 76, 115
Scott, G. B. D., 127*
Scott, George, 195
Selenka, Emil, 143, 157*
Semenova, Iuliyaka, 223
Sheak, Henry, 173, 209*
Sheppard, Raymond, (111)
Shirley, Mary M., 105, 106*
Simpson, George Gaylord, 78, 105*
Smith, Jonothan, 224
Specht, August, 11
Specht, Carl Gottlob, 11
Specht, Friedrich, 11, (14), 22, (108), (116), (118), (147), (155), (160), (254)
Speidel, George, 200, (201)
Steiner, Paul E., 210*
Steinmetz, J., (193)
Stemmler, Carl, (85), 87, 181*, 209
Stirton, Robert A., (20)
Stott, Kenheim, Jr., 183, 197, 210*
Straus, W. L., Jr., 23, 24*, 119, 127*
Suschitsky, W., (92)
Swindler, Daris R., 105*

"Tarzan," 225, (225), (226), 227
Thenius, Erich, 19
Thomas, Warren, 206, (206)
Thoms, W. J., 155, 157*
Todd, T. Wingate, 143, 157*
Tomas, José, (193)
Tratz, Edward, 115, 126*
Traum, J., 210*
Trotter, Mildred, 171*
Tulpius, Nicholas, 110
Turgenev, Ivan, 145
Twiesselmann, F., 34*

Ulmer, Frederisk A., Jr., 127*, 182, 210*
Umlauff, J. F. G., 59
Umlauff, William, (152)
Urbain, A., 116, 127*
Usher-Smith, Jeremy, 210*, 213, 218, 222*

Vanderbilt, George, 64, (64), 105*
Villars-Darasse, M., 60, (60)
Vogel, C., 76, 105*

Wadlow, Robert, 223
Wallace, Alfred Russel, 117, (124), 127*
Ward, Roland, (253)
Warden, Carl J., 157*
Washburn, S. L., 157*
Waterton, Mr., 42, 172

Webster, Daniel, 145
Weidenreich, Franz, 31*, 34*, 143, 145, 157*
Weinert, H., 28
Weissmuller, Johnny, (226)
Wendt, Herbert, 45*
Weyl, Peter K., 231*
Wiley, James, 157*
Wilson, J. Leighton, 38
Willoughby, David P., 34*, 78, 105*, 158*, 166*, 183*, 222*, 255*
Wombwell, Mrs., 42, 172, 173

Wyman, Jeffries, 38, 45*

Yerkes, Ada W., 45*, 56*, 107*
Yerkes, Robert M., 45*, 56*, 89, 107*, 112, 149, 175, 233, 249
Young, J. Z., 18*, 24*
Young, T. E., 155, 158*

Zahl, Paul, 159, 164, 170*
Zehrt, Jack, (200)

Index of Gorillas

An asterisk (*) following a page number indicates citation in a note. A page number in parentheses indicates citation in a caption.

Abraham, 212
Achilla, (85), 87, 181*, 208, (208), 209, 212, 247
Albert, 77, 80, 81, 82, (88), 150, 151, 196, (196), (198), 201, 205, 247, 248
Alvila, 76, 77, (103), 104, (168), 197, 248
Assumbo, 209, 212, 213, 217

Bamboo, 77, 81, 157, 177, (177), 182, 218, 233
Baron, The, 206
Bata, 77, 80, 81, 82, 196, (198), 247
Benoit, 247
Big Man, 202, (205), 212, 247, 251
Bimbo, 77
Binti, 197, (214), (247)
Bobby, 176, (176)
Bobo, 211
Bongo, 206, 248
Bouba, 77, 80, 81, 82, 196, (198)
Bushman, 81, 178, (180), 181*, 211
Bwana, 248

Chey-Gor, 248
Christina, 206
Colo, 206, (206), 248
Colonel, The, 204
Congo, 89, 175, (176), 177

Dinah, 173, (174)
Daniel, 212
Dolly, 197, (214), (247)

Emmy, 206
Empress, 45

Fifi, 211

Gargantua, 188, (188), (190), (191), (192), 248
Goma, 76, 207, (207), 212, 248
Guy, 87, 157, (254)

Hatari, 202, (204)
Helen, 212

Inaki, 248

Jacob, 247
Jambo, (85), 87, 208*
Jambo (#2), 208, (208), 212
Janet Pensarosa, 77
Jim, (138), 197, (247)
Jimmie, 77
John Daniel, 45, 173, (174)
John Daniel II, 45, (175)
Jungle Jeannie, 204

Kenya, 196
King Tut, 202, (203), (204), 233
Kribi Kate, 204, 247, 248

Lilly (=Sumaili), 232
Little John, (104)

Madame Ningo, 173
Mafuka, (110)
Mahari, 202
Makoko, 81, 192, (193), 211, 219, 220, 248
Makula, 208, 212, 247
Mambo, 195, (195), 211
Mamfe, 76, 209, 212, 213, 217, 248
Massa, 157, 178, 181, (182), 233, 247, 248
Massa (#2), (199), (218)
Mbongo, 11, 77, (81), 87, 182, (183), (184), 196, 200*, 217, 220, 222, 248
Meng, (109), (151)
Migger, 208
Mighty Omega, 233
Missus, (248)
Moka, 248
M'Pungu (=Pongo), 42, (43), 173
M'Toto, 157, 181, 189, 191, (192), (193)
Muni, 205
Mzuri, 248

Nandi, 209, 212, 248
Ngagi, 11, (81), 87, 182, (183), (185), (186), 196, 217, 220, 222, 247, 248
Ngi, 77
N'Gi, 178, (179)

Nikumba, 247
N'Pongo, 209

Oka, 81, 192, (195), 211, 219, 220
Oscar, 206

Penelope, 202, 247
Phil, 199, (200), (201), 222
Pongo, 42, (43), 173
Pussi, 45

Quarta, 208

Sally, (109), 147
Samson, 87, 201, (201)
Samson (#2), 202
Shamba, 233
Snowflake, 204, (205)

Souanke, 247
Stefi, 208, 209, 247
Sultan, 45
Sumaili (=Lilly), 232
Susie, 77, 81, 82, 89, 132, 176*, 181, (181), 221, 233

Tara, 202
Terra (=Tara?), 202
Tiffany, 204
Tomoka, 76, 248
Toto. See M'Toto
Trib, 197, (247), 248
Trudy, 212

Vila, (168), 197, (198), (247)

Zaire, 104, 209, 218, 248

Subject Index

An asterisk (*) following a page number indicates citation in a note. A page number in parentheses indicates an illustration.

Age vs. growth in gorillas, 87; (longevity), in man, apes, and various lower mammals, 155–57
Albert National Park, (48)
Anthropometry, of man vs. higher apes, 128–41
Apes, anthropometry of, 128–41; diet of, 18; housing of, 216
Arm length: vs. span, 89; vs. leg length in man and apes, (135); how taken, 90, (130), 133
Australopithecines, 36
Aye-aye, (22)

Bathing, of young gorilla, (217)
Behavior in gorillas, 166–67
Belly girth: how taken, 89; in gorillas, 133
"Bigfoot," or Sasquatch, 25–28, (26), 31, 33, 151*
Bipedal walking in gorillas, (165), 166
Bodily proportions in the gorilla vs. man, 91, (137); shown diagrammatically in man and apes, (132); measurements, how taken, (130)
Body build, formula for, 134*, 136
Body weight in relation to molar size, (30)
Bones, limb, lengths of in gorillas, 70, 71
Bonobo (pygmy chimpanzee), 115, (115)
Brachial index, in adult gorillas, 74
Brain/body index, in primates, 142, 143
Brain, development of, 22; brain weight and intelligence, 142–51

Calories vs. bodyweight, requirements in apes, 221
Capturing of gorillas, 170
Capuchin, brain weight and relative intelligence in, 143, 144
Catarrhini (infraorder), 17, 23
Chest girth, how taken, 89, (130), 132
Chimpanzee: arm length in, (140); brain weight and relative intelligence in, 143, 144; deviations in bodily proportions from lowland gorilla, 136; eruption times of teeth, 99; facial differences from gorilla, (109); life history and description of, 107–16; longevity in, 156; muscular strength of, 154; native names of, 110; physical characteristics of, 109; pygmy, (115); pygmy, brain weight and relative intelligence in, 143, 144; races of, 115; skull, compared with that of orangutan, (127); walking gait in, (111)
Cranial capacity, formulas for, 103, 143
Cranial (skull) cavities in man and gorilla, (146)
Cro-Magnon Man, (33)
Crural (leg/thigh) index, in adult gorillas, 74

Dental age vs. skeletal dimensions, 96, 97
Deviations, bodily, of chimpanzee and mountain gorilla from lowland gorilla, 136

Ears, comparative size of, in man and apes, (137)
Engé-ena (gorilla), 44

Feet, of man, gorilla, and Sasquatch, (33)
Fetus, proportions of, in man and gorilla, (95)
Foot length, how taken, 90, (130), 136

Gestation period: in man, gorilla, and chimpanzee, 104; in gorilla, 168
Gibbons: literature on, 107*; white-handed, (139)
Gigantopithecus, 28–31, (29), 32
Girth measurements, of adult and newborn gorillas, 93
Gorilla and/or gorillas: age vs. growth in, 87; behavior in, 166–67, 213; bodily measurements of, from birth to adult, 76–77, 83–84; body weight, formula for, (79); breeding, pregnancy, and birth in, 211–16; brain weight and relative intelligence in, 143, 144, 175–76; capturing of, 170; chest beating in, 160*; classification of, 24; differentiating characters in three races, 76, 78; early accounts of, 35–45; early zoo, 44–45; fighting prowess of, 227–30; fetus, proportions of, (95); food in captivity, 173, 186, 189, 195, 202, 205, 211–22; foods eaten in wild, 51–53, 160–65;

geographical distribution of, 46–49, (47), (48); gestation period in, 104, 168, 212; growth grid for, (75); hand and foot of, (33); handedness in, 166; head and facial features in, (100–101); heights of, 69–71; height in relation to span, 129–30; height and weight, growth in, 81, 215; infant care in, 167–69; laryngeal air pouches in, 180*; lengths of limb bones in, 70; locomotion in, 166; longevity in, 155–57, 182; mating in, 167–69; migration in, 166; muscular strength of, 151–55, 153*; natural habitats of, (48), 50–55, (54); nesting and climbing of, 163–65; numbers of, 53; origin of name, 36*, 37*, 38; physical distinctions in, 71–76, 78; physique compared with that of man, (72); races of, 46–50; ranges of eastern species, (51); reproduction in, 167–69, 247; sitting height in, 88; skeletal size in relation to dental age, 97; skull, gorilla's compared with man's, 98, (125); skull size in relation to dental age, 102; social relations of, 159–60; span vs. sitting height in, 80; standing height in, 87; survival, chances of, 250–54; swimming in, 216; twin births in, 169, 212; walking gait in, (74), 76; weight and bodily size of, 57–71, 187, 194, 197, 200, 222

Grauer's gorilla (eastern lowland): height of, 71; formulas for height in, 78; limb bone lengths in, 71; skeletal proportions in, 71

Growth curves for man, gorilla, chimpanzee, and horse, (82)

Handedness, in gorillas, 166
Hand length, how taken, 90, (130), 133
Hands, of man, gorilla, and Sasquatch, (33)
Height: standing, in man and the higher apes, 87, 128; in relation to span in gorillas (formula for), 129–30; of adult lowland and mountain gorillas, 69
Horse: growth in, compared with man and apes, (82); pulling power of, 154

Infant care in gorillas, 167–69
Intelligence and brain weight, 142–51; comparative, in primates, scale of, 144
Intermembral index, in adult gorillas, 74

Java ape-man, comparative intelligence in, 143, 144

"King Kong," 224, (224), 224*, 225, 230, 248

Leg length (trochanter height), how taken, 90, (130), 133
Locomotion in gorillas, 166
Longevity: in man, apes, and various lower mammals, 155–57; comparative, formulas for, 156
Lowland gorillas: body measurements in, relative to age, 76–77, 83–84; head and facial features in, (100–101); growth in relation to age, (75); heights of, 69, 70; limb bone lengths in, 70; living girth measurements in, 93; physique of, compared with man, (72); proportions of body in, 86; quadrupedal gait in, (74), 76; skeletal size in relation to dental age, 97; skeleton in, (58–59), (72); skull size in relation to dental age, 102; weights of, 69; weight, formula for, 79

Man: brain weight and relative intelligence in, 143, 144; hand and foot of, (17), (18), (33); longevity in, 156
Mandible, of man, gorilla, and *Gigantopithecus*, (28)

Marmoset, brain weight and relative intelligence in, 143, 144
Mating, in gorillas, 167–69
Measurements of body and limbs in man and apes: adult and newborn, 131; relative, 134
Migration, in gorillas, 166
Monkeys, diet of in the wild, 18, 18*
Mountain gorilla: bodily deviations from lowland gorilla, 136; distinguishing characters in, 78; food in wild of, 161, (162); height of, 69; limb bone lengths in, 71; living girth measurements in, 93; weight in, 69
Muscular anatomy: in gorilla, (95); in man, (94)

Neanderthal Man, (33)
Nesting and climbing of gorillas, 163–65

Orangutan: bodily size of, 117; brain weight and relative intelligence in, 143, 144; eruption times of teeth in, 99; geographical range of, (123); life history and description of, 116–27; longevity in, 156; muscular strength of, 153; names of, 117; nest building of, (126); physical distinctions in, 117, 119, 120; temperament and habits of, 121; throat pouch in, 119, (120)

Platyrrhini (infraorder), 17, 23
Pongo, 37*
Potto, (22)
Pregnancy, in gorillas, 168
Primates: brain weight and relative intelligence in, 143, 144; classification of, 15–24, (23); dentition of, 16; distinguishing characteristics of, 16; evolution of, (19), 22; family tree of, (21), (22), (32); feet of, (18); hands of, (17); heads of, (16); longevity in various, 156; nasal differences in, 17; phylogeny of, (20)
Pygmies (Wambutte), 53
Pygmy chimpanzee, or bonobo, 115, (115)

Rain forest, animals in, 55
Reproduction, in gorillas, 167–69

Sasquatch (or "Bigfoot"), 25–28, (26), 31, 33, 151*
Siamang, brain weight and relative intelligence in, 143, 144
Sitting height, in gorillas, 88, (130)
Skeletal dimensions in relation to dental age, 96, 97
Skeleton: comparison of in man and apes, (129); of immature orangutan, (139)
Skeletons of man and gorilla, (58), (59), (72), (201)
Skull, in man and gorilla, 98, (146)
Span, or horizontal reach, in lowland gorillas, 76–77; formulas for, 85; how taken, 89, (130); in relation to arm length, 89
Stature or standing height in man and the higher apes, 87, 128
Swimming, in gorillas, 216

Tarsier, spectral, (22)
Teeth, molar, correlation with body size, (30), 31, (31)
Tibiohumeral index in adult gorillas, 74
Tooth eruption, age in, 97, 99
Tranquilizing of a gorilla, (169), 212
Tree shrews, 15
Trunk height (anterior), 88, (130), 132; formulas for, in relation to sitting height, 133

Tupaia, 15, (22)
Tupaioidea, 15, 23
Twins, gorilla, 169

Walking gait in gorillas, (74), 76, 166

Weight in gorillas: formula for, (79), 222; and bodily size of, 57–71, 187, 194, 197, 200, 222

Yeti (or "Abominable Snowman"), 151*